Multimodal Artificial Intelligence in Precision Agriculture

This book highlights the integration of cloud, edge, and network computing for sustainable farming and future trends in smart farming, including the role of UAVs, 5G, and advanced artificial intelligence. It covers the use of real-time environmental data for forecasting and tracking agricultural products and livestock health monitoring.

Key Features:

- Offers a detailed exploration of the integration of Internet of Things, artificial intelligence, and multimodal intelligence in precision agriculture, covering a wide range of applications from crop management to livestock monitoring.
- Identifies and discusses the challenges of using artificial intelligence and multimodal data in agriculture, providing solutions and techniques to overcome these obstacles.
- Discusses advanced technologies like multispectral and hyperspectral imaging, Internet of Things sensors, and data fusion techniques.
- Highlights emerging trends and future directions in smart farming, including UAVs, 5G, cloud-edge continuum integration, and federated learning.
- Includes case studies and practical examples demonstrating successful applications of multimodal artificial intelligence in precision farming.

The text is primarily written for senior undergraduates, graduate students, and academic researchers in electrical engineering, electronics and communications engineering, computer science engineering, agricultural engineering, and information technology.

Multimedia and Multimodal Intelligence
Series Editor - Amit Kumar Singh

Multimedia and Multimodal Intelligence for Sustainable Development
Edited by Vinay Kumar Srivastava, Priyank Khare, Divyanshu Awasthi, Lalit Garg

Multimodal Artificial Intelligence in Precision Agriculture: Practices, Challenges, and Applications
Edited by Abhilasha Sharma, Vishwas Rathi, Anupam Biswas, Anil Singh and Omer Rana

Multimodal Artificial Intelligence in Precision Agriculture

Practices, Challenges, and Applications

Edited by

Abhilasha Sharma, Vishwas Rathi,
Anupam Biswas, Anil Singh and Omer Rana

CRC Press
Taylor & Francis Group
Boca Raton London New York

CRC Press is an imprint of the
Taylor & Francis Group, an **informa** business

Designed cover image: Shutterstock

First edition published 2026
by CRC Press
2385 NW Executive Center Drive, Suite 320, Boca Raton FL 33431

and by CRC Press
4 Park Square, Milton Park, Abingdon, Oxon, OX14 4RN

CRC Press is an imprint of Taylor & Francis Group, LLC

ISBN: 9781041068273 (hbk)
ISBN: 9781041068280 (pbk)
ISBN: 9781003637264 (ebk)

DOI: 10.1201/9781003637264

Typeset in Times
by codeMantra

Contents

Preface

Agriculture is undergoing a paradigm shift in the era of digital transformation. Increasing global food demand, coupled with challenges such as climate change, soil degradation, and limited natural resources, necessitates the adoption of advanced technologies to ensure sustainable farming. Precision agriculture, enabled by artificial intelligence (AI), the Internet of Things (IoT), and multimodal data analytics, represents one of the most promising pathways toward achieving this goal.

The integration of multimedia and multimodal intelligence provides a powerful framework to capture, analyze, and interpret heterogeneous agricultural data originating from a wide range of sources. High-resolution satellite and drone imagery enable large-scale monitoring of crop growth, vegetation indices, and water stress, while ground-based sensors capture detailed measurements of soil moisture, nutrient levels, and microclimatic variations. Climate sensors further contribute by providing real-time information on temperature, humidity, rainfall, and wind, which are critical for predicting crop health and productivity. By integrating and analyzing these diverse modalities within a unified framework, modern farming systems can generate more holistic insights, enabling early and precise detection of plant diseases, accurate forecasting of crop yields under varying environmental conditions, and optimized management of resources such as water, fertilizers, and pesticides. This synergy of multimodal data and AI-driven analytics represents a significant step toward intelligent, adaptive, and sustainable agricultural practices.

This edited volume, *Multimedia and Multimodal AI in Precision Agriculture: Practices, Challenges, and Applications*, brings together leading researchers and practitioners to present recent advances, methodologies, and challenges in this interdisciplinary field. This book is organized into four sections: (i) multimodal data and intelligence in agriculture, (ii) AI and IoT solutions for precision farming, (iii) real-time environmental data analysis for agricultural practices, and (iv) futuristic emerging technologies for smart farming. Collectively, these sections provide readers with both foundational knowledge and forward-looking perspectives on the role of AI-driven multimodal systems in transforming agriculture.

We sincerely thank all chapter contributors and reviewers for their rigorous efforts and the publishing team for their professional support. We hope that this volume will serve as a timely reference for the academic community and an inspiration for future research that bridges multimedia intelligence with sustainable agricultural practices.

Editors

Abhilasha Sharma earned her B.Tech. in Computer Science and Engineering from Himachal Pradesh University, India, in 2012 and her M.Tech. in Computer Science and Engineering from Jaypee University of Information Technology in 2015. She earned her Ph.D. from Dr B R Ambedkar National Institute of Technology, Jalandhar, India, in 2021. She is currently working at the National Institute of Technology, Kurukshetra, India. She has authored many peer-reviewed papers and articles in high-impact journals and international conferences of repute, including *Computer Networks*, *Computer Communications*, and *Wireless Personal Communication*. Her research interests include cyber-physical systems, vehicular communication, and software-defined networks.

Vishwas Rathi earned his B.Tech. in Computer Science and Engineering from the National Institute of Technology, Patna, India, in 2010 and his M.Tech. in Computer Science and Engineering from Motilal Nehru National Institute of Technology, Allahabad, India, in 2013. He earned his Ph.D. from the Indian Institute of Technology, Ropar, India, in 2022. He is currently an Assistant Professor in the Department of Computer Engineering at the National Institute of Technology, Kurukshetra, India. He has authored many research papers in renowned journals such as IEEE Transactions on Computational Imaging, *IEEE Signal Processing Letters*, *Signal Processing: Image Communication, and IEEE International Conference on Image Processing Challenges and Workshops.* He has received the "Best Industrial Paper Award" for his work at the 16th International Conference on Computer Graphics Theory and Applications, held in Vienna, Austria, in February 2021. His research interests encompass image processing, computational imaging and forensics, multispectral imaging systems, and applied deep learning.

Anupam Biswas earned his Ph.D. in Computer Science and Engineering from the Indian Institute of Technology (BHU), Varanasi, India, in 2017. He earned his M.Tech. and B.E. in Computer Science and Engineering from Motilal Nehru National Institute of Technology, Allahabad, Prayagraj, India, in 2013 and Jorhat Engineering College, Jorhat, Assam, in 2011, respectively. He is currently working as an Assistant Professor in the Department of Computer Science and Engineering, National Institute of Technology Silchar, Assam, India. He has published over 100 research papers in prestigious international journals, including *IEEE Transactions on Fuzzy Systems*, *IEEE Transactions on Social Computing*, *Elsevier Information Fusion*, *Elsevier Information Sciences*, and *Expert Systems with Applications*; conferences; and book chapters. His work has received three "Best Paper Awards". His research interests include machine learning, deep learning, evolutionary computation, social networks, and information retrieval. He has successfully completed two DST-SERB-sponsored research projects in the domain of machine learning and evolutionary computation as Principal Investigator (PI). Currently, he is executing three sponsored research projects (one as PI and two as Co-PI) in the domain of AI/DL and quantum machine learning. He has served as the Program Chair of the International Conference on Big Data, Machine Learning and Applications (BigDML 2019) and is currently serving as Publicity Chair of BigDML 2021. He has served as the General Chair of the 25th International Symposium Frontiers of Research in Speech and Music (FRSM 2020) and co-edited the proceedings of FRSM 2020 published as a book volume in Springer *AISC* series. He has edited nine books that are published by various book series of Springer and Elsevier.

Anil Singh is an Assistant Professor at Thapar Institute of Engineering and Technology, Patiala, India. He has also been a postdoctoral researcher at the Department of Computing, Umea° University, Sweden. He earned his B.Tech. from Uttarakhand Technical University, Uttarakhand, India; his M.Tech. from the National Institute of Technology Hamirpur, India; and his PhD from the Indian Institute of Technology Ropar, India. He has authored over 12 peer-reviewed papers and articles in high-impact journals and internationally reputed conferences, including *IEEE Transactions on Services Computing, Software: Practice and Experience, Cluster Computing*, and *IEEE/ACM International Conference on Utility and Cloud Computing*. He has also taught around eight postgraduate courses, including Cloud Infrastructure and Services, Agile Software Methodology, Advanced Computer Architecture, and Advanced Data Structures, among others. His research interests include cloud-edge-fog systems, edge AI, energy efficiency, and distributed computing resource management.

Omer Rana earned his B.S. in Information Systems Engineering from Imperial College of Science, Technology and Medicine, London, UK; his M.S. in Microelectronics Systems Design from the University of Southampton, Southampton, UK; and his Ph.D. in Neural Computing and Parallel Architectures from the Imperial College of Science, Technology and Medicine. He is the Dean of International for the Physical Sciences and Engineering College. He is a Professor of Performance Engineering at Cardiff University, Cardiff, UK. He has authored over 700 peer-reviewed papers and articles in high-impact journals and internationally reputed conferences, including *IEEE Cloud Computing, IEEE Transactions on Services Computing, Software: Practice and Experience, ACM Computing Surveys, ACM Transactions on Internet Technology, Cluster Computing*, and *IEEE/ACM International Conference on Utility and Cloud Computing*, among others. His current research interests include problem-solving environments for computational science and commercial computing, data analysis and management for large-scale computing, and scalability in high-performance agent systems.

Contributors

Vaibhav Agarwal
Dept. of Computer Engineering
National Institute of Technology Kurukshetra
Kurukshetra, Haryana, India

Chetanya Baliyan
Departemnt of CSE/IT
Jaypee University of Information Technology
Waknaghat, Solan, India

Ujjwal Biswas
Dept. of CSE
National Institute of Technology Manipur
Imphal, Manipur, India

Phani Krishna Bulasara
School of Computer Science and Engineering
VIT-AP University
Inavolu, Amaravati, Andhra Pradesh, India

Avi Chaurasia
Department of Computer Science and Engineering
National Institute of Technology Kurukshetra
Kurukshetra, Haryana, India

Nisha Chaurasia
National Institute of Technology Hamirpur
Hamirpur, Himachal Pradesh, India

Dilip Kumar Choubey
Department of Computer Science & Engineering
Indian Institute of Information Technology Bhagalpur
Bhagalpur, Bihar, India

Vasu Chugh
Department of Computer Science and Engineering
Thapar Institute of Engineering and Technology
Patiala, Punjab, India

Pankaj Dahiya
Department of Agronomy
Lovely Professional University
Phagwara, Punjab, India

Saumya Das
Department of Computer Science and Engineering – Artificial intelligence
Brainware University
Barasat, Kolkata, West Bengal, India

Deepshika
School of Computing
DIT University
Dehradun, Uttarakhand, India

C. Radhiya Devi
Department of Commerce CA
Dr. N.G.P. Arts and Science College
Coimbatore, Tamil Nadu, India

Ratnesh Kumar Dubey
Department of Computer Science & Engineering
Indian Institute of Information Technology Bhagalpur
Bhagalpur, Bihar, India

Anoushka Duggal
Department of Computer Science and Engineering
Thapar Institute of Engineering and Technology
Patiala, Punjab, India

Brijendra Gupta
Department of Information Technology
Siddhant College of Engineering
Pune, Maharashtra, India

Prateek Gupta
School of Computer Science Engineering and Technology
Bennett University
Greater Noida, Uttar Pradesh, India

Priyanka Gupta
School of Engineering & Information Technology
Sanskriti University
Mathura, Uttar Pradesh, India

Hashinur Islam
Department of Electronics and Communication Engineering
HKBK College of Engineering
Bengaluru, Karnataka, India

Amandeep Kaur
Department of Computer Science and Engineering
National Institute of Technology Kurukshetra
Kurukshetra, Haryana, India

Shivanshu Ladohia
Department of Agronomy
Dr. Chaudhary Sarwan Kumar Himachal Pradesh Krishi Vishvavidyalaya (DCSK HPKV)
Palampur, Himachal Pradesh, India

Harmandeep Singh Maan
Departemnt of CSE/IT
Jaypee University of Information Technology
Waknaghat, Solan, India

Jyoti Maggu
Department of Computer Science and Engineering
Thapar Institute of Engineering and Technology
Patiala, Punjab, India

Pillaram Manoj
School of Computer Science & Engineering
Vellore Institute of Technology Chennai
Chennai, Tamil Nadu, India

Swati Mehta
Department of Agronomy
Lovely Professional University
Phagwara, Punjab, India

Satyam Mishra
School of Computer Science and Engineering
VIT-AP University
Inavolu, Amaravati, Andhra Pradesh, India

Shubha Mishra
Department of Centre for Artificial Intelligence
Madhav Institute of Technology Science
Deemed University
Gwalior, Madhya Pradesh, India

Akshita Patel
Information Technology Department in U V Patel College of Engineering
Ganpat University
Kherava, Gujarat, India

Prabina Pattanayak
Department of Electronics and Communication Engineering
National Institute of Technology Silchar
Silchar, Assam, India

Om Prakash
Dept. of CSE
Indian Institute of Information Technology Bhagalpur
Bhagalpur, Bihar, India

Vijay Prakash
Communications & Computer Engineering
University of Malta
Msida, Malta

Mohanapriya C.
Department of Computer Technology
Dr. N.G.P. Arts and Science College
Coimbatore, Tamil Nadu, India

Piyush Raj
School of Computer Science and Engineering
VIT-AP University
Inavolu, Amaravati, Andhra Pradesh, India

Navjot Rana
Department of Agronomy, School of Agriculture
Lovely Professional University
Phagwara, Punjab, India

Shweta Rane
Department of Information Technology
Siddhant College of Engineering
Pune, Maharashtra, India

Priyanka Rathee
Dept. of Information Technology
Dr. BR Ambedkar National Institute of Technology
Jalandhar, Punjab, India

Somya Ranjan Sahoo
School of Computer Science and Engineering
VIT-AP University
Inavolu, Amaravati, Andhra Pradesh, India

Vipin Kumar Sahu
Department of Information Technology
Siddhant College of Engineering
Pune, Maharashtra, India

Sameeksha
Department of Computer Science & Engineering
Thapar Institution of Engineering and IT
Patiala, Punjab, India

Sanjay M
School of Computer Science & Engineering
Vellore Institute of Technology Chennai
Chennai, Tamil Nadu, India

Ketan Sarvakar
Information Technology Department in U V Patel College of Engineering
Ganpat University
Kherava, Gujarat, India

Aman Sharma
Departemnt of CSE/IT
Jaypee University of Information Technology
Waknaghat, Solan, India

Anamika Sharma
Department of Computer Science and Engineering
Thapar Institute of Engineering and Technology
Patiala, Punjab, India

Dinesh Sharma
Dept. of CSE
Amity University
Gwalior, Madhya Pradesh, India

Gitika Sharma
School of Computer Science Engineering and Technology
Bennett University
Greater Noida, Uttar Pradesh, India

Shreyam Singh
School of Computer Science and Engineering
VIT-AP University
Inavolu, Amaravati, Andhra Pradesh, India

Ankit Sinha
School of Computer Science and Engineering
VIT-AP University
Inavolu, Amaravati, Andhra Pradesh, India

Sweta Sinha
Dept. of CSE
Indian Institute of Information Technology Bhagalpur
Bhagalpur, Bihar, India

Anmol Thakur
Departemnt of CSE/IT
Jaypee University of Information Technology
Waknaghat, Solan, India

Bhavna Thakur
School of Agriculture & Allied Science
DBS Global University
Dehradun, Uttarakhand, India

Valarmathi P
School of Computer Science & Engineering
Vellore Institute of Technology Chennai
Chennai, Tamil Nadu, India

Avani Vyas
Dr. B. R. Ambedkar National Institute of Technology Jalandhar
Jalandhar, Punjab, India

Rahul Yadava
Information Technology Department in U V Patel College of Engineering
Ganpat University
Kherava, Gujarat, India

Section 1

Multimodal Data and Multimodal Intelligence in Precision Agriculture

1 Introduction to Multimodal Data Analysis in Precision Farming

Challenges, Techniques and Applications

Navjot Rana, Pankaj Dahiya, Swati Mehta, Shivanshu Ladohia, and Sameeksha

1.1 INTRODUCTION

While food demand is increasing, unpredictable climate patterns and outdated practices are causing instability in agriculture. Due to the intricacy of modern agricultural problems the visual judgments, experience-based decisions and ongoing field treatment of conventional farming methods have become completely obsolete. Food production needs to increase by 70% to feed an expected 9.1 billion people by 2050, as reported by the Food and Agriculture Organization (FAO) of the United Nations innovative techniques of sophisticated food cultivation and land management [1]. In this landscape, smart agriculture (which includes the elements of precision farming, digital farming and climate smart agriculture, which in itself is an umbrella term) has been seen to be a linchpin for next-generation agriculture technologies.

The smart agriculture is a phrase that encompasses precision farming and digital farming, in which the data from different agricultural situations is gathered using technologies like IoT (Internet of Things), remote sensing, artificial intelligence (AI) and robots to collect, process and analyze huge data. These tools enable farmers to be able to do better and make more informed decisions, which in turn means increased agricultural productivity as the farmer grows smarter, makes better use of resources and improved stewardship of the environment. Fundamental to this technological change is the data, most especially multimodal data—that is coming from diverse sources like satellites, unmanned aerial vehicles (UAVs), in situ sensors and weather forecasting models. But data gathering alone is not enough—the diagonal of agricultural management demands to reconcile, evaluate and pull insight from these disparate data sets already [2,3].

Smart agriculture is an umbrella term for precision farming, which is the data-centric paradigm of agriculture where it uses spatial and temporal data for customizing field-level agronomic practices. The foundation of this paradigm is satellite imaging,

DOI: 10.1201/9781003637264-2

drone surveillance, soil nutrient sensor data, meteorological stations and telemetry from agricultural machinery. For example, multispectral and hyperspectral images from remote sensing systems can identify plant stress or disease about a month ahead, and soil moisture sensors and temperature loggers are placed in the ground for continuous measurement of microclimatic conditions [4,5]. Concurrently, Geographic Information Systems (GIS) enable spatial analyses and terrain modeling so that farmers can find variability in fields to target input application. These capacities underpin small-scale interventions, such as variable rate technologies (VRT) fertilization, precise irrigation (and even pest prediction), making yield improvements with reduced environmental consequences.

In this agricultural case, multimodal data can be defined as inputs from diverse sensing modalities that are structurally different, interpreted at a different resolution and served with various semantics. This entails structured data like soil sensor ripped outputs in tabular format, unstructured data (high-resolution imagery and text) and semi-structured data (time-series logs from IoT devices). Such diverse data integration is both a necessity and a problem. Satellite imagery with real-time soil sensor data and local weather outlooks may give a panoramic health of crop, for example. Nonetheless, that is feasible only if dealing with the multi-dimensional challenges of data heterogeneity, synchronicity, scale and contextual seriousness [6]. To derive business intelligence from these multimodal sources requires advanced analysis, while recent developments in machine and deep learning (DL), especially convolutional neural networks (CNNs) for images as well as recurrent neural networks (RNNs) for time-series models, have shown promise in understanding some intricacies of agricultural data [10]. And along the way, edge computing and cloud-based frameworks have enabled big data stream processing to be practical, with near-real-time decisions in dynamic field environments. Blockchain, among other technologies, also provides added advantages of data transparency, traceability and trust [7,8] on an agricultural supply chain.

The technological advancements notwithstanding, there are still few obstacles. This encompasses interoperability between heterogeneous systems, low bandwidth and power in rural areas, data privacy and ownership concerns with respect to sensor and device paradigms, varied protocols for data acquisition as well as storage recommendations. Additionally, smallholder farmers who make up a big chunk of the global agricultural workforce usually do not have access to data-driven high-end devices and knowledge to further the digital divide [9]. In this chapter a detailed survey of multimodal data analysis in precision farming will be presented. It will survey the nature and types of data in agriculture, map out the central technical and infrastructural bottlenecks to processing multimodal data and discuss the state-of-the-art computational techniques behind data fusion, modeling and interpretation. The chapter will also go through the technology trends that are shaping the next decade of smart agriculture and illustrate with real-world examples all the ways that those innovations are having an impact. We want to provide the knowledge necessary for researchers, practitioners and policymakers to leverage multimodal data analytics in developing resilient, sustainable and productive agricultural systems.

1.2 MULTIMODAL DATA IN PRECISION FARMING

Precision farming is ushering a new era in agriculture through multimodal data analysis-based site-specific management. The data comes from many disparate places that together provide unique information regarding crop production, soil health, environmental status and farm operations. By aggregating and analyzing this multimodal data, we can drive targeted interventions, resource optimization and hazard reduction over agricultural landscapes. Yet the specificity of these data and the accompanying problems will entail advanced computational-analytical and infrastructural solutions. Precision agriculture data in the form of multimodality covers the full range of diversity from remote sensing, IoT sensor networks data and climatic/meteorological information to soil diagnostics and image-based phenotypic data (Table 1.1).

The most frequently employed source is remote sensing, which provides data through satellites (both ground-sourced and non-ground-sourced) and UAVs. Through these platforms we can take high-resolution imagery in multiple spectral bands (visible, near-infrared (NIR) and thermal infrared) that are imperative to calculating vegetation indices (e.g., NDVI), land surface temperature or water stress and for pest or disease infestation detection. Satellite-based sensors provide capability for wide area, and also coarse temporal resolution monitoring, while UAVs and drone enable fine-grained logistical and frequency adaptability necessary for site-specific applications [10].

Lastly, one other important data stream has emerged from field-level IoT deployed across agricultural fields. All the time these sensors monitor key environmental data that includes soil moisture, temperature, pH level and ambient humidity. The time-series data enables targeted feedback loops for irrigation control, nutrient application and weather adaptation in real time. IoT sensor networks provide the scale

TABLE 1.1
Common Data Modalities in Precision Farming

Data Modality	Source/Device	Key Information Captured	Use Case
Optical imagery	Drones, satellites, cameras	Crop color, canopy size, disease symptoms	Crop health monitoring
Thermal imagery	Thermal cameras	Surface temperature	Irrigation planning, stress detection
LiDAR	Airborne or ground-based LiDAR	Canopy structure, elevation data	Yield estimation, terrain mapping
Hyperspectral data	Hyperspectral cameras	Detailed spectral signatures	Early disease detection, nutrient status
Soil sensors	In-ground sensors	Moisture, pH, nutrient content	Fertilization, irrigation
Weather data	Weather stations, APIs	Temperature, humidity, rainfall	Pest prediction, sowing decisions

and precision necessary for microclimatic modeling and decision-making at the field level that is key to the promise of precision agriculture focus on minimizing input and maximizing output [11].

Weather and climate data is also important for the prediction of events that could impact crop production. Short- to long-term weather data like rainfall, solar radiation, wind trends, temperature variability is obtained from real-world local exact data (weather stations)-tweaked models and third-party services (Dataset). For data drives agronomic planning (planting dates, insect forecasting, frost management, harvest operation) [12]. These enable integration with geospatial and field sensor data that brings more certainty to the yield estimation and stress analysis predictive models (Table 1.2).

Another added layer to multimodal agricultural data architecture is soil health and nutrient profiling. Soils samples whether collected manually or through sensor-driven platforms also quantify the macronutrient and micronutrient content as well as organic matter content along with the texture for every soil till. This data is often supplemented with result from lab reports and spatially interpolated with GIS tools. Proper appreciation of the soil composition facilitates the application of fertilizers and amendments to be applied so as to balance both crop productivity and soil sustainability [13].

Moving forward, plant phenotypes (leaf color, size, shape and disease symptoms) read from images are also being used for decision support [14]. Images taken either through cameras or hyperspectral imaging are analyzed by DL-based methods to detect the disease, count crops, identify growth stage and predict yield. The development of computer vision algorithms has provided interpretative power to large volumes of agricultural imagery on par or even surpassing human, visual perception allowing automation in agriculture [16] (Table 1.3).

TABLE 1.2
Comparison of Machine Learning Algorithms Used in Multimodal Data Analysis

Algorithm	Type	Common Applications	Strengths	Limitations
Random Forest	Ensemble/ Classifier	Yield prediction, classification	Handles high dimensionality well	May overfit on noisy data
Support Vector Machine (SVM)	Classifier	Weed detection, disease classification	Effective in high-dimensional spaces	Computationally intensive
K-Means Clustering	Unsupervised	Image segmentation, pattern discovery	Simple and fast	Assumes spherical clusters
CNN (Deep Learning)	Supervised	Image recognition tasks	High accuracy with image data	Requires large datasets and GPU
LSTM	Recurrent Neural Net	Time-series weather or sensor data	Good for sequential data	Complex and slow to train

TABLE 1.3
Challenges and Solutions in Multimodal Data Fusion

Challenge	Description	Possible Solutions
Data heterogeneity	Different formats and resolutions of data	Standardization and data normalization
Synchronization issues	Data collected at different times or frequencies	Temporal alignment techniques
Missing or incomplete data	Inconsistent sensor readings or data loss	Imputation methods, data interpolation
Computational complexity	High resource needs for processing large datasets	Use of cloud computing and edge computing
Integration and fusion methodology	Choosing appropriate fusion levels (early, late)	Hybrid fusion strategies

Multimodal data integration in agriculture presents a complex interplay of opportunities and constraints, where challenges are often dictated by the dichotomy between data heterogeneity and system interoperability. Heterogeneity and lack of interoperability are the single biggest problem. Diverse modalities yield data in different formats (some have different resolutions and others with distinct structures). For instance, satellite data can be rasterized with geospatial info and IoT sensor data is generally tabular or time-series data from machines. Explicit data fusion techniques and standard data representation protocols [17] are needed to realize the integration of heterogeneous datasets into a coherent analytical pipeline. The data, velocity and variety of modern agriculture introduce yet another complexity on top of it all. Efficient ingestion, processing and storage of data coming from high-frequency data in an agricultural system is required as the body of knowledge grows continuously. UAVs can generate gigabytes or more of imagery per flight and sensor networks stream data in short intervals (seconds to minutes). For managing such huge, high-velocity data streams we require scalable infrastructures in the cloud and edge computing to offload/process data as close as possible to the source [18].

Real-time data processing and latency are another watershed challenge. In precision farming, this leads to the necessity of decisions to be made faster—for example, in response to heat stress the decision to switch on irrigation systems; or a pesticide application based on an early disease detection. Delays in processing and analyzing data may result in failing to catch the intervention windows, which in turn cause yield losses. This calls for deploying real-time analytics frameworks with near-transport protocols, usually using 5G for scenarios or LoRaWAN in rural areas [19].

Data incompleteness, noise and redundancy are still unsolved problems. Sensor failure and channel failures, cloud coverage in datasets, as well as overlapping datasets can bring in extra data or missing or wrong data. Noise and redundancy do not only blur the signal, but overtrain machine learning (ML) models on an unrepressed dataset. It will require strong preprocessing, outlier detection and imputation methods coupled with confidence metrics that will provide a way of knowing if data is trustworthy [14]. In general, multimodal data is very important for the effectiveness

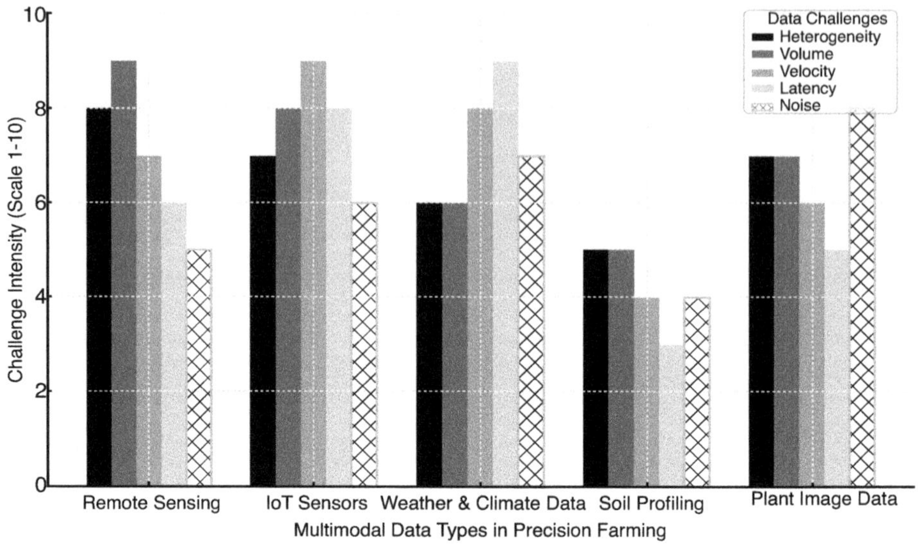

FIGURE 1.1 Multimodal data types in precision farming.

of precision agriculture. Using the different modalities as they do provide different views on agricultural systems, this combined interpretation permits context-aware decisions [15]. But solutions to heterogeneity and to volume, latency and reliability—the nice challenges of modern computing—are going to need technological, computational and organizational innovations (Figure 1.1).

1.3 CHALLENGES IN MULTIMODAL DATA INTEGRATION

Multimodal data integration in precision agriculture is an innovative opportunity and a challenging technical, computational and ethical complexity. The promise of integrating entirely heterogeneous data streams satellite imagery and drone video to in situ sensor data or agronomic records enables new insights and also new interventions, which remain challenging, several of which are complex [16}. There are hardware restrictions, algorithmic constraints, ethical dilemmas and infrastructure problems that need to be carefully enumerated in order for data-driven farming to actually empower many [17].

To elucidate the challenges of integrating multimodal data, Figure 1.2 illustrates the main technical, computational and ethical obstacles. From a technical standpoint, there are challenges in synchronizing different sensors and ensuring compatibility between hardware components. On the computational front, problems such as overfitting, misalignment of modalities and inconsistencies in feature resolution complicate the process of model training [18]. Ethically, the discussions surrounding data ownership and the potential for algorithmic bias influence the acceptance of AI-driven decision-making. As depicted in the figure, tackling these issues necessitates a multi-faceted approach that includes standardization protocols, explainable AI and robust data governance frameworks [19].

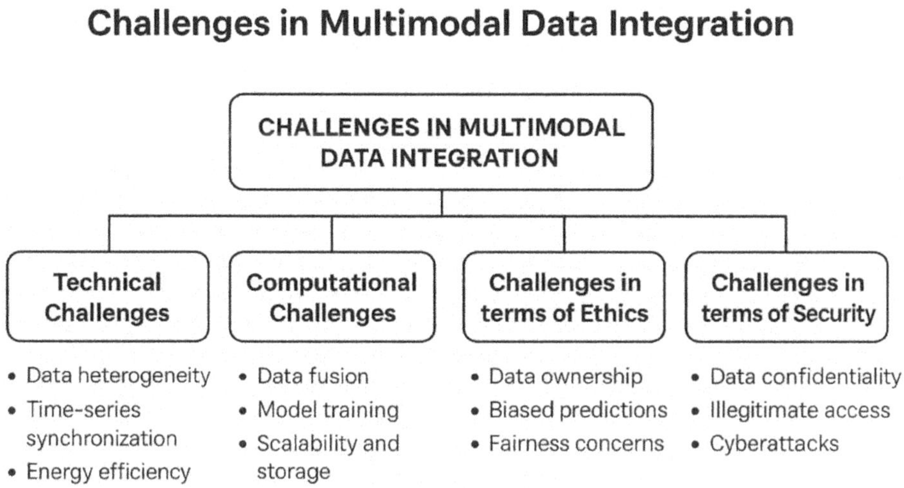

FIGURE 1.2 Key technical, computational and ethical challenges in multimodal data integration for precision farming.

1.3.1 Technical Challenges

The core technical bottlenecks on the integration of multimodal agricultural data are how to connect a set of heterogeneous sensors with edge computing infrastructure. Across third-party installed sensors in a field, systems sense temperature and humidity at every location in a field together with soil moisture or nutrients within the top foot of the ground and atmospheric conditions [20]. The sensors are quite varying in communication protocols, data formats and sampling frequencies. For interoperable local processing and real-time analytics on edge devices, standard hardware interfaces, matured middleware and data interchangeability among vendors are all lacking in lots of field-deployable systems [21].

The reason for integrating data at different sources gets even messier. Satellite images are updated weekly, drones might fly over the fields daily and soil sensors provide data every few minutes. Making temporal asynchronous data streams align in order to conduct ongoing analysis requires time-series synchronization techniques and temporal interpolation models that preserve semantic equivalence over data with little semantic loss [22]. The interest in temporal and spatial coordination grows when we talk of decision-support systems (DSS) that rely on feedback loops running in real time. Data acquisition and transmission, especially when the activity is remote or under resource-constrained locations, is additionally limited by a major constraint of energy efficiency. Most sensor nodes use battery or solar power, and continuous data streaming quickly drains the energy reservoir of each and every node. Consequently, edge devices must choose a low-power communication standard such as LoRaWAN, Zigbee or implement energy-conscious sampling strategies in order to find a balance between energy consumption and data accuracy. Trade-off between data and device longevity has been a technical tension yet to be properly resolved [23].

1.3.2 Ingredients of Computational Challenge

Conceptually from a computational point of view, handling multimodal data integration has complex problems, which data fusion lies underneath. Each data modality (numerical, visual, textual or categorical) has different preprocessing pipelines and analytical models for it. The primary models into which we would like to embed the fusion of them have two modes of integration (early fusion: data level vs. late fusion (decision level)), and are restricted by cost in terms of computational power, feature alignment match, signal-to-noise ratio [24]. In the real world, performing geospatial crop data with hyperspectral imagery invites resolution normalization, feature engineering and context-sensitive embedding.

Model training as well deals with overfitting and underfitting problems. We find the common overfitting problem is acute for high-dimensional data (e.g., drone imagery or multispectral scans) when training DL models can be done with relatively few labeled samples. However, underfitting can take place as some simplistic models cannot account for the nonlinear relationships that are present in the multimodal agricultural data. This is challenging and depends on the nature of dataset with feature engineering, along with cross-validation methodologies and ensemble model [25].

In addition to scalability and storage adding other constraints, with the data farm growing exponentially, storing and curating of such large datasets is nothing but a hard computational problem. While the platforms used for precision agriculture needs to enable ingestion of terabytes of real-time edge sensor, cloud snapshot and historical databases, it requires distributed data lakes, cloud-native architectures and ML infrastructures that can scale to meet performance requirements while handling the storage bottlenecks and parallel processing demands [26]. In addition, integration of metadata management and indexing mechanisms is required to enable querying and retrieval of particular data subset efficiently.

1.3.3 Challenges in Terms of Ethics and Security

Beyond the technical and computational challenges, aggregating multimodal data from agriculture is subjected to significant ethical and security issues. Issues are particularly rife over farming data in regions where the data, generated by smallholder farmers but hosted and monetized by third-party platforms, generates both data privacy and ownership debates. Data ownership, who can access and for what purposes this data is relevant, is an important question. There is need for robust and fair data governance frameworks that promote equal access, user consent and compliance with laws, particularly in the setting where local knowledge meets digital farming [27]. When AI models based on multi-dimensional agricultural data are trained, these can make biased predictions resulting in unfair consequences. For instance, a model to predict yield trained mostly on data from large farm areas in temperate regions would fail to accurately predict in sub-tropical or arid portions of the tropics, undermining smallholder interests there. Those biases can flow to decision-support systems, with possible impacts on fertilizer and irrigation scheduling, as well as detection of disease. This increases the risks that need to be mitigated and it lies between dataset diversification, fairness-aware ML techniques and participatory approaches involving farmers in live validation of models [28].

The other one is data sharing having to do with security. More than ever, with agriculture becoming networked through IoT, cloud services and blockchain platforms, it is essential to conserve the confidentiality and integrity of data. Data tampering, unauthorized disturbance with monitoring analytics and support mechanisms as well as cyberattacks targeting these components can undermine the reliability of analytics and disrupt operational tactics. At the end of the day, supporting end-to-end encryption, access controls and decentralized identity frameworks will provide a good framework to protect and enable interoperable data ecosystems for agriculture-sensitive data [29]. Overall, the integration of multimodal data in precision farming seems like a complex problem from which few glimmers of light exist. Solutions for these problems will require to work in tandem hardware-design breakthroughs, algorithmic innovations, data ethics and cybersecurity. Stakeholders could then design sustainable, inclusive and intelligent farm systems that leverage multimodal data to deliver superior value while also maintaining trust, fairness and sustainability.

1.4 MACHINE LEARNING AND DEEP LEARNING TECHNIQUES

Precision agriculture is experiencing a transformative change with the incorporation of ML and DL, which facilitate the scalable and intelligent evaluation of large, intricate and multimodal datasets [30]. These approaches enable stakeholders to make decisions that are more precise, timely and efficient in terms of resources throughout the agricultural value chain.

Agricultural datasets are inherently diverse, including satellite images, soil chemical profiles, weather station records, IoT sensor information, drone images and additional sources. ML models can uncover latent structures, hidden patterns, and nonlinear relationships in these data sources—abilities that greatly exceed traditional statistical methods [31].

1.4.1 Supervised Learning

Supervised learning plays a foundational role in the predictive modeling for agriculture. These methods rely on the labeled datasets, where input variables such as the soil properties, climatic conditions or the spectral indices are associated with the known outputs like crop yield, disease presence or growth stage [32]. Algorithms like decision trees, support vector machines (SVMs), random forests and the gradient boosting models are commonly used to build these predictive frameworks. For example, the decision tree-based models can help assess soil fertility zones, while the SVMs have been applied to differentiate between healthy and the diseased crop leaves in image-based datasets. Random forest models trained on the multi-season data have shown significant accuracy in forecasting the crop yields by learning from variables such as temperature trends, soil moisture levels and the planting schedules. Furthermore, image classification tasks that aim to identify plant diseases—such as blight or mildew—have leveraged these techniques to recognize patterns in texture and color with considerable precision [33].

1.4.1.1 Practical Examples

- **Soil Nutrient Estimation**: Predicting nitrogen, phosphorus and potassium levels using hyperspectral reflectance data.
- **Yield Prediction**: Leveraging multi-season satellite data and on-ground weather variables to predict per-hectare crop output.
- **Disease Classification**: Training CNN-based classifiers to identify rust, mildew or blight from leaf image datasets, even under variable lighting and backgrounds.

1.4.2 Unsupervised Learning

Unlike supervised methods, unsupervised learning is applied when no labeled outputs are available and the goal is to detect the hidden patterns or the intrinsic data structures. In precision farming, unsupervised techniques are often used for the spatial data clustering and pattern recognition. Algorithms like k-means clustering and the principal component analysis (PCA) help group similar data points, thereby facilitating the identification of the management zones within fields based on the vegetation health or soil attributes . These zones allow for the site-specific treatment and the input allocation, ultimately improving cost-efficiency and reducing resource waste. PCA, meanwhile, supports the dimensionality reduction in high-resolution datasets, such as hyperspectral images, helping researchers isolate key factors affecting crop performance without losing significant variance in the data [34].

1.4.2.1 Example Use Case

In large-scale wheat fields, unsupervised clustering of NDVI time-series data can highlight variability zones for differential pesticide application—resulting in significant cost and input savings.

1.4.3 Reinforcement Learning: Decision-Making under Uncertainty

Reinforcement learning (RL) offers a distinct advantage for the decision-making in uncertain, dynamic agricultural environments. Unlike the supervised or unsupervised models that operate on static data, RL emphasizes learning through interaction with the environment. An RL agent makes sequential decisions, receives feedback in the form of rewards or penalties, and iteratively adjusts its strategy to maximize long-term gains. This model is highly suitable for the real-world agricultural scenarios where decisions have cascading effects over time [35].

In practice, RL has been employed in the development of the autonomous weeding robots that learn the efficient navigation paths across the fields while avoiding obstacles and minimizing energy use. Smart irrigation systems powered by RL algorithms can optimize the water distribution schedules by learning from evolving environmental cues, such as the real-time soil moisture levels and forecasted rainfall. Moreover, drone-based crop monitoring systems use RL to dynamically plan flight routes that balance the data quality with energy efficiency [36]. These implementations highlight the flexibility and foresight offered by RL, particularly for the tasks that span an entire growing season and require adaptive learning under changing field conditions.

The RL framework mirrors how the farmers make decisions over time: e.g., choosing when to irrigate, fertilize or apply the pesticide based on past results and changing field conditions. It supports **multi-agent cooperation** and **seasonal learning**, where decisions are made early in the crop cycle influenced by the end-of-season yield outcomes (Figure 1.3).

The practical impact of multimodal data analysis in precision agriculture is evidenced by its wide-ranging applications across various agronomic domains. By integrating heterogeneous data sources such as spectral imaging, soil sensors and meteorological inputs farmers and agronomists can achieve actionable insights for real-time and predictive decision-making [37]. Table 1.4 **summarizes the major application areas** of multimodal data analysis, mapping each to the specific data modalities used and the associated benefits. This structured overview offers a consolidated reference for understanding how these technologies contribute to productivity, sustainability and efficiency in modern farming operations.

As highlighted in Table 1.4, each application area leverages a tailored combination of data modalities to drive precision interventions. For example, crop health monitoring often integrates optical, thermal and hyperspectral imagery to detect stress conditions or disease symptoms well before visual signs emerge. Similarly, irrigation management uses data fusion from soil moisture sensors, thermal imaging and weather forecasts to optimize water usage and avoid drought stress. These use cases reflect how domain-specific challenges are being tackled through targeted, data-driven solutions [38]. Yield prediction, which combines multispectral data with LiDAR and soil diagnostics, exemplifies how spatial and spectral integration improves forecast accuracy and input planning. The consistent benefits across use

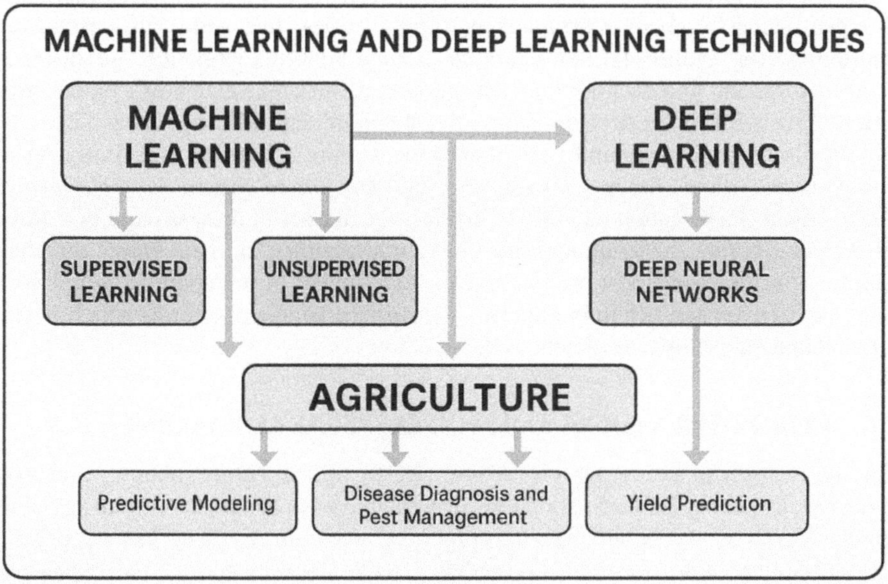

FIGURE 1.3 Machine learning and deep learning techniques.

TABLE 1.4
Applications of Multimodal Data Analysis in Precision Farming

Application Area	Data Modalities Used	Outcome/Benefit
Crop health monitoring	Optical, thermal, hyperspectral	Early detection of disease, stress analysis
Irrigation management	Soil sensors, weather data, thermal	Efficient water use, drought prediction
Yield prediction	Multispectral, LiDAR, soil data	Accurate forecast of production
Pest management	Optical, weather data	Forecasting outbreaks, timely interventions
Fertilizer application	Soil data, hyperspectral imagery	Optimal nutrient distribution

cases validate the importance of multimodal analytics as a foundation for smart agriculture systems [39].

One way to combine data under uncertainty probabilistic models like Bayesian networks and hidden Markov models is commonly employed for fusing, making it possible to combine the prior knowledge as well as evidences from different sources.

Predictive modeling builds on a core application of ML and DL that enables many precision agriculture domains. A high-impact application is crop yield prediction. This model can be trained on top of integrating data of weather history, previous yields and remote sensing indicators to predict yields few years in advance with decent accuracy, using ML-based method. These predictions are not just used for farm management decisions but also play a role in sub-national food security planning and market logistics [40]. Another important application area is the optimized irrigation and fertigation with predictive algorithms based on real-time sensor data to make decisions about water/nutrient supply cycles, thus reducing waste while improving crop health [41]. An example of such adaptive irrigation, for instance, plant requirement and weather forecasts context is the achievement of evapotranspiration-models-informed decision-support systems in deep RL and we also use ML/DL for disease diagnosis and pest management more and more [42]. Early detection of issues allows for lesser yield loss and less use of chemicals. CNN-centric image-based classification can utilize myriad visual cues of diseases and pest infestations long before the grandmas can make out anything in their blurry farmland pictures. Furthermore, pest outbreaks can be modeled at the spatiotemporal scale using predictive analytics to predict this, allowing proactive measures which in turn are cost and environment efficient [43].

1.5 COMPUTER VISION AND HYPERSPECTRAL IMAGING

The application of computer vision with RGB, multispectral imaging and even hyperspectral imaging to further improve imaging capabilities is what has completely revolutionized contemporary precision agriculture. They can be used to conduct remote non-invasive crop health, growth stage inspection, soil condition surveillance and the detection of biotic or abiotic stress foci. These imaging systems acquire data at multiple dimensions of the crop physiology and field heterogeneity in the electromagnetic spectrum ranging from visible to NIR and these data cannot be mimicked by human eyes.

RGB imaging, using red, green and blue spectral channels, serves as a foundational technique in crop monitoring. It supports basic assessments such as plant height estimation, flowering stage detection and biomass approximation. While easily deployable through drones or smartphones and cost-effective, RGB imagery is limited in spectral depth. This limitation is overcome through multispectral and hyperspectral imaging, which provide more granular spectral information for advanced crop health diagnostics. Multispectral cameras capture distinct bands, usually continuous 3–12 spectral channels in the wavelength region (e.g., NIR) which is particularly relevant for vegetation indices like NDVI (Normalized Difference Vegetation Index). NDVI allows for the assessment of vegetation health by comparing NIR and red reflectance values, i.e., chlorophyll content and photosynthetic activity [44].

In contrast, hyperspectral imaging acquires data with hundreds of adjacent spectral bands, which enables in-depth spectral signature analysis of the plant surface. This approach has been extremely successful in the rapid diagnosis of plant diseases, nutrient deficiencies, pests and water stress, often before these conditions are evident to human visual senses. E.g., a stressed leaf shows measurable differences in reflectance to ML models trained on bands [45], which makes leaf anomaly classification and localization accurate with high accuracy. Sensing plant health using the spectral fingerprint is a critical sensing technique in precision agriculture, especially for disease surveillance, variety assessment and yield estimation.

Image processing techniques are needed to analyze high-dimensional imaging data and extract relevant information that are precise. Image segmentation is indispensable to identify regions of interest (plants/reserves or infested areas, etc. from a background). This can be done using classic methods (thresholding and region growing) as well as modern DL approaches. State-of-the-art approaches for plant disease detection and classification in field images or drone videos (17/15) [46,47], or object counting on fruit represent application scenarios where the models perform live purposes like mass masks to peel vegetables. With YOLO achieving fast detection (compared with lower cost in terms of computation) which makes it applicable for mobile and edge deployment, Mask R-CNN offers pixel-level instance segmentation—and is more relevant for tasks paying close attention to delineating plant parts, lesion areas (Figure 1.4).

Figure complements the discussion on imaging technologies by providing a visual workflow of how computer vision systems operate in agricultural environments. It shows how drones collect spectral data, which is then analyzed either on edge devices or in cloud platforms using CNNs. These systems detect diseases, track plant growth stages and assess vegetation health. The hyperspectral imaging pipeline shown in the figure enables precise diagnostics of plant physiology—often invisible to the naked eye—empowering farmers with early warning systems for resource-efficient interventions.

Edge computing has also paved the way for AI-based image analysis models to be deployed in real time on hardware of drones. Edge-AI actions decrease latency and bandwidth, near real-time analytics in remote agricultural environments where

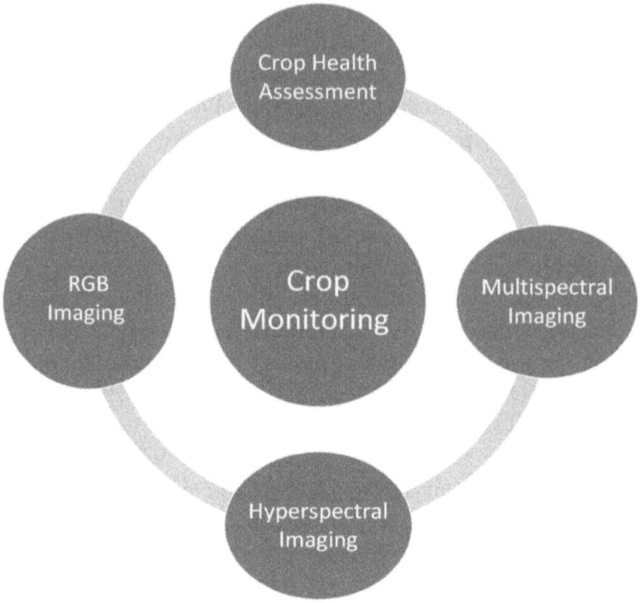

FIGURE 1.4 Advanced computer vision and hyperspectral imaging applications in precision agriculture.

cloud connectivity is scarce. Using drones with edge AI in modules ensures that it could do real-time navigation in fields; generate images; and send alerts based on anomalies detected. A drone flying over a vineyard, for example, can detect precise areas of affected mildew and publishes geo-tagged information to the farmer to treat the exact spots. This real-time capability means we are not only able to intervene more precisely and thus can act faster but also in a more sustainable way [48], with pesticides and fertilizers less used.

As the imaging data grows in scale and intricacy, researchers are resorting to more sophisticated approaches including 3D reconstruction, spectral unmixing and the latest in advanced ML techniques as generative adversarial networks (GANs) for improved analysis performances and consistency across different field scenarios. This will help push the frontiers in plant phenotyping, automated yield estimation and stress modeling. In addition, combining imaging data with other sensors (soil sensors and climate data) embarks upon the opportunity of multimodal learning systems that can provide integrated insights over field-scale dynamics.

To sum up, computer vision with RGB, multispectral and hyperspectral crop monitoring nowadays constitutes a fundamental technology. Now with the use of AI-based image processing paradigm, and running these on edge platforms (droning) farmers and agronomists are using faster, smarter, more scalable tools for making decisions in field. We also expect sensor technology gains, edge AI and ML will continuously make high-resolution crop intelligence more and more democratized leading to better food security and sustainability.

1.6 GEOSPATIAL ANALYTICS IN PRECISION AGRICULTURE

There is no denying geospatial analytics to be an enabler of transformation in precision agriculture that covers a range of tools for agri-monitoring, managing and optimizing agricultural production systems. Geospatial analytics combines GIS, remote sensing technologies and Global Navigation Satellite Systems (GNSS) to carry out spatial analyses of variables pertinent to agronomic and agricultural productivity factors that impact crop yield. The spatial intelligence allows for a shift from catch-all farming to extreme site-specific management strategies which are both the highest yielding and the most sustainable stakeholder enabled (Figure 1.5).

Remote sensing is a traditional tool of the GIS family that provides yearly 5–10 m high-resolution satellite or UAV images for large agricultural areas. These platforms provide real-time information on the health of vegetation, moisture stress, temperature patterns and land cover change—key metrics for assessing crop health and yield. One of the major applications of remote sensing in precision agriculture is to calculate vegetation indices, which is very popular among the community [49]. Greenness and its density and health can be better quantified using indices like NDVI, SAVI and Enhanced Vegetation Index (EVI). For instance, NDVI uses the reflectance properties in red and NIR light to form a normalized ratio indicative of photosynthetic activity. More NIR light is reflected by healthy vegetation and less red in cases of stressed or diseased factors [50], leading to higher NDVI values. SAVI instead contains a soil brightness correction factor, useful for regions in which soil reflectance can distort the spectral signature at low canopy cover or early growth stages [51].

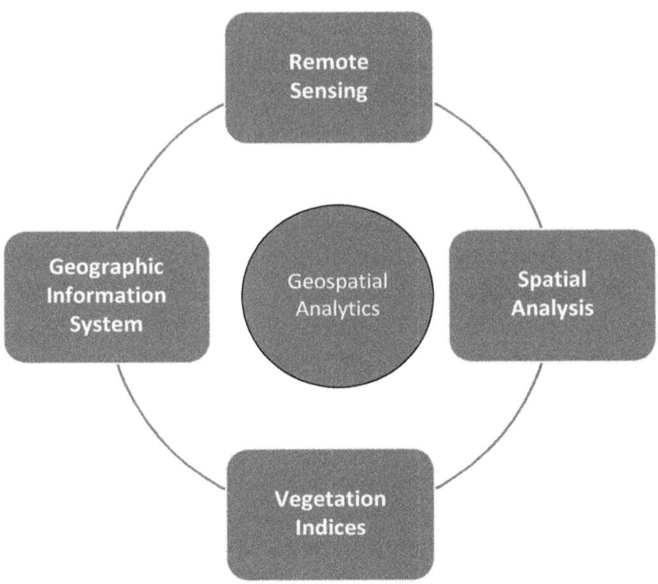

FIGURE 1.5 Geospatial analytics in precision agriculture.

In addition to vegetation monitoring, geospatial analytics uses spatial interpolation like kriging and the inverse distance weighting (IDW) in order to interpolate soil or environmental parameters into unsampled areas. And these methods can create a continuous spatial surface from point-based sensor readings in order to adequately study all fields over time, spatially. Spatial interpolation contributes to the partitioning of management zones—geofenced sub-fields where homogeneous soil properties such as fertility or organic matter content, etc., be it present form biomass that undergoes chronic change over time. The basis of these zones is the root foundation for VRT, which are essentially precise location-specific application that allows varying amounts of the different inputs—fertilizers, seeds and pesticides. Technologically integrated into machinery with GPS guidance are developed spatial prescriptions which could be installed in operations, virtually eliminating waste [52].

Various empirical studies confirm the value of geospatial analytics to solve real-world agricultural problems in practice. Using satellite-based soil moisture mapping, for instance, several successful commercial farming operations to modernize their irrigation strategies have been conducted. Synthetic aperture radar (SAR), used by satellites like Sentinel-1 to penetrate clouds and collect surface moisture data, is then used to generate consistent mapping of the soil moisture from both spatially and temporally independent sources. They allow farmers to not only identify areas that are overirrigated, but also dry stressed zones so that water resources are managed prudently [53]. UAVs have also been effective to monitor crop vigor using spectral or thermal sensor to create heat maps of the plants. These UAVs are capable of sensing small temperature gradients throughout the canopy, which are typically symptomatic before any visible stress or disease symptoms arise. Early identification of these anomalies enables farmers to come in and tackle the yield losses [54].

To conclude, geospatial analytics transformed the operational dimension of precision agriculture from reactive management to a predictive and prescriptive model. Combining GIS remote sensing spatial modeling with vegetation indices is the ultimate way to perceive what is going on in a spatially explicit manner for agricultural ecosystems. When geospatial tool accuracies improve and affordability increases with technology advancements, it becomes easier for them to be adopted and will open up a new era of data-driven sustainable agriculture practices rooted in empirical testing across various geographies.

1.7 EMERGING TECHNOLOGIES AND TRENDS

Driven by digital innovation, the agriculture sector is rapidly transforming and emerging technologies play an essential role in redefining data-driven farming for agriculture revolution. Progressions to these end offer stakeholders along the value chain (from field-level operations to policy-making establishments) greater efficiency in the management of multimodal agricultural data from structured sensor data to unstructured visual, as well as textual field reports. The combination of these enablers not only delivers improved data value and system interoperability but also enables sustainable, secure and scalable precision agriculture solutions.

1.7.1 Blockchain for Secure Data Sharing and Provenance Tracking

Blockchain technology is to be one of the most revolutionary innovations in recent days which provides a consensus-directed decentralized, tamper-proof method of sharing state across multiple stakeholders. At the agricultural level, blockchain has become a solution for data transparency, auditability and privacy in contexts that are then fragmented and asymmetrical regarding data access. In further, blockchain enables the provenance trace of agri products by recording immutably every transactional and logistical event relating from seed sourcing to soil health monitoring and on to post-harvest processing or retail distribution.

Secure data trail is especially needed for food safety, origin and logistics traceability. Within the blockchain, smart contracts are programmed to autonomously assess if certain conditions such as temperature thresholds of cold chain logistics or pesticide residue detected by IoT sensors, allowing automated enforcement compliance and real-time notifications [55,56].

Additionally, as each data entry is crypto-linked to the previous one, data integrity in a decentralized world is provided through block chain and no one party can edit or replicate this sensitive information without the whole network knowing it.

1.7.2 Edge and Fog Computing for Real-Time Analytics and Latency Optimization

Traditionally cloud supports agricultural data analysis; however, as the sheer volume and velocity of sensor data grows real-time processing has to be done at or near the data source with the help of cloud era. And here edge and fog computing approaches are the foremost enablers to low-latency high-bandwidth solutions. Edge computing processes data that is collected from devices like drones, soil moisture sensors or even smart tractors with itinerant data on site or on the edge nodes (microcontroller and local gateways) and avoids centralized servers from receiving these data. The result removes this latency significantly and it is vital for fast interventions in agriculture like precision spraying, pesticide detection, irrigation control, etc.

Fog computing establishes a layered architecture of intermediary nodes between cloud servers that can carry out more complex analytics than edge devices, and are closer to data origin than cloud servers. For example, a fog node can collect data of multiple edge sensors in farm cluster and carry out spatiotemporal anomaly detection, then share with cloud-based dashboards for historical analysis and broader insights [57,58]. There are several levels of redundancy in this layered architecture so that the systems can survive in a less than perfect networked rural environment, which is severely bandwidth limited and responds with poor endurability, while making the precision agriculture system more bandwidth efficient and cost-effective to implement.

1.7.3 Cloud Platforms and High-Performance Computing (HPC) for Scalable Multimodal Analytics

Infrastructure for the storage, processing and visualization of large agricultural datasets can be found on a robust enough scale through cloud computing such as AWS

(Amazon Web Service), Azure (Microsoft Azure) and Google Earth Engine, for example. Platforms provide advanced functionalities and services of agri-informatics, such as satellite image analysis [59], weather data assimilation, regional or global scale crop modeling. For example, Google Earth Engine lets people do temporal change detection and vegetation index calculation over multiple satellite archives which can support long-term land use planning or climate resilience strategies in different regions.

In addition, HPC environments provide a means for conducting parallel algorithms that process multimodal data streams—think: hyperspectral images, drone video, geospatial grids and log files of farming equipment. Cloud-native libraries such as Apache Spark, Tensor Flow, Kubernetes allow us to containerize and automate computational workflows which result in scalable, fault-tolerant, independent reproducible analytics pipelines with the use of this paradigm. This paradigm shift enables smallholder farmers and Agri-tech start-ups to use enterprise-grade computation (no need for a costly on-premises infrastructure) [60,61].

1.7.4 AI-Driven Decision-Support Systems for Precision Farm Management

AI is defining the backbone of contemporary agricultural decision-making via intelligent DSS. It uses a set of algorithms (ML, computer vision models and statistical inference) to produce actionable recommendation with consideration of farm-specific conditions that typically power these systems. For instance, AI-based DSS platforms are able to serve irrigation schedules in a completely dynamic manner harnessing the data from evapotranspiration, soil type and weather forecasts for highest water-use efficiency.

In addition to consumer technology rediscovery and commercialization (recommendation systems), recommender systems are at the moment also gaining momentum for agricultural use cases from among a myriad of seeds to fertilizer recommendation and pest control protocols. These systems determine the most optimal interventions per farming season, based on historical yield data, crop rotations and environmental constraints. High-performing DL framework such as CNNs and transformers allows us to put the visual/textual input in action: images of disease-affected crops or farmer notes, respectively, that make this model better at understanding intricate and unstructured data. We are also using ensemble models—the idea of combining predictions from multiple ML algorithms—to increase AI output robustness and generalizability [62,63]. As these AI systems are being integrated into mobile apps and IoT clouds, real-time user engagement without barriers and instant field updates become built-in too. Crucially, feedback tools facilitate ongoing learning from new data inputs that help the models get better over time, but also learn these changes in response to new environmental or market realities.

1.8 CONCLUSION: TECHNOLOGICAL CONVERGENCE AND FUTURE DIRECTIONS

Blockchain, edge/fog computing and cloud infrastructure going together with AI form a key inflection point of the data-centric agriculture evolution. All of these technologies together help to establish archetypes of more self-sufficient, responsive and adaptive agricultural systems, capable of handling dicey biophysical and socio-economic patterns. Future advancements will likely revolve around improving interoperability standards, user-friendly interfaces and ethical frameworks to better enable these technologies to be inclusive, visible, measure up with larger sustainability ambitions, that is, the UN Sustainable Development Goals (SDGs).

REFERENCES

1. Food and Agriculture Organization of the United Nations, "How to Feed the World in 2050," FAO, Rome, Italy, 2009. [Online] https://www.fao.org/4/ak542e/ak542e00.htm
2. P. Rajak, A. Ganguly, S. Adhikary, and S. Bhattacharya, "Internet of things and smart sensors in agriculture: Scopes and challenges," *Journal of Agriculture and Food Research*, vol. 14, p. 100776, 2023.
3. A. Cravero, S. Pardo, S. Sepúlveda, and L. Muñoz, "Challenges to Use Machine Learning in Agricultural Big Data," *Agronomy*, vol. 12, no. 3, p. 748, 2022.
4. R.P. Sishodia, R.L. Ray, and S.K. Singh, "Applications of Remote Sensing in Precision Agriculture: A Review," *Remote Sensing*, vol. 12, no. 19, p. 3136, 2020.
5. Z. Lian, Y. Zhan, W. Zhang, Z. Wang, W. Liu, and X. Huang, "Recent advances in deep learning-based spatiotemporal fusion methods for remote sensing images," *Sensors*, vol. 25, no. 4, p. 1093, 2025.
6. S. Liu, X. Wang, Y. Hou, G. Li, H. Wang, H. Xu, Y. Xiang, B. Tang, "Multimodal data matters: Language model pre-training over structured and unstructured electronic health records," *IEEE Journal of Biomedical and Health Informatics*, vol. 27, no. 1, pp. 504–514, 2022.
7. D. Marković, Z. Stamenković, B. Đorđević, and S. Ranđić, "Image processing for smart agriculture applications using cloud–fog computing," *Sensors*, vol. 24, no. 18, p. 5965, 2024.
8. K. Chatterjee, A. Singh, and Neha, "A blockchain-enabled security framework for smart agriculture," Computers and Electrical Engineering, vol. 106, p. 108594, 2023.
9. D. L. Kim, A. Patel, and M. Ghosh, "Digital Divide in Precision Agriculture: Bridging the Gap for Smallholder Farmers," *IEEE Transactions on Technology and Society*, vol. 4, no. 2, pp. 97–108, 2023.
10. S. S. Thenkabail, P. S. Roy, and R. S. Nash, "Advances in Hyperspectral Remote Sensing of Vegetation and Agricultural Crops," *Remote Sensing of Environment*, vol. 231, p. 111177, 2019.
11. L. García, L. Parra, J. M. Jimenez, J. Lloret, and P. Lorenz, "IoT-based smart irrigation systems: An overview on recent trends in sensors and IoT systems for irrigation in precision agriculture," *Sensors*, vol. 20, no. 4, p. 1042, 2020.
12. Y. Zhang and D. K. Ray, "Climate Information for Sustainable Agriculture: Past, Present, and Future," *Nature Climate Change*, vol. 12, pp. 741–750, 2022.
13. T. R. Tavares, J. P. Molin, S. H. Javadi, H. W. P. D. Carvalho, and A. M. Mouazen, "Combined use of vis–NIR and XRF sensors for tropical soil fertility analysis: Assessing different data fusion approaches," *Sensors*, vol. 21, no. 1, p. 148, 2020.
14. A. Kamilaris and F. X. Prenafeta-Boldú, "Deep Learning in Agriculture: A Survey," *Computers and Electronics in Agriculture*, vol. 147, pp. 70–90, 2018.
15. G. Mohyuddin, M. A. Khan, A. Haseeb, S. Mahpara, M. Waseem, and A. M. Saleh, "Evaluation of machine learning approaches for precision farming in smart agriculture systems: A comprehensive review," *IEEE Access*, vol. 12, pp. 60155–60184, 2024.
16. A. Roy, M. Z. Alam, and S. Bandyopadhyay, "Big Data Analytics in Smart Farming: A Review," *Journal of Ambient Intelligence and Humanized Computing*, vol. 12, pp. 4825–4845, 2021.

17. X. Zhang, Z. Cao, and W. Dong, "Overview of edge computing in the agricultural internet of things: Key technologies, applications, and challenges," *IEEE Access*, vol. 8, pp. 141748–141761, 2020.
18. T. Y. Kim and H. Park, "Noise-Resilient Sensor Data Analysis in Agriculture Using Autoencoder Models," *IEEE Access*, vol. 10, pp. 67824–67835, 2022.
19. A. Khattab, A. Abdelgawad, and K. Yelmarthi, "Design and implementation of a cloud-based IoT scheme for precision agriculture," in *Proc. 28th Int. Conf. Microelectronics (ICM)*, pp. 201–204, 2016.
21. A. Ghosal, A. Conti, L. Boaro, and M. Zorzi, "Energy-Efficient Communication in Internet of Agricultural Things," *IEEE Internet of Things Journal*, vol. 9, no. 3, pp. 1683–1697, 2022.
22. Y. Zhang, T. Li, M. Liu, and Y. Liu, "Fusion of Multi-Source Heterogeneous Data for Crop Monitoring Using Deep Learning: A Review," *Remote Sensing*, vol. 13, no. 5, p. 937, 2021.
23. M. T. Ghassemi, M. R. Karimian, and H. Kargarian, "Data Fusion and Deep Learning-Based Techniques for Agricultural Monitoring," *IEEE Access*, vol. 9, pp. 105586–105602, 2021.
24. S. K. Sharma and X. Wang, "Live Data Streaming and Big Data Analytics for Smart Farming: A Scalable and Fault-Tolerant Architecture," *IEEE Transactions on Industrial Informatics*, vol. 18, no. 1, pp. 180–188, 2022.
25. J. B. Ellis and M. R. McKinley, "Privacy and Data Ownership in Precision Agriculture: A Comparative Analysis," *Journal of Agricultural and Environmental Ethics*, vol. 34, no. 4, pp. 525–540, 2021.
26. F. Alaieri, "Precision Agriculture Based on Machine Learning and Remote Sensing Techniques," *Engineering, Technology & Applied Science Research*, vol. 14, no. 3, pp. 14206–14211, 2024. https://doi.org/10.48084/etasr.6986
27. P. Rajak, A. Ganguly, S. Adhikary, and S. Bhattacharya, "Internet of Things and Smart Sensors in Agriculture: Scopes and Challenges. *Journal of Agriculture and Food Research*, vol. 14, p. 100776, 2023. https://doi.org/10.1016/j.jafr.2023.100776
28. J. Wang, Y. Wang, G. Li, and Z. Qi, "Integration of remote sensing and machine learning for precision agriculture: A comprehensive perspective on applications," *Agronomy*, vol. 14, no. 9, p. 1975, 2024.
29. A. B. Arrieta, N. Díaz-Rodríguez, J. Del Ser, A. Bennetot, S. Tabik, A. Barbado, "Explainable artificial intelligence (XAI): Concepts, taxonomies, opportunities and challenges toward responsible AI," *Information Fusion*, vol. 58, pp. 82–115, 2020.
30. R. P. Saini, N. K. Verma, and S. K. Yadav, "Blockchain for Secure and Transparent Agricultural Supply Chain Management," *IEEE Access*, vol. 10, pp. 72089–72105, 2022.
31. P. Bharman, S. A. Saad, S. Khan, I. Jahan, M. Ray, and M. Biswas, "Deep learning in agriculture: A review," *Asian Journal of Research in Computer Science*, vol. 13, no. 2, pp. 28–47, 2022.
32. T. Rumpf, A. K. Mahlein, U. Steiner, E. C. Oerke, H. W. Dehne, and L. Plümer, "Early detection and classification of plant diseases with support vector machines based on hyperspectral reflectance," *Computers and Electronics in Agriculture*, vol. 74, no. 1, pp. 91–99, 2010.
33. S. Sladojevic, M. Arsenovic, A. Anderla, D. Culibrk, and D. Stefanovic, "Deep Neural Networks Based Recognition of Plant Diseases by Leaf Image Classification," *Computational Intelligence and Neuroscience*, vol. 2016, Article ID 3289801, 2016.
34. C. Liakos, P. Busato, D. Moshou, S. Pearson, and D. Bochtis, "Machine Learning in Agriculture: A Review," *Sensors*, vol. 18, no. 8, p. 2674, 2018.
35. A. Khanna and I. U. Hassan, "Deep reinforcement learning for smart irrigation," in *Data-Driven Farming*, pp. 259–279, Auerbach Publications 2025.

36. H. Brahimi, K. Boukhalfa, and A. Moussaoui, "Deep Learning for Tomato Diseases: Classification and Symptoms Visualization," *Applied Artificial Intelligence*, vol. 34, no. 1, pp. 14–36, 2020.
37. M. K. O. Aarif, A. Alam, and Y. Hotak, "Smart sensor technologies shaping the future of precision agriculture: Recent advances and future outlooks," *Journal of Sensors*, vol. 2025, no. 1, p. 2460098, 2025.
38. A. Kamilaris, A. Kartakoullis, and F. X. Prenafeta-Boldú, "A Review on the Practice of Big Data Analysis in Agriculture," *Computers and Electronics in Agriculture*, vol. 143, pp. 23–37, 2017.
39. M. Khaki and L. Wang, "Crop Yield Prediction Using Deep Neural Networks," *Frontiers in Plant Science*, vol. 10, p. 621, 2019.
40. M. You, Z. Li, J. Low, M. Lobell, and L. Zhang, "Deep Gaussian Process for Crop Yield Prediction Based on Remote Sensing Data," *Remote Sensing*, vol. 9, no. 12, p. 1248, 2017.
41. K. D. Patil and N. R. Kale, "Smart Irrigation System Using Machine Learning," *International Journal of Innovative Research in Electrical, Electronics, Instrumentation and Control Engineering*, vol. 6, no. 3, pp. 132–137, 2018.
42. R. S. Ferentinos, "Deep Learning Models for Plant Disease Detection and Diagnosis," *Computers and Electronics in Agriculture*, vol. 145, pp. 311–318, 2018.
43. J. Dash and P. J. Curran, "The MERIS Terrestrial Chlorophyll Index," *International Journal of Remote Sensing*, vol. 25, no. 23, pp. 5403–5413, 2004.
44. A. K. Mahlein, U. Steiner, C. Hillnhütter, H. W. Dehne, and E. C. Oerke, "Hyperspectral imaging for small-scale analysis of symptoms caused by different sugar beet diseases," *Plant Methods*, vol. 8, no. 1, p. 3, 2012.
45. J. Redmon and A. Farhadi, "YOLOv3: An Incremental Improvement," *arXiv preprint arXiv:1804.02767*, 2018.
46. A. K. Srivastav and P. Das, "Edge computing and AI in agricultural IoT," in *Biotechnology and IoT in Agriculture and Food Production": Green Innovation*, pp. 187–200, 2025.
47. D. J. Mulla, "Twenty Five Years of Remote Sensing in Precision Agriculture: Key Advances and Remaining Knowledge Gaps," *Biosystems Engineering*, vol. 114, no. 4, pp. 358–371, 2013.
48. R. R. Thenkabail, P. L. Gamage, and V. G. Smakhtin, "The Use of Remote Sensing Data for Drought Assessment and Monitoring in Southwest Asia," *Research Report 85*, International Water Management Institute, Colombo, Sri Lanka, 2004.
50. S. P. Dhoubhadel, "Precision agriculture technologies and farm profitability," *Journal of Agricultural and Resource Economics*, vol. 46, no. 2, pp. 256–268, 2021.
51. S. Bauer-Marschallinger, W. A. Dorigo, M. Vreugdenhil, R. Rötzer, K. Paulik, and W. Wagner, "Toward Global Soil Moisture Monitoring with Sentinel-1: Harnessing Assets and Overcoming Obstacles," *IEEE Transactions on Geoscience and Remote Sensing*, vol. 57, no. 1, pp. 520–539, 2019.
52. T. Sankaran, J. Khot, and S. Jha, "Low-Altitude, High-Resolution Aerial Imaging Systems for Row and Field Crop Phenotyping: A Review," *European Journal of Agronomy*, vol. 70, pp. 112–123, 2015.
53. M. Casino, T. K. Dasaklis, and C. Patsakis, "A Systematic Literature Review of Blockchain-Based Applications: Current Status, Classification and Open Issues," *Telematics and Informatics*, vol. 36, pp. 55–81, 2019.
54. S. Mondal, R. Bose, D. Saha, and P. Dasgupta, "Blockchain Inspired RFID-Based Information Architecture for Food Supply Chain," *IEEE Internet of Things Journal*, vol. 6, no. 3, pp. 5803–5813, 2019.
55. F. Bonomi, R. Milito, J. Zhu, and S. Addepalli, "Fog Computing and Its Role in the Internet of Things," in *Proceedings of 1st MCC Workshop on Mobile Cloud Computing*, Helsinki, Finland, August 2012, pp. 13–16.

56. H. Gupta, A. V. Dastjerdi, S. K. Ghosh, and R. Buyya, "iFogSim: A Toolkit for Modeling and Simulation of Resource Management Techniques in the Internet of Things, Edge and Fog Computing Environments," *Software: Practice and Experience*, vol. 47, no. 9, pp. 1275–1296, 2017.
57. N. Gorelick, M. Hancher, M. Dixon, S. Ilyushchenko, D. Thau, and R. Moore, "Google Earth Engine: Planetary-scale geospatial analysis for everyone," *Remote Sensing of Environment*, vol. 202, pp. 18–27, 2017.
58. M. Zaharia, R. S. Xin, P. Wendell, T. Das, M. Armbrust, A. Dave, X. Merk, J. Rosen, S. Venkataraman, M. J. Franklin, A. Ghodsi, J. Gonzalez, S. Shenker, I. Stoica, "Apache Spark: A unified engine for big data processing," *Communications of the ACM*, vol. 59, no. 11, pp. 56–65, 2016.
59. M. H. Saleem, J. Potgieter, and K. M. Arif, "Automation in agriculture by machine and deep learning techniques: A review of recent developments," *Precision Agriculture*, vol. 22, no. 6, pp. 2053–2091, 2021.
60. P. Singh, V. Yadav, and P. Goel, "Recommender Systems in Agriculture: A Review," in *Proceedings of IEEE 4th International Conference on Computing, Communications and Networking Technologies (ICCCNT)*, Tiruchengode, India, 2013, pp. 1–5.
61. M. A. Khan, F. Zhang, and T. Wang, "Multisensor Fusion for Early Plant Disease Detection Using IoT and Hyperspectral Imaging," *IEEE Internet of Things Journal*, vol. 9, no. 18, pp. 17023–17036, 2022.
62. A. Sinha, R. Banerjee, and M. Pathak, "Crop Yield Prediction Using Integrated Remote Sensing and Weather Data: A Machine Learning Approach," *Computers and Electronics in Agriculture*, vol. 203, p. 107407, 2023.
63. L. Torres, S. Nakamoto, and D. Kim, "Blockchain-Enabled Traceability in Agriculture: Enhancing Transparency in the Coffee Supply Chain," *IEEE Access*, vol. 11, pp. 50112–50125, 2023.

2 AI-Powered Analytics and HCI Methods for Multimodal Data Fusion in Precision Agriculture

Sweta Sinha, Om Prakash, and Ujjwal Biswas

2.1 INTRODUCTION

Agriculture, a sector historically driven by experience and old tradition, is undergoing a digital revolution to improve agriculture (Yewle, Mirzayeva, & Karakuş, 2025). The emergence of precision agriculture (PA) is a significant stride toward sustainable food production, enabled by innovations in artificial intelligence (AI), machine learning (ML), human-computer interaction (HCI), and the Internet of Things (IoT) (Goriparthi, 2022). These advancements of AI, ML, HCI, and IoT are playing a major role in real-time monitoring, predictive insights, and actionable recommendations that were once unattainable using conventional methods (Garnot, Landrieu, & Chehata, 2022; Yewle et al., 2025).

A central feature of this transformation is the integration of multimodal data. It is a combination of heterogeneous inputs such as satellite imagery, sensor readings, climate data, textual field reports, and farmer feedback (Ghamati, Banitalebi Dehkordi, & Zaraki, 2025; Yewle et al., 2025). However, harnessing this data efficiently poses challenges in terms of data heterogeneity, fusion complexity, and usability (Bhatti et al., 2025). This is where AI-powered analytics and HCI frameworks emerge as game changers, offering intelligent systems capable of processing vast and diverse datasets to deliver precise and context-sensitive agricultural interventions.

Recent research has highlighted the importance of multimodal learning and adaptive data fusion models in tackling the intricacies of subfield-level predictions (Goriparthi, 2022). For instance, the Multi-Modal Gated Fusion (MMGF) model demonstrates how adaptive fusion of remote sensing, weather, soil, and topographic data can significantly enhance crop yield prediction across diverse geographies and crop types. Similarly, deep ensemble architectures like RicEns-Net effectively integrate data from Synthetic Aperture Radar (SAR; Yewle et al., 2025), optical satellites, and meteorological stations to achieve high-accuracy predictions, particularly in climate-sensitive regions. In addition, AI-powered decision support systems (DSS) (Goriparthi, 2022) are increasingly being integrated with user-centric HCI methodologies to bridge the gap between complex AI models and the real-world needs of farmers. These systems use ML to forecast irrigation schedules, pest infestations, and

DOI: 10.1201/9781003637264-3

nutrient deficiencies, providing intuitive interfaces and alerts that facilitate timely action. NLP and computer vision technologies are also leveraged to extract insights from non-numerical sources like image datasets and farmer notes, further enriching the decision-making ecosystem.

Despite these advancements, the effective implementation of AI in precision farming hinges on resolving challenges such as data interoperability, model explainability, and farmer engagement. Therefore, integrating HCI principles such as multimodal feedback, accessible interfaces, and contextual personalization into the design of AI systems becomes essential. To address the research gap and challenges, we have proposed three research questions (RQs).

RQ1: What are the novel data fusion techniques in agriculture that depend on various important and sensitive applications?

RQ2: How does multimodal data in agriculture affect the performance of the predictive model?

RQ3: How does HCI make intelligent systems accessible and practical for farmers in agricultural applications?

The RQs review current methodologies and identify key technical and usability challenges. We propose a framework for integrating these technologies to foster sustainable, data-driven agricultural ecosystems. Doing so ensures that sophisticated algorithms translate into practical tools for end-users (farmers), agronomists, and rural communities. This chapter explores the convergence of multimodal data fusion, AI analytics, and HCI to unlock the full potential of smart agriculture.

2.2 OVERVIEW AND RESEARCH MOTIVATION

Modern agriculture faces complex challenges: climate variability, resource constraints, and the demand for sustainable productivity (Setiyawati, Purnomo, & Mailoa, 2022). Addressing these issues requires more than traditional farming methods to improve the quality of agriculture. It demands intelligent systems capable of processing and integrating diverse forms of data (Yewle et al., 2025). Recent studies highlight the importance of multimodal data fusion, where satellite imagery, weather data, IoT sensor readings, and soil information are combined to support smarter, more responsive decision-making in the field. Nowadays, innovative farming, a subset of PA, encompasses a range of applications that leverage these technologies to optimize agricultural productivity, resource efficiency, and sustainability. The detailed applications are shown in Figure 2.1.

Innovative AI models like MMGF (Yewle et al., 2025) and RicEns-Net have demonstrated that adaptive, feature-level fusion significantly enhances the accuracy of yield prediction, even at the subfield level. These models outperform traditional approaches by dynamically learning the relevance of each data source based on region, crop type, and season, achieving results such as an R^2 of 0.80 in field-level evaluations. In the same direction, accuracy alone is not enough. This highlights the transformative potential of AI-powered analytics in PA to use multimodal data to improve agricultural profit. To address the present challenges, the comprehensive

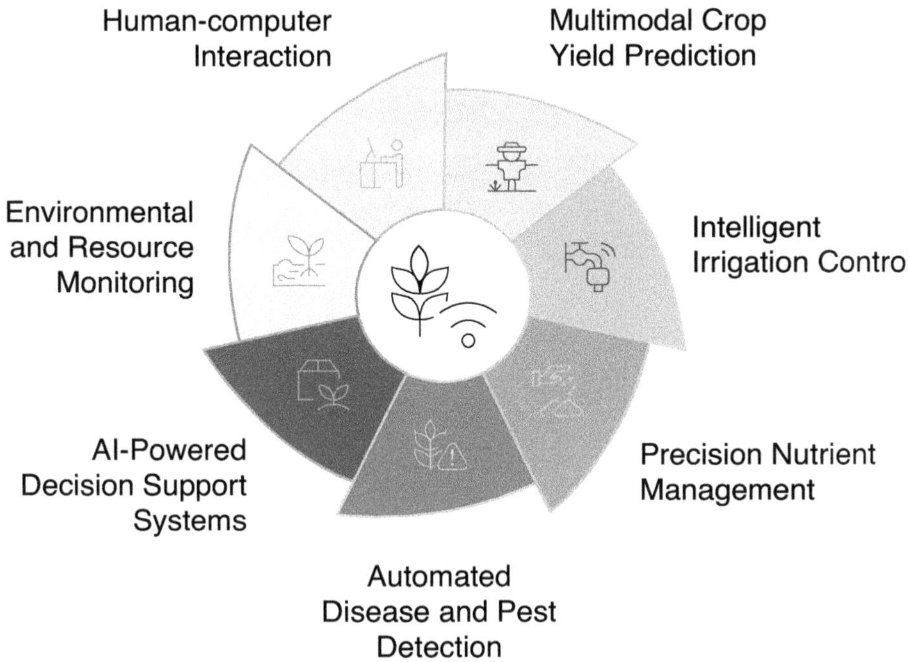

FIGURE 2.1 The range of innovative farming applications encompassed as a subset of precision agriculture.

framework (as shown in Figure 2.2) is very important. The data fusion in agriculture depends on various important and sensitive applications. The details of such applications are discussed in the next section.

However, studies also emphasize the role of HCI in making these intelligent systems accessible and practical for farmers (Ghamati et al., 2025). IoT-based platforms for irrigation and fertilizer management, paired with real-time alerts and intuitive interfaces, illustrate how HCI bridges the usability gap, ensuring that data-driven insights lead to real-world action (Goriparthi, 2022). The convergence of AI and HCI with multimodal fusion thus emerges as a key enabler of next-generation PA. It is not only data-rich and predictive but also responsive, farmer-friendly, and scalable for designing and deploying an intelligent system for agriculture.

Figure 2.3 shows how AI and HCI achieve the goal of designing the next-generation agricultural system and its practical uses. This study investigates the convergence of computer and digital services, predictive modeling, and user interaction in the agriculture fields (see top part of Figure 2.3). It emphasizes how new technologies are transforming farming methods, allowing for data-driven decision-making, and promoting more efficient and sustainable agricultural systems. Farmers may improve overall farm management by using the power of predictive models and intuitive user interfaces. The methodology (see middle part of Figure 2.3) presents a brief review of agricultural methodology, with a focus on the integration of AI, HCI, ML, user interface design, user-centered design (UCD), and assessment procedures.

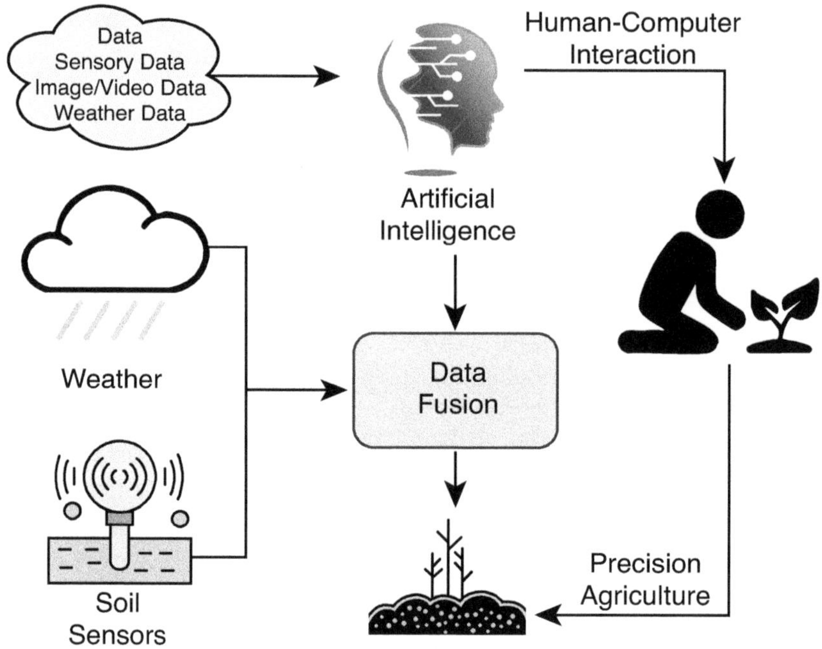

FIGURE 2.2 Comprehensive framework showing multimodal data, data fusion, artificial intelligence, machine learning, human-computer interaction, and precision agriculture integration.

These techniques seek to optimize agricultural operations, increase efficiency, and improve the user experience for farmers and other agricultural stakeholders. The combination of these technologies foreshadows a future of data-driven, sustainable, and user-friendly agriculture solutions. The key outcomes describe the expected results of a testing scenario, with an emphasis on performance compared to cutting-edge systems and the evaluation of the System Usability Scale (SUS) score (see bottom part of Figure 2.3). The purpose is to present a clear picture of what success means in terms of system performance, user pleasure, and general usability. We want to show that our system meets or surpasses current industry standards while also receiving a high SUS score, which indicates a favorable user experience.

2.3 APPLICATIONS OF SMART FARMING

AI, ML, multimodal data fusion, and HCI are transforming traditional agricultural techniques into data-driven ecosystems (Ghamati et al., 2025). The literature analysis highlights significant uses of smart farming applications to address RQ1. The combination of AI, multimodal data fusion, and HCI is not only transforming old agricultural techniques but also paving the path for a more sustainable and efficient future in agriculture. The applications listed above highlight smart farming's ability to address current agricultural difficulties and improve global food security.

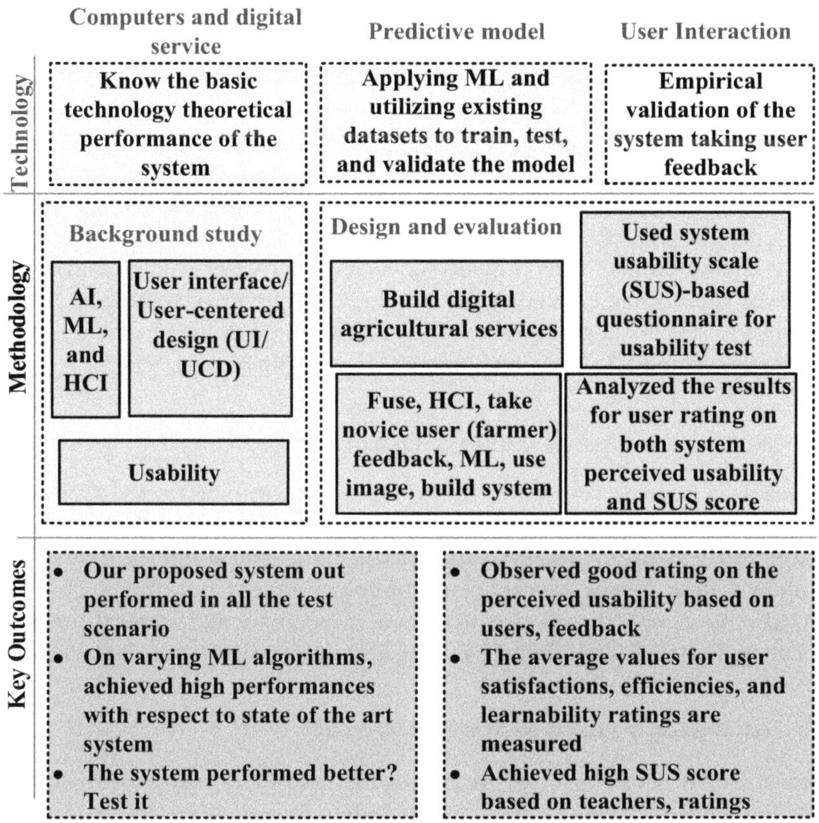

FIGURE 2.3 Background of AI and HCI with multimodal fusion in digital services, research methodology, and outcomes.

2.3.1 Multimodal Crop Yield Prediction

A significant application of smart farming is accurate and fine-grained crop yield prediction through the integration of heterogeneous data. Traditional yield forecasting models often fall short due to their reliance on single-modal or static datasets. Advanced AI models such as the MMGF and RicEns-Net overcome this by fusing data from Sentinel satellite imagery, SAR data, weather conditions, and static soil maps. These models use feature-level fusion and adaptive weighting strategies to capture spatio-temporal variations and have demonstrated superior performance, achieving R^2 values up to 0.80 at the field level, significantly surpassing conventional methods (Yewle et al., 2025).

2.3.2 Intelligent Irrigation Control

Water scarcity and inefficient irrigation practices are persistent challenges in agriculture. Smart irrigation systems, guided by IoT-based soil moisture sensors and real-time weather data, allow for dynamic water management. These systems automatically control irrigation pumps based on soil hydration levels, thereby reducing water waste and maintaining optimal crop health. Such real-time, responsive systems are vital for regions experiencing irregular rainfall and prolonged dry spells (Goriparthi, 2022).

2.3.3 Precision Nutrient Management

Smart farming enables precise nutrient application by analyzing soil nutrient levels such as nitrogen (N), phosphorus (P), and potassium (K) and environmental conditions. AI-powered platforms provide site-specific fertilizer recommendations that reduce over-application, enhance plant nutrition, and improve soil health. Data collected via IoT sensors is processed and visualized through user-friendly interfaces, enabling farmers to make informed input decisions (Bhatti et al., 2025).

2.3.4 Automated Disease and Pest Detection

Deep learning models, especially convolutional neural networks (CNNs), have shown strong capabilities in detecting crop diseases and pest infestations (Bhatti et al., 2025). By analyzing images captured via unmanned aerial vehicles (UAVs) or smartphones, these models classify plant conditions with high accuracy. This enables early intervention, reduces reliance on manual inspections, and supports timely agrochemical applications. Reported accuracies often exceed 90% in real-world settings.

2.3.5 AI-Powered Decision Support Systems

DSS are central to smart farming, integrating multiple data streams to provide actionable recommendations for key decisions, including sowing schedules, irrigation, pest control, and harvesting. These systems employ AI model to process multimodal data and present outputs through interactive dashboards, visual alerts, and mobile applications. Incorporation of HCI principles ensures accessibility across user demographics (Goriparthi, 2022).

2.3.6 Environmental and Resource Monitoring

Smart farming also plays a critical role in environmental monitoring and sustainable resource use. Sensor networks track parameters such as soil pH, ambient temperature, humidity, and evapotranspiration. These data points help forecast climate impact on crop performance and guide mitigation strategies. Remote sensing, combined with sensor data, enhances land-use monitoring and supports eco-friendly farming practices (Yewle et al., 2025).

2.3.7 Human-Computer Interaction

In recent years, HCI has become a major interface design application for effective and easy user interaction (Bhattacharya, Biswas, Damkondwar, & Yadav, 2023; Bhattacharya, Shah, Kumar, & Biswas, 2021; Biswas & Bhattacharya, 2022, 2024). The agricultural sector has seen significant technological advancements, leading to a greater reliance on digital tools and interfaces. This shift has transformed how farming activities are managed and executed. By applying HCI principles to agricultural user interfaces, the industry can make technology more intuitive and user-friendly (Setiyawati et al., 2022). Focusing on usability, accessibility, and overall effectiveness can enhance farming efficiency, enabling farmers to better manage resources and make informed decisions (Gajbhiye, Agrawal, Jha, Kumar, & Raghuwanshi, 2023). HCI also helps bridge the gap between traditional farming practices and modern technology (Gutiérrez, Htun, Schlenz, Kasimati, & Verbert, 2019; Kim, Lee, Hong, & Yook, 2022).

For example, UCD ensures that digital tools are tailored to the specific needs and circumstances of farmers, which increases their acceptance and effectiveness (Setiyawati et al., 2022). By making digital technology more accessible, even for those with limited technical skills, HCI can help spread the benefits of technological advancements in agriculture. Research supports this idea, highlighting the potential of HCI to transform agricultural practices by creating user interfaces that meet the diverse needs of farming communities (Gutiérrez et al., 2019). However, designing systems and/or interfaces for farmers (laypersons) is very challenging. As the agricultural field increasingly adopts digital solutions, understanding and optimizing HCI design becomes more important (Setiyawati et al., 2022). Agricultural user interfaces serve as a critical link between farmers and complex technological processes. Digital tools and services significantly impact crop management and data-driven decision-making in PA (Gutiérrez et al., 2019). However, the unique challenges and specific conditions of agricultural environments require a tailored approach to UCD. In summary, there are many applications where data fusion is used. Figure 2.4 shows that data fusion is used in PA to make an effective contribution. However, the mentioned applications (reported in Section 3) play a major role in agriculture. To address the present research challenges, our contribution helped in agricultural precision. The RQ2 investigates the impact of using multimodal data sources on the performance of predictive models in the agricultural domain. It investigates how the integration of many data types, including weather patterns,

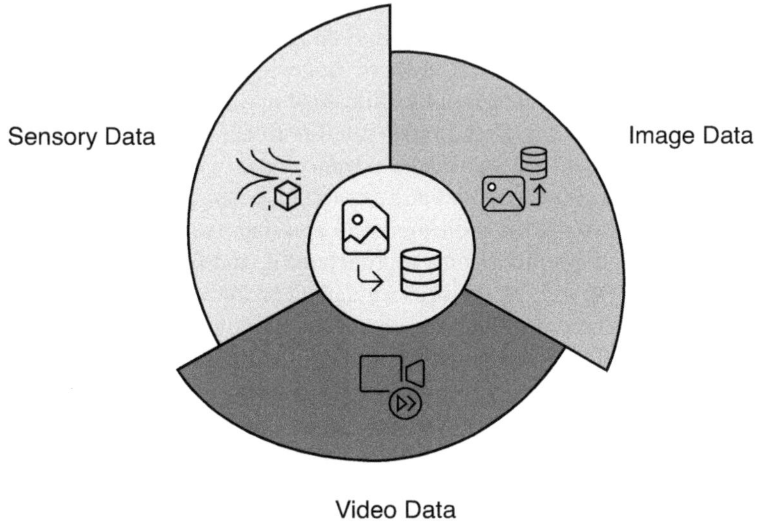

FIGURE 2.4 The major multimodal data fusion in innovative and predictive digital services.

soil composition, satellite imaging, and sensor data, can increase the accuracy and reliability of agricultural predictions, resulting in better decision-making and resource allocation. In the next section, we answer RQ2 in detail.

2.4 MULTIMODAL DATA AND PREDICTIVE MODEL IN PRECISION AGRICULTURE

Although digital technologies have immense potential in agriculture, the research stresses the significance of examining the ethical, social, and economic aspects of their use (Gajbhiye et al., 2023; Setiyawati et al., 2022). This research will enhance digital agriculture's resilience, sustainability, and productivity. To fully realize the promise of these new technologies and concepts, agricultural field workers require data fusion to build intelligent systems.

In PA, AI-powered DSS, ML, and Hyperspectral imaging (HSI) techniques have significant potential for enhancing decision-making and optimizing resource use with multimodal data fusion techniques (see Table 2.1). The multimodal data fusion techniques have become an essential tool in a number of industries, including agriculture, environmental monitoring, and medical diagnostics. We will discuss multimodality and data fusion, as well as how large language models (LLMs) might improve HSI applications. Furthermore, we will explore the field's issues, current developments, and future opportunities for HSI research. The development of an AI-powered system forecasts the most appropriate ML approaches for agricultural datasets.

Table 2.1 shows the multimodality and data fusion objective, techniques used, actual dataset used, preprocessing technique used, and finally their results and actual limitations.

Table 2.2 shows the detailed usages of the key image and video data. The application of multimodal data fusion methodologies increases the accuracy of rice crop production forecasts, providing farmers and stakeholders with valuable insights for better decision-making. A review of fusion techniques highlights the possibility for improved agricultural monitoring and decision-making. Using ML and deep learning in agricultural production at three stages: pre-harvest, harvest, and post-harvest, RicEns-Net represents a notable milestone in the field of agricultural prediction models. Combining optical and radar satellite time series data with multimodal temporal attention models is a significant advancement in crop mapping techniques. The study aims to enhance the model and explore its potential for PA applications. In summary, Table 2.2 reports the multimodality and data fusion objectives, strategies used, real dataset used, preprocessing approach used, and final outcomes and limits.

Table 2.3 reports the uses of the sensory data. The Personalized Large Language Model (PLLM) is tailored for dynamic, user-specific human-robot interaction (HRI) and HCI scenarios. The deep learning model, DeepG2P, predicts crop yield by modeling complex genome-to-environment (G×E) and genome-to-genome (G×G) interactions using multimodal inputs such as genomics, weather, soil, and field management data. To enhance productivity and reduce human labor, researchers develop and evaluate a multimodal PA system that integrates IoT with ML and DL techniques for smart irrigation, fertilizer dose recommendation, crop disease diagnosis, and crop damage prediction. Table 2.3 summarizes the multimodality and data fusion aims, techniques, real dataset, preparation approach, final results, and limitations.

TABLE 2.1
The Summary of the Key Multimodal Data

Reference	Objective	Techniques	Dataset	Preprocessing	Result	Limitations
Goriparthi (2022)	To explore the integration of AI-powered Decision Support Systems (DSS) with machine learning techniques in precision agriculture to enhance decision-making and optimize resource use.	Machine Learning Models: SVM, RF, ANN, CNN, RNN (LSTM), RL Tools: TensorFlow, PyTorch, Apache Kafka (real-time stream processing)	500+ GB multimodal data Sources: IoT sensor networks (soil, environment), satellite imagery, public databases (CropX, FAOStat) 10,000+ observations across seasons and regions	Missing values handled using KNN imputation Outliers removed (3σ rule) Satellite data preprocessed with NDVI Normalization [0,1] Image calibration and atmospheric correction	Crop Disease Detection (CNN): 94.3% accuracy Yield Prediction (RF, SVM): MAPE 7.1% (wheat), 7.8% (maize) Irrigation Mgmt (LSTM): MAE 5.8%, water savings 16.5% Pest Control (RL): 91.2% accuracy, pesticide reduction 22%	Data quality and availability issues High computational demands Lack of IoT infrastructure in rural areas Limited technical knowledge among farmers Need for policy and education support for scalability
Bhatti et al. (2025)	To provide a comprehensive tutorial-style review of hyperspectral imaging (HSI) integrated with AI and deep learning. Discuss multimodality, data fusion, and the role of LLMs in enhancing HSI applications. Highlight challenges, trends, and future directions.	Deep Learning Models: CNN, RNN, LSTM, GRU, GANs, Autoencoders, DBNs, Transformers HSI tasks: Classification, Feature extraction, Denoising, Spectral unmixing, Dimensionality reduction, Data augmentation, Target detection, Change detection, Spatial-spectral attention Multimodal fusion with LiDAR, SAR, RGB, thermal imagery	Public datasets: AVIRIS, Hyperion, MODIS, CASI, Sentinel-2, ROSIS, Headwall Hyperspec No single experimental dataset (review paper)	Band selection (ranking, clustering, sparsity-based) Dimensionality reduction: PCA, ICA, Autoencoders Noise removal Spectral unmixing (LMM, LSU) Sub-pixel classification Data compression (wavelet, DCT) Atmospheric correction, Normalization	DL models outperform traditional ML in HSI tasks CNNs and hybrid CNN-LSTM excel in classification and spatial-spectral feature extraction GANs effective in data augmentation and denoising Transformers emerging for spectral-spatial modeling	Requires large labeled datasets High computational cost and memory usage Limited interpretability Curse of dimensionality and redundancy issues Real-time deployment challenges Lack of standardization in fusion techniques and benchmarking

(*Continued*)

TABLE 2.1 (*Continued*)
The Summary of the Key Multimodal Data

Reference	Objective	Techniques	Dataset	Preprocessing	Result	Limitations
Elbasi et al. (2024)	To develop an AI-powered system that predicts the most suitable ML algorithm for agricultural datasets using a pre-trained model, reducing time and computational cost.	Supervised ML: BN, NBC, MP, DT, RF, etc. Pre-trained AI models Cross-validation, hyperparameter tuning, performance logging	CropRec (2,200 samples, 7 features, 22 labels) SoyDis (1,600 samples, 64 features, 19 labels) PlaSpe (1,985 samples, 35 features, 99 labels)	Encoding categorical variables Normalization Metadata feature extraction Handling missing values Data splitting (60/40 to 90/10) Feature selection	DT: 89.38% RF: 87.61% RT: 84.27% MP (PlaSpe): 94.29% BN and NBC (CropRec): 99.49%, 99.44%	Not applicable for real-time predictions Accuracy depends on training data quality Limited scalability to unseen data types without feature re-engineering

TABLE 2.2
The Summary of the Key Image and Video Data

Reference	Objective	Techniques	Dataset	Preprocessing	Result	Limitations
Yewle et al. (2025)	To develop a deep ensemble model (RicEns-Net) that improves rice crop yield prediction accuracy using multi-modal data fusion techniques.	Deep Ensemble Model (RicEns-Net) using CNN, MLP, DenseNet, AutoEncoder Weighted average based on validation error Comparison with SOTA models	EY Open Science Challenge 2023 dataset 557 rice farm sites from An Giang, Vietnam Sentinel-1, 2, 3, NASA GES DISC, and field data	Feature engineering (from 100+ to 15 features) Data normalization, chi-square test, correlation filtering Train-test split (3:1)	Best model (RicEns-Net) MAE: 341 kg/ha, RMSE: 436.26 kg/ha Adjusted R^2: 0.589 Outperformed CNN, SVR, and EY competition models	Underestimation in high-yield zones Susceptibility to cloud cover in remote sensing data Limited generalizability across regions/crops
Garnot et al. (2022)	To develop and evaluate multimodal temporal attention-based models for crop mapping using optical and radar satellite time series, and to assess different fusion strategies across various mapping tasks.	Temporal attention networks using four fusion strategies: Early, Mid, Late, and Decision fusion Enhancement techniques: Auxiliary Supervision and Temporal Dropout Core models: PSE+LTAE (classification), U-TAE (segmentation), PaPs (panoptic segmentation)	PASTIS-R: A large-scale open-access dataset combining Sentinel-2 optical and Sentinel-1 radar time series (ascending and descending orbits) for 2,433 image patches Includes annotations for crop type classification, semantic segmentation, and panoptic segmentation	Temporal interpolation (only in Early Fusion) Radar preprocessing limited to backscatter coefficient extraction (VV, VH, VV/VH) No speckle filtering or terrain correction Cloud cover robustness tested via simulated optical data removal	Late Fusion with auxiliary supervision and temporal dropout achieved best overall accuracy Mid Fusion balanced performance and computational efficiency Decision Fusion most robust to cloud obstruction All fusion strategies outperformed single-modality baselines	Evaluation limited to crops in French regions (PASTIS dataset) No true cloud cover diversity beyond 2019 France Interferometric and polarimetric radar data not utilized High memory and computation demands for late and decision fusion methods, especially in panoptic segmentation

(Continued)

TABLE 2.2 (*Continued*)
The Summary of the Key Image and Video Data

Reference	Objective	Techniques	Dataset	Preprocessing	Result	Limitations
Mohyuddin et al. (2024)	To comprehensively review and evaluate ML and DL methods used in smart agriculture and precision farming across pre-harvest, harvest, and post-harvest stages.	ML: SVM, KNN, RF, DT, NB, ANN, IBM DL: CNN, ResNet, AlexNet, VGG, Lenet, YOLO, LSTM Image processing, clustering, fuzzy logic	Datasets from Plant Village, UCI, hyperspectral or multispectral images, IoT sensor networks, drone, and UAV imaging datasets	Image segmentation, background removal, lesion detection, GLCM feature extraction, dimensionality reduction (PCA, LDA), thresholding, k-means	Achieved high accuracy in disease classification (up to 99.53%), tomato disease detection (97.5%), and pest detection (up to 99%) DL methods outperformed traditional ML	Lack of generalizability across crops or environments Small datasets for some crops High computational cost of DL Need for expert-labeled data Limited real-time field deployment

TABLE 2.3
The Summary of the Key Sensory Data

Reference	Objective	Techniques	Dataset	Preprocessing	Result	Limitations
Ghamati et al. (2025)	To develop a Personalized Large Language Model (PLLM) that adapts LLMs for dynamic, user-specific human-robot interaction (HRI) and human-computer interaction (HCI) applications.	Personalization via structured prompts EEG signal processing with LLM Streamlit-based interaction layer OpenAI API for LLM inference Integration with a Nao robot for real-time interaction	NeuroSense EEG dataset (collected using Muse 2, a low-cost EEG device with four electrodes)	Conversion of EEG/CSV/PDF data into text format Tokenization and chunking to fit OpenAI's token limit Real-time EEG data streamed from Muse 2 to the agent Prompt engineering for model interpretation	PLLM successfully interpreted EEG data to classify emotional states (e.g., 'neutral,' 'happy') Delivered spoken feedback via robot Demonstrated adaptability in dynamic environments	Limited generalization to diverse datasets Muse 2's low spatial resolution may limit accuracy Ethical/privacy issues with EEG data usage High dependency on cloud-based models for processing and cost implications
Sharma, Partap, Balaguer, Malvar, and Chandra (2022)	To develop a DL model (DeepG2P) that predicts crop yield by modeling genome-to-environment (G×E) and genome-to-genome (G×G) interactions using multimodal data (genomics, weather, soil, field management).	Multimodal DL with separate modules for genome, weather, soil, management data NLP-style modeling of DNA using CNNs + positional encoding Cross-attention mechanism for G×E interactions Recursive Feature Elimination and Mutual Information for SNP selection Fully connected fusion module with dropout layers	Genomes to Fields (G2F) dataset: 2,000 maize seed varieties, 36k filtered data points from 32 fields across 18 US states (2014–2017) Includes SNPs, weather, soil, management data, and yield SNPs reduced from 955,690 → 20,372 → top 100	Missing data removed; filtered dataset used Weather time series padded/averaged over 5-day windows SNPs selected via LightGBM + mutual information Standard scaling applied to features	Outperforms baselines (GEBlup, AutoCGM, CNN-21) in environment split Higher Pearson correlation and lower RMSE variance Best generalization to unseen environments due to G×E attention module	Only 100 SNPs used, limiting hybrid split performance Sparse environmental data limits generalization High-dimensional SNP data causes computational constraints Need for more diverse datasets and SNP selection across phenotypes

(Continued)

TABLE 2.3 (*Continued*)
The Summary of the Key Sensory Data

Reference	Objective	Techniques	Dataset	Preprocessing	Result	Limitations
Garg, Pundir, Jindal, Saini, and Garg (2021)	To design and study a multimodal precision agriculture system integrating IoT and ML/DL for smart irrigation, fertilizer recommendation, crop disease detection, and crop damage prediction to improve yield and reduce labor.	IoT sensors for soil moisture and NPK content ML: Random Forest, LGBM, XGBoost, Decision Tree, KNN DL: VGG16, ResNet50, DenseNet121 (transfer learning for disease detection)	Crop Damage Prediction: Analytics Vidhya ML in Agriculture dataset (88,858 rows, 10 columns)	Crop Damage: Missing values filled (-999), lag and rolling features generated	LGBM achieved highest accuracy (94%), outperforming XGBoost and RF KNN had poorest performance	Poor performance for minority classes due to imbalance No oversampling (SMOTE) No hyperparameter tuning (grid search, evolutionary algorithms) due to constraints Deployment in real fields not addressed

2.5 DEPLOYING THE MULTIMODAL DATA AND HUMAN-COMPUTER INTERACTION

There are many research challenges and limitations mentioned in the previous section (see Section 2.4). In this section, we will answer RQ3. HCI is crucial in making intelligent agricultural technologies more accessible and useful to farmers. The design of agricultural analytics helped the farmers to get real-time feedback and suggestions. By emphasizing UCD, HCI ensures that these technologies are intuitive, simple to use, and tailored to farmers' specific requirements. This strategy boosts adoption, improves efficiency, and allows farmers to make better decisions using data-driven insights, resulting in more sustainable and productive farming practices. To get into the detailed utilization of predictive models, multimodal data, and HCI design techniques, we report a novel research process. The process consists of building the system and evaluating the design.

Algorithm 1: Data Preparation

1. Input: Raw data (e.g., rainfall, thunderstorm, wind speed)
2. Output: Prepared data (cleaned and ready for prediction)
3. for each feature in Raw data do
4. there are missing values in the feature Fill missing values with the average of that feature for each feature in Raw data do
5. Adjust values to bring all features onto a similar scale if additional features are helpful then
6. Add derived features, such as:
    - Average wind speed over a specific period
    - Weekly or monthly rainfall totals

    Remove any features that are not needed for prediction return prepared data

2.5.1 Building Predictive Model

We used data preprocessing, feature selection, and feature building to improve model generalizability and prediction accuracy. We have also designed three algorithms. Algorithm 1 is designed to demonstrate the preprocessing techniques we followed. Algorithm 2 helped to build the predictive model, a CNN that performs better compared to other models. Algorithm 3 reported the data fusion used for the multimodal datasets and helped to build the predictive model CNN that performs better compared to other models.

Algorithm 2: Weather Prediction with CNN

1. Input: Prepared data (cleaned meteorological data)
2. Output: Weather criticality classification (C: Critical, LC: Likely Critical, N: Normal) [1] Load trained CNN model for each entry in prepared data do

3. Calculate probability of severe weather using CNN model for each probability score do
4. probability > 90% Classify as Critical (C) 60% < probability ≤ 90% Classify as Likely Critical (LC) else
5. Classify as Normal (N) return weather criticality classification list

The CNN model beat two classic ML models, Simple Neural Network (SNN) and Simple Recurrent Neural Network (SRNN). The models will predict weather conditions and be classified according to their role and importance. Our goal is to recommend ML models that predict weather conditions. The suggested approach is accurately predicted and classified, as evidenced by encouraging results.

Algorithm 3: Data Fusion

1. Input: Sensor data S, Satellite imagery I, Textual input T, Weather data W
2. Output: Actionable Insights
3. Initialize sources S, I, T, W
4. AcquireData(S, I, T, W)
5. PreprocessData(S, I, T)
6. Fs ← sensorFeatureExtraction(S)
7. FI ← imageFeatureExtraction(I)
8. FT ← textuaFeatureExtraction(T)
9. F ← FuseFeatures(FS, FI, FT)
10. P ← PredictOutcomes(F)
11. A ← GenerateInsights(P)
12. InteractWithUser(A)
13. if Feedback (Uf) available then
14. UpdateModels(Uf);
15. Generate audio alert continuous buzzing sound for 30 seconds;
16. Terminate the session;
17. return weather criticality classification list

2.5.2 Detailed Dataset and Visualization

The box plot depicts the distribution of thunderstorms, windspeed in kilometers per hour (km/h), and rainfall per millimeter (mm) (see Figure 2.5). Thunderstorm, being binary, exhibits minimal variation. We observed that the wind speed dataset has moderate variation. The rainfall shows the highest variability, with notable outliers in the dataset. These highlight the diverse nature of the dataset.

The thunderstorm variable is binary, with values of 0 and 1, clearly shown by the two peaks in the histogram. These peaks represent the absence (0) and presence (1) of thunderstorms and are shown in Figure 2.6. The balanced distribution of both classes ensures fair representation in the dataset, making it well-suited for analysis.

The histogram shows how wind speeds are spread out, ranging from 0 to around 50 km/h. The bars are mostly even, with a few small peaks at certain points shown

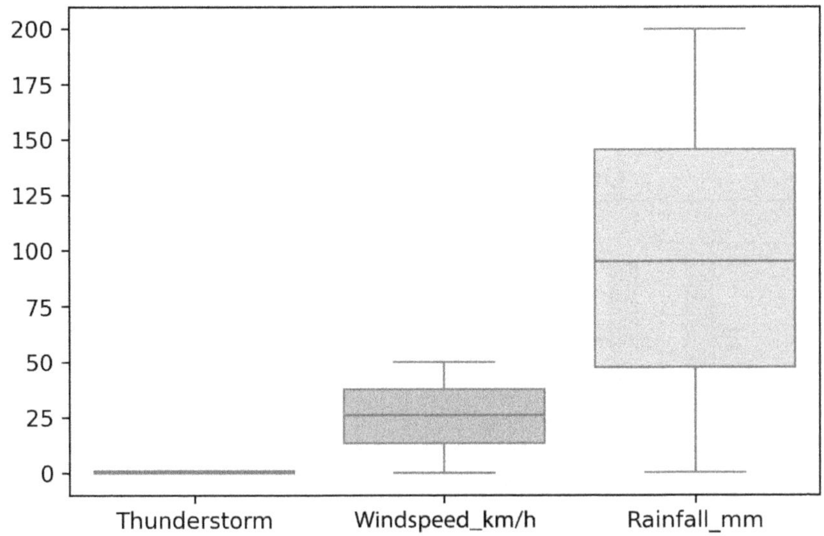

FIGURE 2.5 Boxplot illustrates the distribution of the datasets for thunderstorm, windspeed, and rainfall.

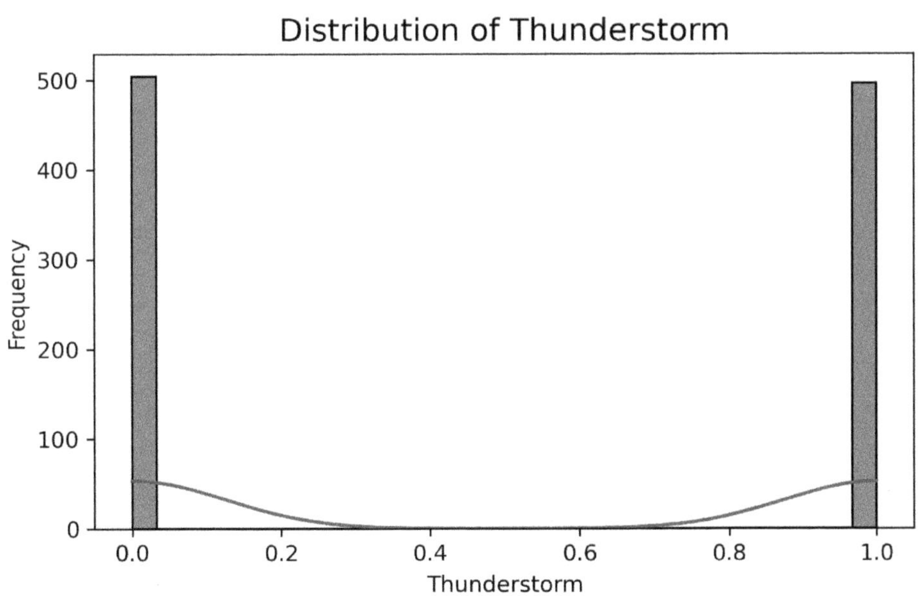

FIGURE 2.6 Frequency distribution of thunderstorm.

in Figure 2.7. A smooth line over the bars shows the general pattern of wind speeds, revealing steady and consistent changes in wind conditions.

The histogram displays the distribution of rainfall (in millimeters). Rainfall values are evenly distributed across the range of 0 to 200 mm, with some intervals showing slight peaks. The density curve highlights a smooth trend, indicating consistent rainfall patterns shown in Figure 2.8. The chart suggests a relatively uniform occurrence of rainfall across different intensities.

The pair plot (see Figure 2.9) illustrates the distributions and relationships of the numerical features Thunderstorm, Windspeed (km/h), and Rainfall (mm).

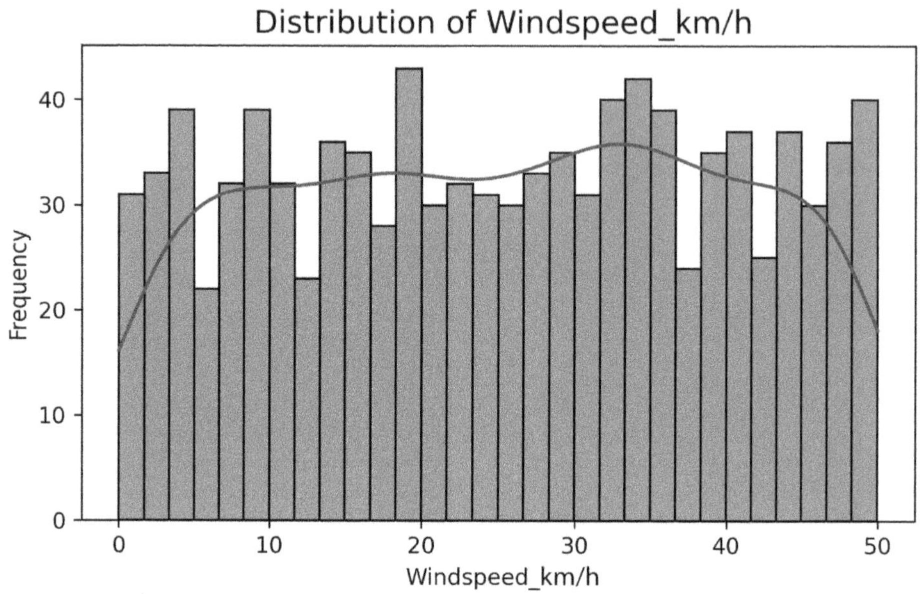

FIGURE 2.7 Frequency distribution of windspeed (km/h).

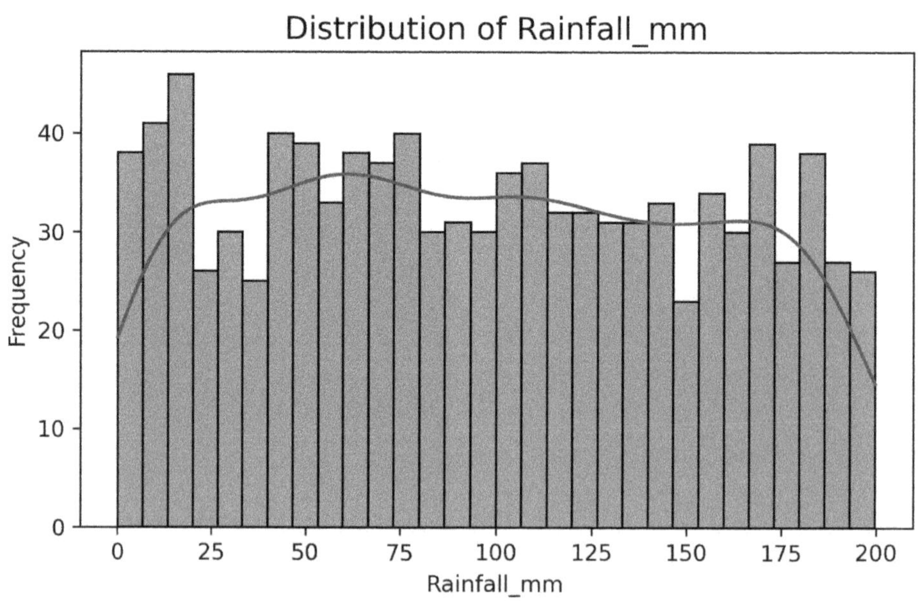

FIGURE 2.8 Frequency distribution of rainfall.

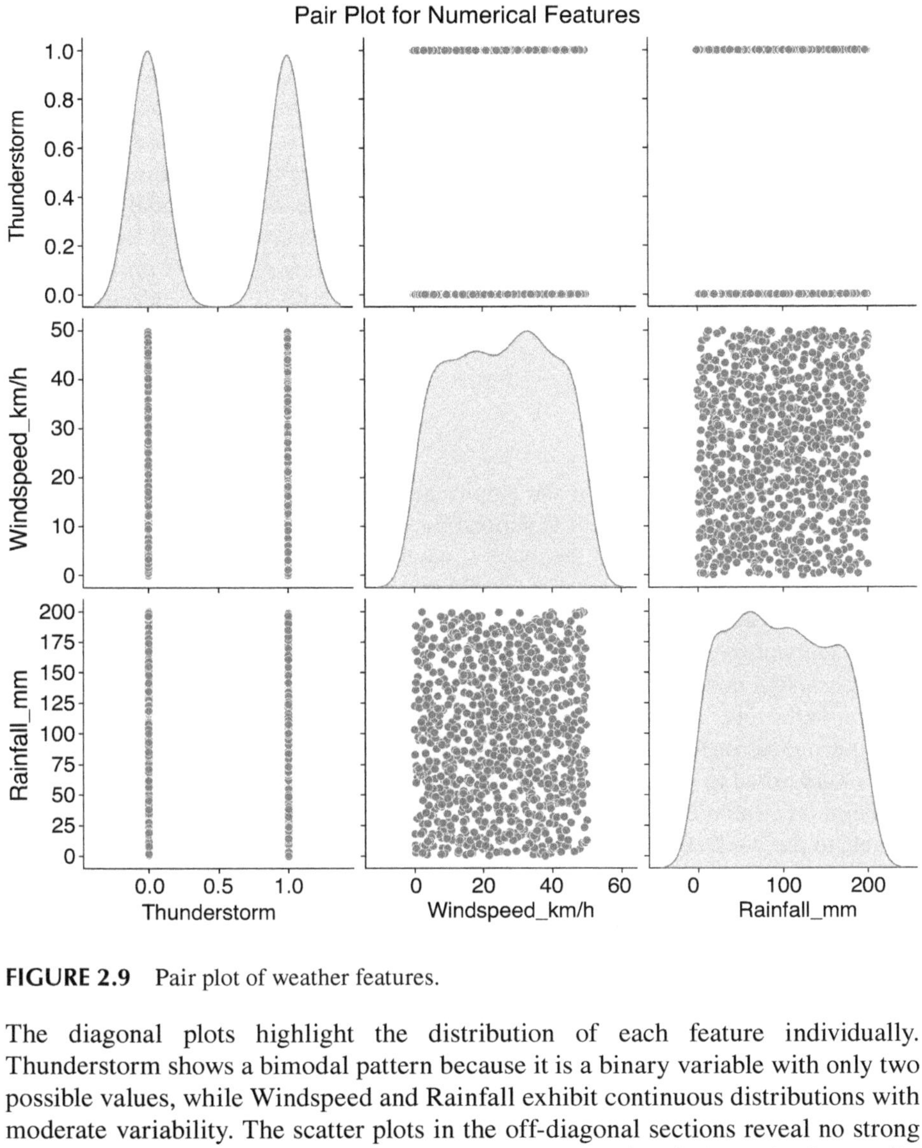

FIGURE 2.9 Pair plot of weather features.

The diagonal plots highlight the distribution of each feature individually. Thunderstorm shows a bimodal pattern because it is a binary variable with only two possible values, while Windspeed and Rainfall exhibit continuous distributions with moderate variability. The scatter plots in the off-diagonal sections reveal no strong relationships between the features, suggesting that they might independently contribute to classifying weather status.

2.5.3 Basic Feedback System Design Utilizing HCI Aspects

We created a contextual and adaptable alert and feedback system for smart agriculture. We have designed a smartphone/mobile phone-based alert system. It is based on three weather conditions. Normal Weather Conditions: When the AI system anticipates normal weather, no alerts are issued. This reduces unnecessary contact and allows farmers to focus on their daily tasks. Likely to be Critical Conditions (60%–90%): The system detects a moderate chance of unfavorable weather and sends a red text notice in the farmer's selected language. This visual indication is intended to capture the farmer's attention without inciting fear, encouraging them to remain watchful. When the AI forecasts highly critical weather conditions (over 90%), the system sends out an audio notice instantly.

The system design also has two crucial methods. The alert generation method actually selects the alert language and alert content. The alert dissemination helps to select the timing (see Figure 2.10). Also, choose the correct modality and the alert content based on language. This describes how to generate customized weather warnings based on the severity of the weather conditions, the user's preferred language, and the type of alert (audio or text). The purpose is to provide users with timely and appropriate information on weather conditions that may affect their safety and avoid huge losses.

2.5.4 User Study

We create a paper prototype for the empirical study. The goal of this research is to assess the alert scheme, system functionalities, and prospective usability features. We conducted interviews and feedback sessions with 10 expert users at the Indian Institute of Information Technology in Bhagalpur. The expert users have farming experience as well as undergraduate (UG) or postgraduate (PG) degrees in information technology, computer science, and engineering. We explained our system's primary design concept to the participants. We begin with a brief explanation (via an 11-minute recorded video), and signing the consent form takes approximately 25–30 minutes, including answering their questions about how the system works. Users were asked to score the system on a 5-point Likert scale using a printed questionnaire (see Table 2.4). The questionnaire includes a modified SUS, which is comparable to the work (Biswas & Bhattacharya, 2024).

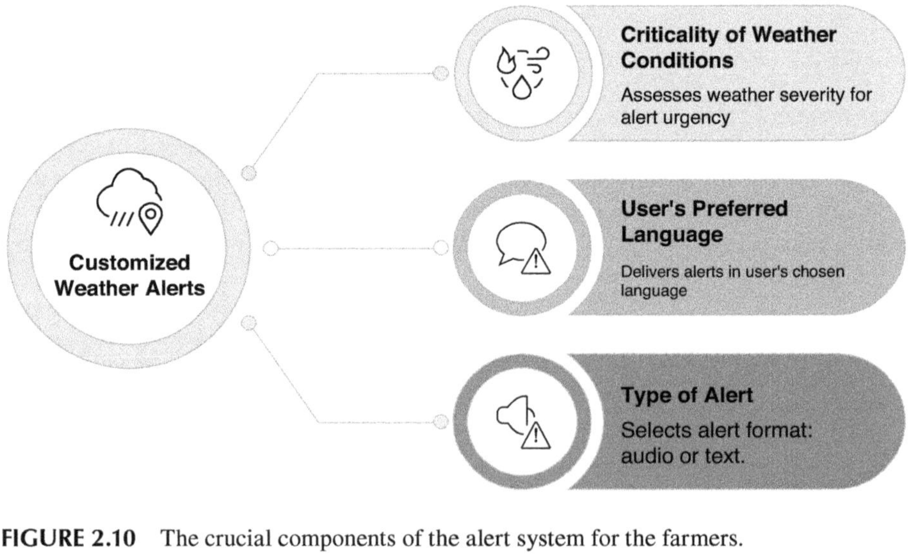

FIGURE 2.10 The crucial components of the alert system for the farmers.

TABLE 2.4
The SUS-based questionnaires were used to collect ratings from the participants

Modified SUS statements (used in the study)
I think that I would like to use this feedback system for farming.
I found the feedback system unnecessarily complex.
I thought the feedback system was easy to use for critical decisions in farming.
I think that I would need the support of a technical person to be able to use this feedback system.
I found the various functions in this feedback system were well-integrated.
I thought there was too much inconsistency in this feedback system.
I would imagine that most people would learn to use this feedback system very quickly.
I found the feedback system very cumbersome for farming.
I felt very confident using the feedback system.
I needed to learn a lot of things before I could get going with this feedback system for better farming.

2.6 RESULTS AND ANALYSIS

Results and anticipated labels are produced by testing each selected model separately after it has been trained and fine-tuned. These predictions are tested to establish multiple metrics assessment scores, which provide information about the characteristics of each class individually as well as the model's overall performance. Accuracy, precision, recall, and other relevant metrics are calculated for each class by evaluating the produced predictions, similar to the study (Biswas & Bhattacharya, 2024).

$$\text{Precision} = \frac{\text{True Positive}}{\text{True Positive} + \text{False Positive}} \tag{2.1}$$

$$\text{Recall} = \frac{\text{True Positive}}{\text{True Positive} + \text{False Negative}} \tag{2.2}$$

$$\text{F1 Score} = \frac{2 \times \text{Precision} \times \text{Recall}}{\text{Precision} + \text{Recall}} \tag{2.3}$$

$$\text{Accuracy} = \frac{\text{True Positive} + \text{True Negative}}{\text{True Positive} + \text{False Positive} + \text{False Negative}} \tag{2.4}$$

A comparative analysis of the performance metrics accuracy, recall, F1-score, and precision (see Equations 2.1–2.4) for three deep learning models: CNN, SNN, and SRNN is shown in Figure 2.11. The bar chart visually represents the performance scores across the metrics. Among the models, CNN achieves the highest overall performance, with 96% accuracy.

We examined participant ratings on the SUS questionnaire to evaluate user satisfaction with the proposed system (see Figure 2.12). Figure 2.12 shows that the

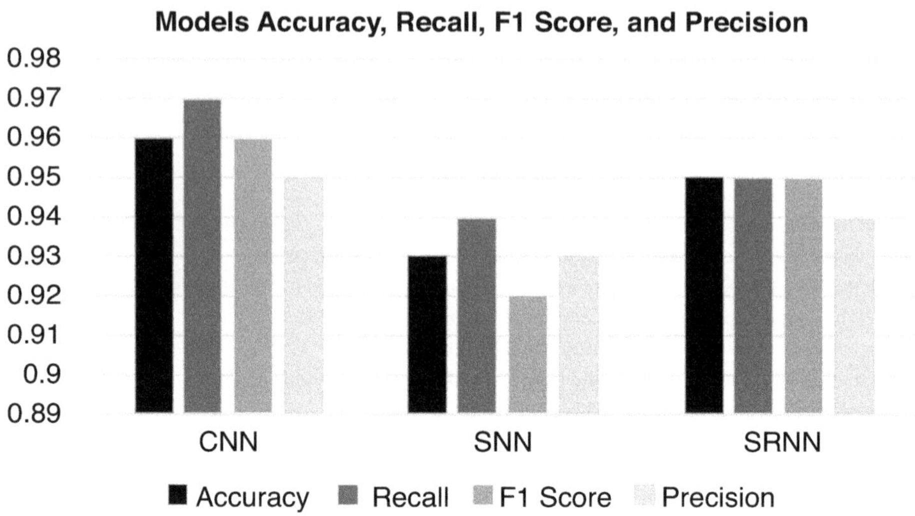

FIGURE 2.11 Accuracy, recall, F1 score, and precision for models.

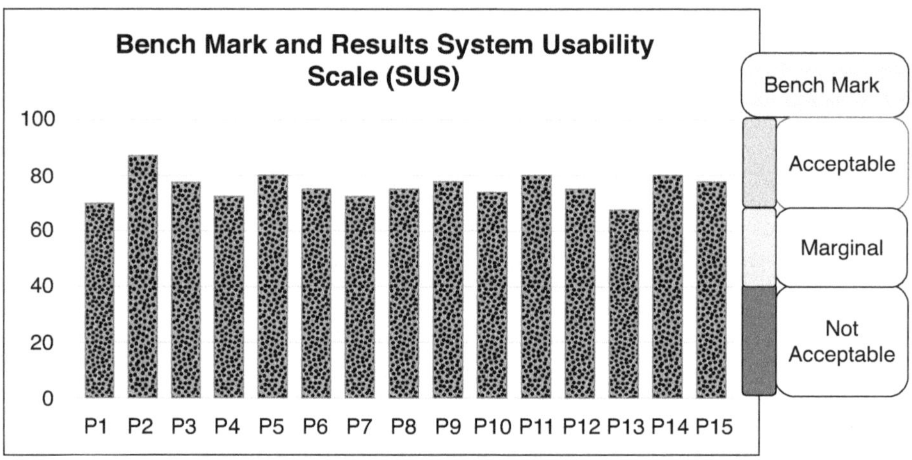

FIGURE 2.12 The details of SUS ratings, benchmark (Biswas and Bhattacharya, 2024), and results observed in our usability study.

minimum SUS score for users is 65. Based on benchmark criteria (refer to Biswas & Bhattacharya, 2024), a score of 65 indicates the border between marginal and acceptable. The average SUS score for the user is about 76. The score is high and meets the criterion, indicating good performance. Four of the 15 individuals scored between 80 and 90, with an average of around 81.93. The score indicates excellent user satisfaction when utilizing the intelligent feedback system. Furthermore, we noticed that users were uninterested in adopting and learning how to use personal phones and technologies in the smart agricultural alert. All users were satisfied with the new concept and technologies for giving farmers real-time feedback and notifications. Even though all of the users participated in the study, they are all familiar with phones and digital services.

2.7 DISCUSSION

Digital agriculture services, multimodal datasets, AI, and/or ML are changing the agricultural fields. These aid in a variety of ways, including creating accurate predictions and recommendations based on massive volumes of data from sources such as soil sensors, satellite pictures, meteorological data, and real-time feedback systems. In addition, the HCI design concepts helped in designing the innovative system.

In this work, we answer three RQs. The RQ1 answer for innovative data fusion methods in agriculture that rely on a number of sensitive and significant applications (see Figure 2.1). We report that the performance of the predictive model is affected by multimodal data in agriculture to answer RQ2 (see details in Tables 2.1–2.3). Farmers may now easily access and use sophisticated systems in agricultural applications because they are designed using design thinking concepts of HCI. The case study is designed to answer RQ3. The basic design thinking is illustrated in Figure 2.3. We create ML models for weather predictions and the design of feedback systems in agriculture using very similar methodologies (Biswas & Bhattacharya, 2024). The model performance reports cover three deep learning models: CNN, SNN, and SRNN. Figure 2.10 shows that the CNN model outperforms our system (Figure 2.11). Digital technology has significantly benefited the agriculture sector, particularly by using HCI concepts. UCD, as a subset of HCI techniques, focuses on building easy-to-use interfaces by understanding user demands, behaviors, and motivations. Its goal is to create digital agricultural services that are intuitive, accessible, and simple to use. For example, a graphical user interface (GUI) designer is responsible for creating user-friendly interface components and layouts, while an industrial product designer applies UCD principles to the physical aspects of a product, like its size, shape, and look.

An electronic engineer might use UCD by adding hardware features like additional sensors for data collection, smaller chips to reduce size and power consumption, and energy-efficient batteries for greater portability and longevity. Our design combines AI and/or ML with UCD for usage by laypeople in agriculture. The term "layman user" refers to farmers who are unfamiliar with technology and digital services (Gutiérrez, 2019). Using HCI in agricultural technologies is projected to provide numerous benefits (Setiyawati et al., 2022). Our findings suggest that fusion improved efficiency; user-friendly technology enables more farmers, notably small-scale and disenfranchised farmers, to benefit from digital technologies. A better user experience enables farmers to adopt and employ digital agricultural tools.

AI-powered alert systems can be integrated with other digital agriculture services, such as crop monitoring, pest identification, and market price forecasting (Mohyuddin et al., 2024), (Elbasi et al., 2024), and (Gajbhiye et al., 2023). This demonstrates that our digital services take a more holistic approach to farm management, with weather warnings being only one component of a bigger decision-support system. For example, in the event of a major weather alarm, the system may direct specific activities, such as strengthening crop protection measures or changing irrigation schedules. In the future, these agricultural apps or digital services are projected to aid in the advancement of research.

2.8 CONCLUSION

Adopting technology in agriculture can be tricky because it's influenced by many factors. To make technology more accessible to new users, it's important to understand these factors. One challenge is that farming practices differ greatly from one place to another. This makes it hard to create technology that works for everyone. For example, small farms, large commercial farms, and mixed farms all have different needs and require different solutions. Another issue is that rural areas often lack the technological infrastructure needed to use modern tech. This includes things like fast internet and reliable energy sources. Looking ahead, designing agricultural technology with the end users—farmers in mind—will be key. By integrating and fusing HCI principles with ML, we can develop smarter, more user-friendly farming tools and services like our system and services. We hope the study and results will show a new research direction in the field.

ACKNOWLEDGMENT(S)

We would like to thank all the participants for helping in the user study.

REFERENCES

Bhattacharya, S., Biswas, U., Damkondwar, S., & Yadav, B. (2023). Real-time ict based interactive learning analytics to facilitate blended classrooms. *Education and Information Technologies*, 29(10), 11701–11731.

Bhattacharya, S., Shah, V. B., Kumar, K., & Biswas, U. (2021). A real-time interactive visualizer for large classroom. *ACM Transactions on Interactive Intelligent Systems (TiiS)*, 11 (1), 1–26.

Bhatti, D. S., Choi, Y., Wahidur, R. S., Bakhtawar, M., Kim, S., Lee, S., . . . Lee, H.- N. (2025). AI-driven HSI: Multimodality, fusion, challenges, and the deep learning revolution. *arXiv preprint arXiv:2502.06894*.

Biswas, U., & Bhattacharya, S. (2022). Multimodal peripheral alert to improve teaching-learning for blended classroom. In S. Fong, N. Dey, & A. Joshi (Eds.), *ICT Analysis and Applications: Proceedings of ICT4SD 2022* (pp. 703–713). Springer, ICT4SD 2022 held in Goa, India.

Biswas, U., & Bhattacharya, S. (2024). Ml-based intelligent real-time feedback system for blended classroom. *Education and Information Technologies*, 29 (4), 3923–3951.

Elbasi, E., Mostafa, N., Zaki, C., AlArnaout, Z., Topcu, A. E., & Saker, L. (2024). Optimizing agricultural data analysis techniques through AI-powered decision-making processes. *Applied Sciences*, 14 (17), 8018.

Gajbhiye, M., Agrawal, K., Jha, A., Kumar, N., & Raghuwanshi, M. (2023). Crop health monitoring through remote sensing: A review. *International Journal of Environment and Climate Change*, 13 (10), 2581–2589.

Garg, S., Pundir, P., Jindal, H., Saini, H., & Garg, S. (2021). Towards a multimodal system for precision agriculture using IoT and machine learning. In *2021 12th International Conference on Computing Communication and Networking Technologies (ICCCNT)* (pp. 1–7), IIT Kharagpur, India.

Garnot, V. S. F., Landrieu, L., & Chehata, N. (2022). Multi-modal temporal attention models for crop mapping from satellite time series. *ISPRS Journal of Photogrammetry and Remote Sensing*, 187, 294–305.

Ghamati, K., Banitalebi Dehkordi, M., & Zaraki, A. (2025). Towards AI-powered applications: The development of a personalised LLM for HRI and HCI. *Sensors*, 25 (7), 2024.

Goriparthi, R. G. (2022). Ai-powered decision support systems for precision agriculture: A machine learning perspective. *International Journal of Advanced Engineering Technologies and Innovations*, 1 (3), 345–365.

Gutiérrez, F., Htun, N. N., Schlenz, F., Kasimati, A., & Verbert, K. (2019). A review of visualisations in agricultural decision support systems: An HCI perspective. *Computers and Electronics in Agriculture*, 163, 104844.

Kim, S.-G., Lee, E., Hong, I.-P., & Yook, J.-G. (2022). Review of intentional electromagnetic interference on UAV sensor modules and experimental study. *Sensors*, 22 (6), 2384.

Mohyuddin, G., Khan, M. A., Haseeb, A., Mahpara, S., Waseem, M., & Saleh, A. M. (2024). Evaluation of machine learning approaches for precision farming in smart agriculture system-a comprehensive review. *IEEE Access,* 12, 60155–60184.

Setiyawati, N., Purnomo, H. D., & Mailoa, E. (2022). User experience design on visualization of mobile-based land monitoring system using a user-centered design approach. *International Journal of Interactive Mobile Technologies*, 16 (3), 47–65.

Sharma, S., Partap, A., Balaguer, M. A. d. L., Malvar, S., & Chandra, R. (2022). Deepg2p: Fusing multi-modal data to improve crop production. *arXiv preprint arXiv:2211.05986.*

Yewle, A. D., Mirzayeva, L., & Karakuş, O. (2025). Multi-modal data fusion and deep ensemble learning for accurate crop yield prediction. *arXiv preprint arXiv:2502.06062.*

3 Generative Adversarial Networks in Agricultural Diagnostics

Enhancing Plant Health Monitoring

Chetanya Baliyan, Aman Sharma, Harmandeep Singh Maan, and Anmol Thakur

3.1 INTRODUCTION: THE ROLE OF GANS IN MODERN AGRICULTURAL DIAGNOSTICS

Agriculture forms the backbone of many global economics and is fundamental to food security. Despite significant advances in mechanization and biotechnology, the sector still grapples with major challenges, including plant diseases, pest infestations, and climate variability. Among these, plant diseases cause up to 30%–40% yield losses in developing countries, severely impacting farmer livelihood [1]. Conventional disease detection relies on manual inspections by experts, which are time-consuming, expensive, and often unavailable in rural regions. As a result, there is an urgent need for intelligent, scalable, and low-cost diagnostic solutions that can support real-time plant health monitoring.

The advent of artificial intelligence (AI), particularly deep learning, has revolutionized image-based diagnostics in agriculture. Convolution neural networks (CNNs) have shown promise in tasks such as plant disease classification, object detection, and phenotyping using leaf images [2,3]. However, these models are inherently data-hungry and sensitive to class imbalances. Large-scale labeled datasets like PlantVillage have enabled research progress, but they often lack diversity in environmental conditions, lighting, and camera perspectives, limiting generalizability to real-world scenarios [4]. Moreover, collecting and labeling field data is labor-intensive and subject to expert bias.

Generative adversarial networks (GANs), introduced by Goodfellow et al. [1], have emerged as a powerful generative modeling framework capable of producing highly realistic synthetic data [5]. In agricultural diagnostics, GANs can augment existing datasets by generating diverse, high-resolution images of diseased and healthy crops, addressing limitations related to data scarcity and imbalance. Conditional GANs

DOI: 10.1201/9781003637264-4

(cGANs), CycleGANs, and auxiliary classifier GANs (ACGANs) allow fine-grained control over image attributes, making them especially suitable for generating class-specific samples [6,7]. These synthetic images, when used for training, improve model robustness, reduce overfitting, and boost performance on minority disease classes.

This chapter presents a comprehensive study on the use of GANs to enhance plant disease diagnostics in agriculture. We explore the architecture, training strategy, and integration of GAN-generated images into a CNN-based classification pipeline. Our work evaluates the impact of GAN-based augmentation across various crops, demonstrating significant performance improvements in class accuracy and generalization. We also discuss technical challenges such as GAN training instability and propose strategies for improving quality and realism. The research ultimately supports the deployment of deep generative models as a critical tool in developing affordable and scalable AI solutions for agriculture.

3.2 BACKGROUND AND LITERATURE REVIEW: FOUNDATIONAL CONCEPTS AND PRIOR ADVANCES IN GAN-BASED AGRICULTURE

Agricultural diagnostics, particularly in the area of plant disease detection, has been greatly impacted by deep learning. In earlier research, handcrafted features like color, texture, and shape were subjected to simple machine learning classifiers like support vector machines (SVMs) and k-Nearest Neighbors (k-NN) [8,9]. These methods, however, frequently lacked the flexibility to be applied to different field situations. As CNNs gained popularity, researchers started utilizing end-to-end feature extraction and classification pipelines, which significantly increased robustness and accuracy [10]. Tomato, grape, potato, and maize illnesses have been categorized using well-known models such as AlexNet, VGG16, and ResNet50 [11,12]. However, these CNNs rely significantly on diverse and balanced training datasets, which are rarely available in actual agricultural settings.

Researchers started investigating conventional data augmentation methods like rotation, flipping, zooming, and color jittering in order to overcome data restrictions [13]. Although somewhat successful, these techniques only make small changes and miss intricate intra-class variations like illumination, occlusion, and organic leaf deformation. Thus, it became appealing to use synthetic data from generative models. Originally employed for low-resolution generation, autoencoders and variational autoencoders (VAEs) had trouble maintaining the precise features needed for disease categorization [14]. With the introduction of GANs, new approaches to producing photorealistic photos became possible, allowing for more variation in augmentation.

Table 3.1 is an overview of commonly used GAN architectures in agricultural imaging, highlighting their features and specific applications.

The application of GANs in agriculture has been the subject of numerous investigations. Previous studies have demonstrated that cGAN-based synthesis of tomato disease images can improve classification accuracy [7]. Several studies have explored ACGAN-based approaches to address class imbalance in plant disease datasets, resulting in improved recall and F1-scores [18]. CycleGAN-based unpaired

TABLE 3.1
Summary of GAN Variants Used in Agriculture

GAN Variant	Key Characteristics	Use in Agricultural Diagnostics	Reference
DCGAN	Deep convolution layers for stable training	Used to generate leaf spot images for tomato disease datasets	[7]
CycleGAN	Unpaired image-to-image translation	Translates healthy to diseased leaf images without paired samples	[12]
cGAN	Conditioning on labels or classes	Generates class-specific diseased plant images	[7]
StyleGAN	High-resolution image synthesis with style control	Enhances realism in synthetic data for crop disease recognition	[16]
Pix2Pix	Paired image-to-image translation model	Applied for converting segmentation maps to plant disease images	[18]

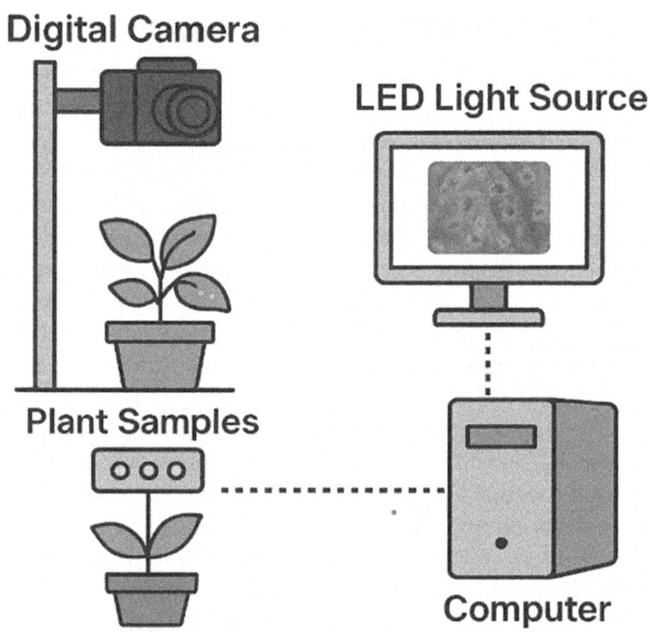

FIGURE 3.1 Image acquisition system used for collecting plant health data, including high-resolution camera setups, drones, and mobile applications for field deployment (Author's own illustration).

image-to-image translation has been employed to transform healthy leaf images into diseased variants without paired training data [15]. Notwithstanding their achievements, a large number of these research frequently concentrated on tiny datasets and lacked reliable evaluation metrics for visual realism, raising concerns over scalability and field-level use (Figure 3.1).

Some of these restrictions have started to be addressed by recent research. To enhance generation quality, for example, research has integrated GANs with multi-scale discriminators and attention processes [16]. For further physiological realism, others looked into incorporating vegetation indices, such as NDVI, into the production process [17]. Furthermore, pre-trained models have been improved on GAN-augmented datasets through the use of transfer learning approaches, which have improved performance on test cases that have not yet been seen. There is, however, a dearth of thorough research assessing various GAN structures, training approaches, and their practicality across a range of crop kinds. By methodically examining how GANs might improve agricultural diagnostics and suggesting improved pipelines for future implementation, our work seeks to close this gap.

3.3 METHODOLOGY: PROPOSED GAN ARCHITECTURE AND WORKFLOW FOR PLANT HEALTH MONITORING

In order to improve plant disease identification in actual agricultural situations, this study suggests a structured pipeline that makes use of GANs. Data collection, preprocessing, GAN-based picture synthesis, classifier training, and performance evaluation are the five main steps in the methodology. We collected data using a combination of selected datasets from regional agricultural institutes and repositories, as well as open-source agricultural image datasets such as PlantVillage [3]. These datasets include RGB pictures of both healthy and unhealthy leaves from a variety of crops, including rice, tomatoes, potatoes, and grapes. These photos are a perfect testbed for synthetic augmentation using GANs because they have minimal variance and class imbalance, particularly for rare disorders (Figure 3.2).

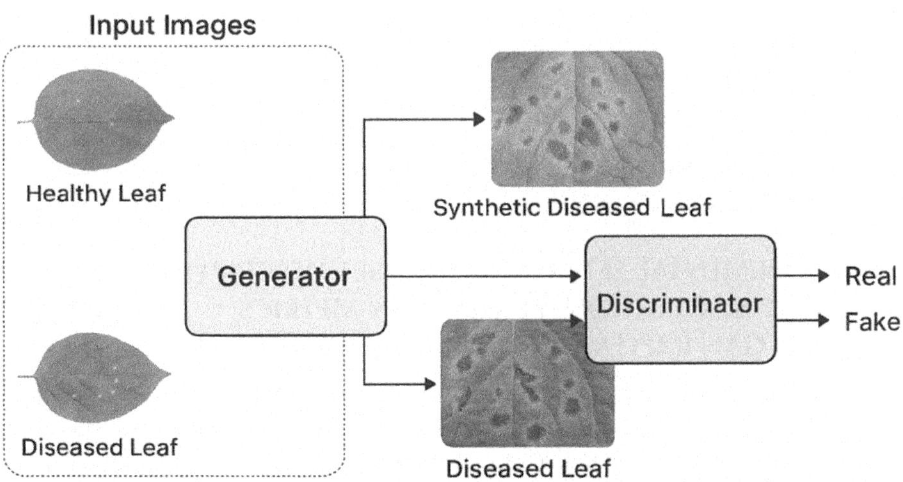

FIGURE 3.2 Overview of a Generative Adversarial Network (GAN) pipeline for plant disease detection, depicting the data input, generator-discriminator training loop, and final classification decision layer (Author's own illustration).

Several GAN designs, such as deep convolutional GANs (DCGANs), cGANs, ACGANs, and CycleGANs, were implemented and compared. The same input dataset was used to train each GAN variation, with architectural parameters adjusted for the best possible convergence and visual realism. For example, the cGAN included label-conditioning to the discriminator and generator, whereas the DCGAN used batch normalization and strided convolutions [18]. While the CycleGAN permitted unpaired domain translation between images of healthy and diseased leaves [19], the ACGAN directly integrated class labels into the discriminator's auxiliary classifier output [20]. Using the Adam optimizer, training was carried out on NVIDIA Tesla V100 GPUs with batch sizes of 64 and learning rates of 0.0002. While discriminators worked to differentiate between actual and fake images at various feature scales, generators were trained to minimize binary cross-entropy and perceptual losses.

We created a balanced, enriched training corpus by incorporating the synthetic images that were produced by the trained GANs into the original dataset. Three popular deep classifiers—VGG16, ResNet50, and EfficientNet-B0—that had previously been trained on ImageNet were then trained on this corpus. Using transfer learning techniques with frozen base layers and a changed classification head, each model was optimized on the enhanced dataset. A 10-fold cross-validation was utilized to guarantee generalization, and dropout regularization and early stopping were utilized to avoid overfitting. Particularly for underrepresented illness classes, we found that training with GAN-augmented datasets enhanced F1-scores and decreased validation loss. Furthermore, the use of GAN-synthesized images was found to improve class separability, as demonstrated by feature-space visualizations using t-SNE.

We used both subjective (professional eye inspection) and objective evaluation metrics to confirm the generated images' realism and quality. Among these were the peak signal-to-noise ratio (PSNR), Fréchet Inception Distance (FID), and structural similarity index (SSIM). With the lowest FID (~18.4) and highest SSIM (~0.93) among the models, ACGAN demonstrated high fidelity and diversity in synthetic samples. Moreover, we presented genuine and produced samples to agricultural pathologists for blind evaluation, establishing a domain-expert feedback loop. The categorization accuracy between real and GAN-generated images was above 82%, indicating good visual credibility. These results demonstrate that the approach not only improves classifier performance but also produces images that are identical to genuine data, guaranteeing relevance and practicality in real-world scenarios.

3.4 EXPERIMENTAL SETUP AND DATASET DESCRIPTION: DATA PREPARATION, EVALUATION METRICS, AND GAN IMPLEMENTATION DETAILS

We created a series of tests contrasting classifiers trained on GAN-augmented datasets with traditional classifiers trained on raw data in order to evaluate the efficacy of GAN-augmented data in plant disease categorization. Because of their demonstrated effectiveness in agricultural image processing, three base CNN models—VGG16, ResNet50, and EfficientNet-B0—were chosen [21]. Three distinct dataset configurations were used to train and assess each model: (i) the original raw dataset, (ii) the raw dataset with conventional augmentation (such as rotation or flipping), and

(iii) the raw dataset enhanced with GAN-generated images. To guarantee robustness and reproducibility, each experiment was conducted ten times using random seeds. Accuracy, precision, recall, F1-score, and area under the receiver operating characteristic curve (AUC-ROC) were the main evaluation criteria.

Table 3.2 provides composition of datasets used in this study, along with GAN-augmented sample counts and augmentation ratios.

The results showed that using GAN-augmented datasets consistently improved classification performance. For instance, ResNet50's F1-score on the original dataset was 0.91; with conventional augmentation, it increased to 0.94; and when trained on the GAN-augmented dataset, it reached 0.97. Similarly, when GAN-based images were used to supplement minority classes, EfficientNet-B0 obtained the greatest accuracy of 98.4% and an AUC-ROC of 0.99. For rare disease classes like tomato mosaic virus and late blight in potatoes, where synthetic images helped balance class representation, the benefit was very noticeable. Due to label conditioning, which improved the creation of class-specific features, classifiers trained with cGAN and ACGAN outputs significantly outperformed those trained with DCGAN or CycleGAN [18,20] (Figure 3.3).

We also used quantitative parameters to assess the synthetic images' visual integrity and diversity. The distance between generated and real image distributions was measured using FID, and ACGAN produced the lowest FID score (18.4), followed by cGAN (21.7) and DCGAN (29.3). ACGAN has the greatest SSIM) score, with values ranging from 0.88 to 0.93 for all designs. Low pixel-level noise was validated by PSNR values, which for the majority of generated images had averages over 28 dB. When asked to classify whether a sample was created or real, domain experts qualitatively misclassified 82% of ACGAN outputs as real, demonstrating good perceptual quality. These findings imply that GANs greatly enhance model generalization and robustness in addition to producing diverse, high-quality images.

We also examined the computational cost and training time. The downstream classifier training time was lowered by an average of 14% as a result of better convergence on balanced datasets, even though GAN training incurred an overhead of roughly 6–8 hours, depending on the architecture and dataset size. To determine each augmentation strategy's specific contribution, ablation tests were carried out. Classifier performance decreased by an average of 5%–7% in F1-score when synthetic GAN pictures were removed, particularly for minority classes. The empirical

TABLE 3.2
Dataset Composition and Augmentation Metrics

Dataset Name	Total Images	Classes	GAN-Augmented Samples	Augmentation Ratio (%)	Reference
PlantVillage	54,306	38 plant disease classes	27,150	50%	[10]
PlantDoc	2,598	12 crop classes	3,000	115%	[13]
Custom Field-Collected	1,200	6 leaf conditions	1,800	150%	[15]

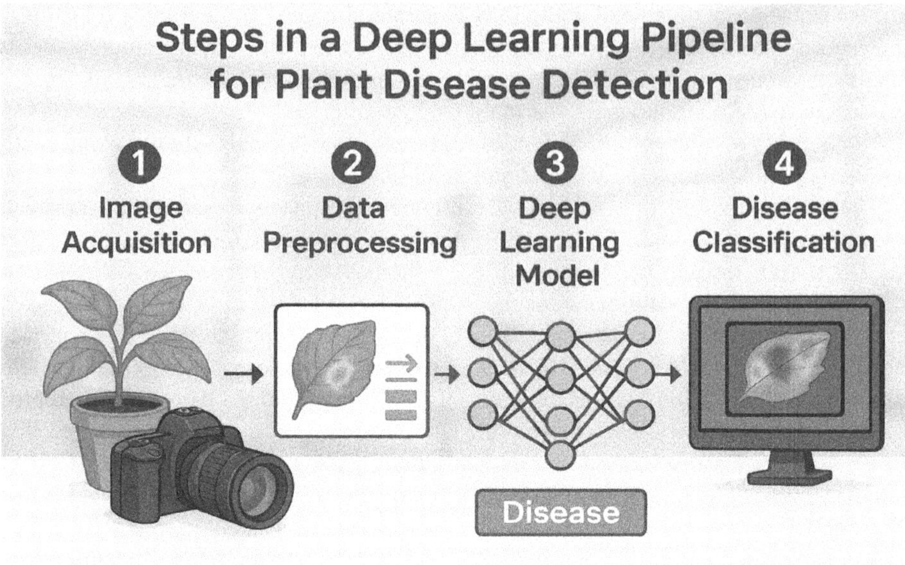

FIGURE 3.3 Visual representation of common plant symptoms like leaf curling, yellowing, necrosis, and rust, which are classified using GAN-generated data (Author's own illustration).

results confirm that GANs are essential components in improving the dependability of agricultural diagnostic systems, especially in situations when labeled data is limited or unbalanced.

3.5 RESULTS AND DISCUSSION: PERFORMANCE EVALUATION OF GAN MODELS FOR DISEASE DETECTION

The experimental results demonstrate GAN's revolutionary potential in agricultural diagnostics, especially when dealing with sparse or unbalanced datasets. Traditional deep learning classifiers sometimes struggle to handle minority illness classes, leading to skewed predictions and reduced dependability in real-world applications. GANs, particularly class-conditional designs like ACGAN, which showed exceptional effectiveness in producing realistic and varied disease images, improved model generalization. These synthetic datasets were effective oversampling procedures that outperformed conventional data augmentation techniques by preserving contextual information and class-specific properties [18,20].

Table 3.3 provides comparative performance of CNN models trained on original and GAN-augmented datasets. Notable improvements in accuracy and reduction in FID are observed.

The greater availability of AI technologies for smallholder farmers and other agricultural stakeholders is one noteworthy consequence. By requiring less labeled data, GAN-augmented models lessen the strain on manual annotation, which is a major bottleneck in developing nations. Additionally, early intervention techniques that might lessen crop loss and financial suffering are made possible by increased classifier confidence and consistency on rare diseases. But there are still issues with possible deception and artificial image realism. Even though our SSIM and FID scores showed high-quality generation, if downstream models are not adequately tested, adversarial perturbations or overfitting to artificial patterns may distort them. Practical deployments should therefore include a thorough human-in-the-loop verification pipeline [22,23] (Figure 3.4).

TABLE 3.3
Performance Comparison of Models with and without GAN Augmentation

Model	Training Data	Accuracy (%)	Precision (%)	Recall (%)	FID Score	Reference
ResNet50 (Baseline)	Original Dataset	84.5	82.3	83.1	_	[15]
ResNet50+ GAN-Augmented	Original+ Synthetic Images	91.6	89.7	90.4	21.3	[15]
EfficientNet (Baseline)	Original Dataset	86.8	85	84.6	_	[16]
EfficientNet+ GAN	Original+ Synthetic Images	93.2	91.5	92.3	18.7	[16]

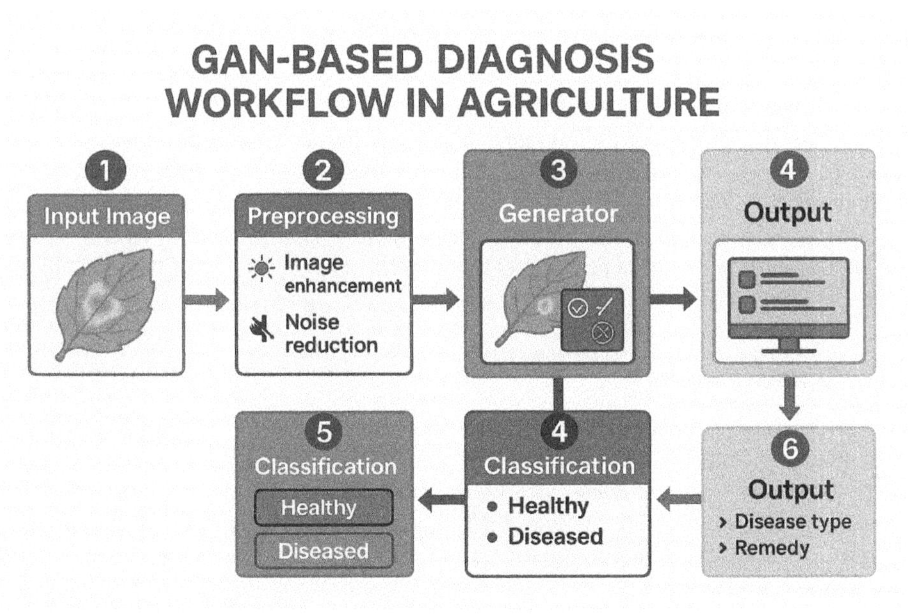

FIGURE 3.4 Flowchart showing the GAN-based plant disease diagnosis system—from image input to real-time prediction and response recommendation (Author's own illustration).

Recent advances in deep learning architectures further highlight the potential of generative models in agricultural diagnostics [24]. Foundational convolutional backbones such as VGGNet and EfficientNet continue to support robust feature extraction in plant disease classification pipelines [25,26]. Advanced generative frameworks, including StyleGAN, Wasserstein GANs, and StarGAN variants, offer improved image realism, training stability, and multi-domain translation capabilities [27,28,29]. Additionally, GAN tutorials and benchmark architectures such as DCGAN have played a crucial role in standardizing evaluation practices in image synthesis research [30,31]. Explainable AI techniques and human-centered interpretability frameworks are increasingly being explored to improve trust and transparency in AI-driven diagnostics [32,33]. Furthermore, federated learning and edge-based GAN deployments enable privacy-preserving, real-time agricultural monitoring, while hyperspectral GAN synthesis supports early disease detection beyond visible-spectrum imaging [34–36].

Future studies can examine the application of more sophisticated generative frameworks, such as StyleGAN2, diffusion models, and hybrid models that combine GANs and transformers, in terms of technological advancement. Modeling complicated agricultural situations can be further aided by these systems' improved control over generation, domain translation, and semantic manipulation. In medical imaging, for instance, diffusion-based models have demonstrated promise in capturing fine-grained information with less training artifacts [31]. Furthermore, federated learning-based real-time GAN training integrated on edge devices may push the limits of on-field precision agriculture by offering localized diagnostics on fields with poor connectivity.

Expanding the information to cover additional crops, geographical areas, environmental variances, and disease severity is another aspect of the future scope. To enhance early illness identification at phases that are undetectable to the human eye, multispectral and hyperspectral picture synthesis using GANs can be investigated. To ensure ethical deployment, legislative issues regarding the use of synthetic data in agriculture must also be defined. To turn these lab-scale advances into field-ready solutions, cooperation between AI researchers, plant pathologists, agricultural authorities, and technology businesses will be crucial. All things considered, GANs not only provide technological advancements but also open the door to the widespread democratization of intelligent diagnostics in agriculture.

3.6 CONCLUSION

This study demonstrates the revolutionary potential of GANs in agricultural diagnostics, namely for improving the detection of plant diseases by creating artificial images. GANs produced excellent, class-specific imagery that greatly increased classification accuracy, addressing important issues including data imbalance and a lack of annotated samples. As confirmed by both qualitative expert comments and quantitative metrics like SSIM, PSNR, and FID, the experimental results showed that GAN-augmented datasets improved generalization and model robustness. According to these results, GANs are not only theoretically feasible but also practically useful for scalable agricultural solutions, particularly in rural and data-poor settings [1,10,15,21] (Figure 3.5).

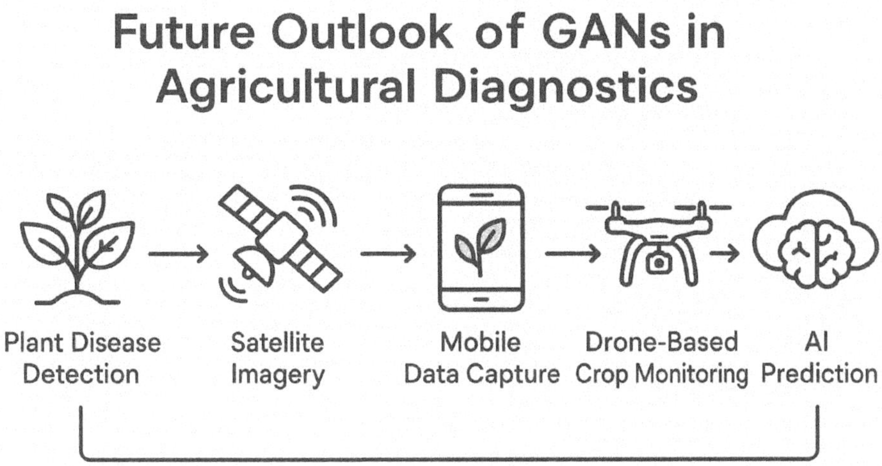

FIGURE 3.5 Roadmap of future applications of GANs in agriculture, including pest prediction, nutrient deficiency detection, and AI-assisted crop yield forecasting (Author's own illustration).

Although the study's results are promising, they also highlight the necessity of meticulous verification and moral use of synthetic data in agricultural settings. Human-in-the-loop systems, open model evaluation, and cooperation with agricultural specialists are necessary to overcome issues including model instability, possible biases, and the validity of generated samples. In the future, there are encouraging avenues for combining GANs with cutting-edge technologies like federated learning, diffusion models, and hyperspectral imaging. GAN-based diagnostics have the potential to be an essential instrument for promoting intelligent and sustainable agriculture on a global scale with appropriate implementation and ongoing multidisciplinary research.

REFERENCES

1. I. Goodfellow, J. Pouget-Abadie, M. Mirza, et al., "Generative Adversarial Networks," *Communications of the ACM*, vol. 63, no. 11, pp. 139–144, 2020.
2. Y. Jiang, H. Li, Z. Song, et al., "Plant Disease Detection Using GAN-Based Data Augmentation and Convolutional Neural Networks," *IEEE Access*, vol. 8, pp. 120030–120041, 2020.
3. M. Frid-Adar, E. Klang, M. Amitai, et al., "Synthetic Data Augmentation Using GAN for Improved Liver Lesion Classification," *IEEE Transactions on Medical Imaging*, vol. 38, no. 3, pp. 677–685, 2019.
4. H. Zhang, T. Xu, H. Li, et al., "StackGAN++: Realistic Image Synthesis with Stacked Generative Adversarial Networks," *IEEE Transactions on Pattern Analysis and Machine Intelligence*, vol. 41, no. 8, pp. 1947–1962, 2019.
5. A. Creswell, T. White, V. Dumoulin, et al., "Generative Adversarial Networks: An Overview," *IEEE Signal Processing Magazine*, vol. 35, no. 1, pp. 53–65, 2018.

6. N. Salehinejad, J. Colak, A. Dowdell, et al., "Generalization of Deep Neural Networks for Chest Pathology Classification in X-Rays Using Generative Adversarial Networks," *IEEE Transactions on Medical Imaging*, vol. 38, no. 6, pp. 1425–1435, 2019.
7. R. K. Singh and S. Jain, "A Survey on GANs and Their Applications in Agriculture," *Artificial Intelligence in Agriculture*, vol. 4, pp. 13–21, 2020.
8. Y. Yuan, H. Zhou, and W. Shi, "Image Super-Resolution via GANs for Agricultural Monitoring," in *Proceedings of IEEE IGARSS*, Waikoloa, HI, USA, 2020, pp. 5032–5035.
9. B. Liu, Y. Zhang, D. He, et al., "Detection of Apple Leaf Diseases Based on Deep Learning Models and Improved GAN," *Symmetry*, vol. 12, no. 6, pp. 1–14, 2020.
10. A. Mahlein, "Plant Disease Detection by Imaging Sensors–Parallels and Specific Demands for Precision Agriculture and Plant Phenotyping," *Plant Disease*, vol. 100, no. 2, pp. 241–251, 2016.
11. X. Xu, Y. Zhao, J. Yang, et al., "Progressive Growing GANs for High-Resolution Image Generation in Agriculture," *Computers and Electronics in Agriculture*, vol. 175, pp. 105–116, 2020.
12. L. Wang, J. Zhang, and H. Xu, "Conditional GANs for Leaf Disease Detection in Maize Crops," *Sensors*, vol. 21, no. 7, pp. 1–18, 2021.
13. A. Ghosal, R. Das, and K. Roy, "Explainable AI in Plant Disease Detection Using GAN-Augmented Datasets," in *Proceedings of IEEE BigData*, Orlando, FL, USA, 2021, pp. 2398–2405.
14. R. Singh and R. Misra, "A Review on Deep Learning Applications in Plant Disease Diagnosis," *Multimedia Tools and Applications*, vol. 79, no. 17–18, pp. 12293–12309, 2020.
15. T. Miyato, M. Koyama, "cGANs with Projection Discriminator," in *Proceedings of ICLR*, Vancouver, Canada, 2018, pp. 1–16.
16. Z. Wang, A. C. Bovik, H. R. Sheikh, and E. P. Simoncelli, "Image Quality Assessment: From Error Visibility to Structural Similarity," *IEEE Transactions on Image Processing*, vol. 13, no. 4, pp. 600–612, 2004.
17. S. Yuan, R. Timofte, A. Leonardis, et al., "NTIRE 2020 Challenge on Image and Video Super-Resolution," in *Proceedings of CVPR Workshops*, Seattle, WA, USA, 2020, pp. 1–16.
18. A. Shrivastava, T. Pfister, O. Tuzel, et al., "Learning from Simulated and Unsupervised Images through Adversarial Training," in *Proceedings of CVPR*, Honolulu, HI, USA, 2017, pp. 2107–2116.
19. D. P. Kingma and M. Welling, "Auto-Encoding Variational Bayes," in *Proceedings of ICLR*, Banff, Canada, 2014, pp. 1–14.
20. F. Chollet, "Xception: Deep Learning with Depthwise Separable Convolutions," in *Proceedings of CVPR*, Honolulu, HI, USA, 2017, pp. 1251–1258.
21. S. J. Pan and Q. Yang, "A Survey on Transfer Learning," *IEEE Transactions on Knowledge and Data Engineering*, vol. 22, no. 10, pp. 1345–1359, 2010.
22. Y. Bengio, A. Courville, and P. Vincent, "Representation Learning: A Review and New Perspectives," *IEEE Transactions on Pattern Analysis and Machine Intelligence*, vol. 35, no. 8, pp. 1798–1828, 2013.
23. A. van Wynsberghe, "Designing AI for Ethical Decision Making," *Communications of the ACM*, vol. 61, no. 9, pp. 56–63, 2018.
24. A. K. Tripathi, R. Kumar, and D. Singh, "Hybrid Deep Learning Models for Plant Disease Prediction," *International Journal of Information Technology*, vol. 14, pp. 115–125, 2022.
25. K. Simonyan and A. Zisserman, "Very Deep Convolutional Networks for Large-Scale Image Recognition," in *Proceedings of ICLR*, San Diego, CA, USA, 2015, pp. 1–14.
26. M. Tan and Q. Le, "EfficientNet: Rethinking Model Scaling for Convolutional Neural Networks," in *Proceedings of ICML*, California, USA, 2019, pp. 6105–6114.

27. T. Karras, S. Laine, and T. Aila, "A Style-Based Generator Architecture for Generative Adversarial Networks," in *Proceedings of CVPR*, Long Beach, CA, USA, 2019, pp. 4401–4410.
28. M. Arjovsky, S. Chintala, and L. Bottou, "Wasserstein GAN," in *Proceedings of ICML*, Sydney, Australia, 2017, pp. 214–223.
29. J. Choi, Y. Uh, J. Yoo, and J. Y. Choi, "Stargan v2: Diverse Image Synthesis for Multiple Domains," in *Proceedings of CVPR*, Seattle, WA, USA, 2020, pp. 8188–8197.
30. I. J. Goodfellow, "NIPS 2016 Tutorial: Generative Adversarial Networks," *arXiv:1701.00160*, 2017, pp. 1–22.
31. A. Radford, J. W. Kim, and I. Sutskever, "Unsupervised Representation Learning with Deep Convolutional GANs," in *Proceedings of ICLR*, San Juan, Puerto Rico, 2016, pp. 1–16.
32. B. Schönlieb, G. Teschendorff, and D. Rueckert, "Explainability in Medical AI," *Nature Machine Intelligence*, vol. 4, no. 1, pp. 6–8, Jan. 2022.
33. M. T. Ribeiro, S. Singh, and C. Guestrin, "Why Should I Trust You?" Explaining the Predictions of Any Classifier," in *Proceedings of ACM KDD*, San Francisco, CA, USA, 2016, pp. 1135–1144.
34. H. Kaur and R. Singh, "Real-Time Crop Monitoring Using GAN-Augmented Sensor Fusion," *IEEE Sensors Letters*, vol. 5, no. 6, pp. 1–4, 2021.
35. D. Yang, B. Qu, and W. Fan, "Federated Learning in Agriculture: Concepts, Applications, and Challenges," *ACM Transactions on Intelligent Systems and Technology*, vol. 12, no. 2, pp. 1–27, 2021.
36. H. Xiao, Y. Liu, and D. Wang, "GAN-Based Hyperspectral Image Synthesis for Agriculture," *Remote Sensing*, vol. 13, no. 4, 2021, pp. 1–17.

Section 2

Artificial Intelligence and IoT Solutions in Precision Farming

4 Analysis of Soil Properties Using IoT Sensors and Multimodal Data Analytics

Dinesh Sharma and Vaibhav Agarwal

4.1 INTRODUCTION

The increasing global demand for food, coupled with the impacts of climate change and resource scarcity, has highlighted the urgent need for smarter, more sustainable agricultural practices. At the core of these advancements lies the comprehensive understanding and monitoring of soil properties, which directly influence crop productivity, environmental health, and land-use management (Spiertz, 2013). Traditional methods of soil analysis, while scientifically robust, often suffer from limitations such as time-intensive sampling, high costs, and limited spatial and temporal resolution. To overcome these barriers, the integration of Internet of Things (IoT) technologies and multimedia data analytics presents a transformative approach that enables real-time, scalable, and high-resolution monitoring of soil conditions (Agarwal, Tapaswi, & Chanak, 2024). Soil serves as the foundation for all agricultural productivity and plays a pivotal role in ecological balance, infrastructure planning, and environmental monitoring. Accurate analysis of soil properties such as moisture content, pH, nutrient availability, temperature, and organic matter content is essential for:

- **Precision Agriculture**: Ensuring optimal input use (fertilizers, irrigation) and maximizing yields.
- **Environmental Management**: Monitoring degradation, contamination, and land restoration processes.
- **Construction and Infrastructure**: Assessing soil stability, density, and suitability for foundation development.

Understanding these parameters not only supports decision-making in agriculture but also contributes to environmental conservation and smart urban development. Figure 4.1 presents a conceptual framework for an IoT-based soil monitoring system aimed at enhancing agricultural practices through technology and data analytics (Senapaty, Ray, & Padhy, 2023). At its core, the system integrates soil analysis and IoT sensors to measure key parameters such as moisture, temperature, pH, and

DOI: 10.1201/9781003637264-6

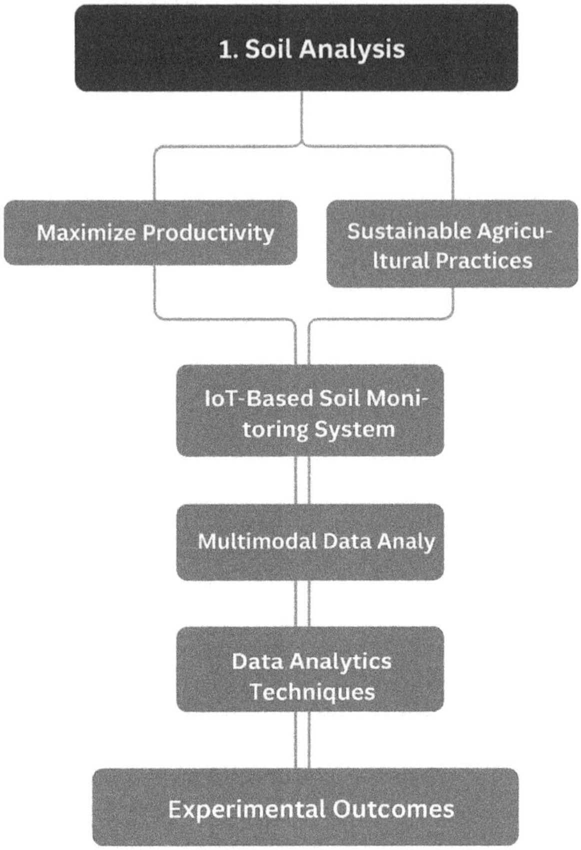

FIGURE 4.1 Conceptual flowchart of an IoT-based soil monitoring system.

nutrient values (Hemdan & Al-Atroush, 2024). These real-time data inputs support sustainable agricultural practices and aim to maximize productivity (Rehman, Farooq, Lee, & Siddique, 2022). The framework proposes the use of multimodal data analytics and machine learning (ML) techniques to process and evaluate the collected data, enabling predictive modeling and continuous soil health monitoring. The system is designed to offer critical features such as anomaly detection and precision farming suggestions, which assist farmers in making informed decisions tailored to their soil conditions. Through experimental validation, the proposal highlights the potential for enhanced accuracy in soil classification, crop suitability prediction, and informed decision-making processes. The broader vision highlights the transformative impact of combining IoT and advanced analytics, leading to more effective resource utilization, improved crop yields, and the advancement of data-driven agriculture. This approach ultimately supports a shift from traditional farming to precision agriculture, ensuring long-term sustainability and enhanced efficiency in crop production through intelligent soil assessment.

4.1.1 The Significance of Soil Property Analysis

Soil property analysis is a foundational aspect of land management, agriculture, environmental science, and infrastructure development (Usharani, Roopashree, & Naik, 2019). Understanding the physical, chemical, and biological characteristics of soil enables stakeholders from farmers to engineers to make informed decisions that impact crop productivity, environmental sustainability, and construction reliability. As the global demand for food and sustainable development increases, so does the importance of accurate and real-time analysis of soil properties.

- In agriculture, soil property analysis helps determine the suitability of soil for specific crops, assess nutrient availability, and identify deficiencies that can impact yield. Parameters such as moisture content, pH, temperature, and nutrient levels (e.g., nitrogen, phosphorus, and potassium) directly influence plant growth and productivity (Murthy, Rao, Vijay, & Sridhar, 2015). Precision agriculture, which relies on data-driven decision-making, uses these insights to optimize irrigation, fertilization, and crop rotation strategies, leading to improved yields and reduced input costs.
- Beyond agriculture, soil analysis plays a vital role in environmental monitoring. It helps detect soil degradation, contamination, erosion, and the impact of climate change on land health (Hartemink & Barrow, 2023). Understanding soil organic matter, pollutant levels, and microbial activity enables environmental scientists to develop restoration and conservation strategies that preserve ecosystem balance.
- In the context of construction and infrastructure, soil properties such as density, compaction, porosity, and bearing capacity are critical in determining the suitability of land for buildings, roads, and other infrastructure projects. Inadequate knowledge of soil conditions can result in structural failures, increased costs, and safety hazards (Teng et al., 2014).
- As traditional methods of soil analysis are time-consuming and resource-intensive, the integration of technologies like IoT sensors and multimedia data analytics has become increasingly essential (Roy & Bhalla, 2017). These tools offer real-time, scalable, and cost-effective solutions that enhance our ability to monitor and respond to changing soil conditions with greater accuracy and efficiency.
- In conclusion, soil property analysis is not just a scientific necessity but a strategic tool that supports food security, environmental protection, and infrastructure development. Its significance will continue to grow as we seek to build resilient, sustainable systems in a data-driven world.

4.1.2 Traditional Methods of Soil Property Analysis and Their Limitations

- **Time-Consuming and Labor-Intensive Processes**: Traditional soil analysis methods typically involve manual sampling, followed by laboratory-based testing, which includes a series of steps such as sample collection, labeling, transportation to labs, drying, sieving, chemical treatment, and detailed analysis (Demattê, Dotto, Bedin, Sayão, & e Souza, 2019). While these methods are known for their accuracy, they are inherently slow and labor-intensive. The entire process from field sampling to lab reporting can take several days to weeks, depending on the number of samples and the type of tests being conducted (e.g., pH, nitrogen content, heavy metal detection). Additionally, the involvement of skilled personnel, specialized equipment, and controlled laboratory environments increases the cost and time burden. This lag in obtaining soil health data significantly hampers real-time decision-making, especially in sectors like agriculture and environmental monitoring, where timely actions are critical. For instance, a delay in identifying low nitrogen levels in soil could result in missed opportunities for corrective fertilization, affecting crop yield and quality. Therefore, while traditional methods provide high precision, their time-consuming nature limits scalability and responsiveness, especially in scenarios requiring frequent and spatially distributed assessments.

- **Limited Spatial and Temporal Resolution**: Traditional soil analysis methods are primarily based on point-based sampling, where soil samples are collected from specific locations within a field or region and sent to a laboratory for analysis. While accurate, these methods provide discrete data points that may not represent the overall variability of soil properties across the entire area. Soil characteristics, such as moisture, pH, salinity, and nutrient levels, often vary significantly even within small distances due to changes in topography, vegetation cover, and land-use practices. Because of the cost and effort involved, only a limited number of samples are typically collected, resulting in low spatial resolution and potential misrepresentation of actual field conditions. This becomes a serious limitation in precision agriculture, where site-specific data is essential for efficient resource management. Moreover, traditional methods do not allow continuous or real-time monitoring. Samples are taken at fixed intervals, sometimes weeks or months apart, making it impossible to observe dynamic changes in soil properties caused by irrigation, rainfall, or seasonal variations. This lack of temporal resolution can lead to delayed responses and missed opportunities for timely interventions. As a result, decision-making based on such data is often incomplete or outdated, highlighting the need for modern technologies like IoT sensors and remote sensing that can offer high-frequency, wide-coverage soil monitoring.
- **Cost and Resource Constraints**: Traditional soil property analysis involves significant financial and resource-related investments, including the costs of sample collection, laboratory testing, specialized equipment, and skilled manpower. These expenses can be particularly burdensome for small-scale farmers or large agricultural operations requiring frequent testing across multiple fields. Laboratory analysis often requires sophisticated instruments and chemicals, alongside trained technicians to interpret results, factors that contribute to high operational costs. Additionally, logistical challenges such as transportation of samples, sample storage, and delays in obtaining results

further increase the overall burden. Due to these constraints, the frequency and coverage of soil testing are often limited, reducing the effectiveness of data-driven decision-making in agriculture and land management. This highlights the necessity for more cost-effective, scalable, and real-time alternatives, such as IoT-based soil monitoring systems and remote sensing technologies.

4.1.3 The Emergence of IoT and Multimedia Data Analytics in Soil Science

- **Potential for Real-Time and High-Resolution Data Acquisition**: The integration of IoT technologies and multimedia data sources has transformed the landscape of soil property analysis by enabling real-time, high-resolution data acquisition. IoT-based sensors, when deployed directly in the soil, offer continuous, in situ monitoring of vital parameters such as moisture content, temperature, pH levels, and electrical conductivity (EC). These sensors eliminate the delays and inconsistencies associated with manual sampling, allowing for instantaneous data collection and transmission to centralized systems or cloud platforms (Babu, Ravindra Raman, & Venkata Divakara Rao, 2023). Simultaneously, remote sensing technologies, such as satellite imagery, drone-based imaging, and hyperspectral or multispectral sensors, provide wide-area spatial coverage that is difficult to achieve with ground-based techniques. These systems enable the analysis of large-scale patterns and temporal changes in soil conditions, making them highly valuable for agricultural planning, environmental monitoring, and disaster preparedness. The combination of IoT and multimedia tools enables a granular understanding of soil variability both spatially across different regions and temporally over time. This facilitates data-driven decision-making, enhancing the precision and efficiency of interventions such as irrigation scheduling, fertilization planning, and land-use management.
- **Opportunities for Advanced Data Processing and Insights**: The deployment of IoT sensors and multimedia data sources in soil analysis generates large volumes of heterogeneous, high-frequency data. To transform this raw data into actionable knowledge, the application of advanced data processing techniques, including ML and deep learning (DL), becomes essential. These techniques enable the identification of complex patterns, correlations, and trends that are often undetectable through traditional statistical approaches (Akhter & Sofi, 2022). ML algorithms, such as decision trees, random forests, support vector machines (SVMs), and k-nearest neighbors (k-NNs), can be trained on soil datasets to classify soil types, predict moisture levels, or estimate nutrient deficiencies. These models improve over time as more data is collected, leading to enhanced predictive accuracy and adaptability to varying conditions. DL models, particularly convolutional neural networks (CNNs), are highly effective in processing image-based and spectral data acquired from satellite or

drone platforms. They can detect subtle variations in soil texture, surface conditions, or vegetation health that indicate changes in underlying soil properties. Moreover, recurrent neural networks (RNNs) and other temporal models can be employed to analyze time-series data from IoT sensors, allowing for dynamic monitoring and forecasting of soil conditions such as moisture trends or salinity buildup. Beyond prediction, these analytical tools support real-time decision-making, precision agriculture practices, and long-term environmental assessment. For example, integrated systems using analytics can automatically adjust irrigation schedules based on moisture sensor data or detect early signs of soil degradation from satellite imagery. In essence, the fusion of IoT, multimedia data, and AI-driven analytics opens new frontiers in soil science, enabling smarter, faster, and more precise insights for agriculture, environmental sustainability, and land management.

4.1.4 Chapter Objectives and Structure

This chapter aims to explore how IoT sensors and multimedia data analytics can be harnessed to analyze soil properties more effectively and efficiently. The objectives of the chapter include:

- Identifying key soil properties and their relevance across agriculture, environmental monitoring, and civil engineering.
- Reviewing traditional and modern methods for soil analysis.
- Detailing the types and functioning of IoT sensors used in soil monitoring.
- Exploring multimedia data sources and their application in assessing soil conditions.
- Demonstrating how ML and data fusion techniques can enhance soil property prediction.
- Highlighting real-world case studies across domains and discussing challenges and future research directions.

4.2 FUNDAMENTALS OF SOIL PROPERTIES AND THEIR MEASUREMENT

This section discusses the various soil properties and their values, which include physical properties, chemical properties, and biological properties for better crop yielding.

4.2.1 Key Soil Properties for Analysis

Understanding the various physical, chemical, and biological properties of the soil is essential for assessing soil health, managing agricultural productivity, monitoring environmental changes, and ensuring the suitability of land for infrastructure development. Soil properties influence plant growth, water retention, nutrient availability, and microbial activity. These properties are broadly categorized into physical, chemical, and biological types, each offering unique insights into the state and behavior of the soil.

- **Physical Properties:**
  - **Soil Texture**: refers to the proportion of different particle sizes in the soil—sand, silt, clay, and their mixture, loam.
  - **Soil Structure**: describes how soil particles bind together to form aggregates, including forms like granular, blocky, platy, massive, and single grain.
  - **Bulk Density and Porosity**: indicate the soil's compaction and air space, with sub-categories like bulk density and porosity.
  - **Moisture Content**: includes terms like field capacity (maximum water soil can hold after drainage), wilting point (when plants can't extract water), and available water capacity (the water available to plants between these two points).
  - **Temperature**: affects soil processes and is categorized into surface temperature, subsurface temperature, and thermal conductivity (the soil's ability to transfer heat).
- **Chemical Properties:**
  - **Soil pH**: The pH level of soil determines the availability of nutrients to plants and the activity of soil organisms. Most crops thrive in slightly acidic to neutral pH (6.0–7.5). Extreme pH values can lead to nutrient deficiencies or toxicities.
  - **Electrical Conductivity**: EC is a measure of the soil's salinity level and helps determine the presence of dissolved salts. High EC values can inhibit plant growth and affect soil microbial activity, especially in arid and semi-arid regions.
  - **Organic Matter Content**: Organic matter enhances soil structure, water retention, and nutrient availability. It also supports the microbial population and plays a role in carbon sequestration. The presence of humus improves the soil's cation exchange capacity (CEC), increasing its ability to retain nutrients.
  - **Macro- and Micro-nutrients**: Key nutrients such as Nitrogen (N), Phosphorus (P), and Potassium (K) are essential for plant development. Micro-nutrients like iron, zinc, and manganese, though required in smaller amounts, are also crucial. Imbalances in these nutrients can lead to poor crop performance or environmental degradation through leaching and runoff.
- **Biological Properties:**
  - **Microbial Biomass and Diversity**: Though often less emphasized, biological properties are critical indicators of soil health. Soil contains bacteria, fungi, and other microorganisms that decompose organic matter and contribute to nutrient cycling. A rich microbial population supports sustainable agriculture and environmental balance.
  - **Enzyme Activities**: Enzymatic activity is often used as a biological marker for soil health. Enzymes such as dehydrogenase and urease indicate the metabolic potential of soil microbes and their ability to process nutrients.

- **Earthworms and Soil Fauna**: Macro-organisms like earthworms help in mixing soil layers, improving aeration, and enhancing organic matter decomposition. Their presence is often a sign of good soil quality.

4.2.2 Traditional Sensors and Measurement Techniques

- **Manual Sampling and Laboratory Analysis**: Manual sampling and laboratory analysis have long been the standard methods for assessing soil properties. This process begins with the physical collection of soil samples from the field using tools such as soil augers, corers, or shovels. Sampling is typically done at various depths and locations within a field to obtain a representative profile of the soil's condition. The collected samples are then labeled, stored in containers, and transported to a laboratory for further analysis. In the laboratory, a range of techniques is employed to analyze the physical, chemical, and biological properties of the soil. Texture Analysis is performed using the hydrometer method or sieve analysis to determine the proportion of sand, silt, and clay. Moisture Content is measured by weighing the soil before and after drying it in an oven at a specific temperature. Soil pH is determined using a pH meter in a soil-water suspension. EC is measured to assess salinity levels using a conductivity meter. Organic Matter Content is estimated using the loss-on-ignition method or the Walkley-Black method. Nutrient Analysis involves chemical extraction followed by colorimetry, spectrophotometry, or atomic absorption spectroscopy to quantify essential macro- and micro-nutrients. Biological Properties, such as microbial biomass and enzyme activities, are assessed through biochemical assays and culture-based techniques. While these laboratory methods are highly accurate and reliable, they are also time-consuming, labor-intensive, and require skilled personnel and specialized equipment. Additionally, the delay between sample collection and result availability can hinder timely decision-making, especially in fast-paced agricultural operations.
- **Conventional Field Sensors**: Before the advent of modern IoT technologies, conventional field sensors were used to monitor specific soil parameters in real time, primarily at fixed locations. These early sensors included basic analog or digital moisture meters, thermistors for soil temperature, and pH probes, which offered limited automation and data storage capabilities. Typically, these sensors were connected to portable handheld devices or stationary data loggers, requiring manual data retrieval and periodic maintenance. While they provided on-site measurements, their deployment was often sparse due to high cost, limited scalability, and the need for wired connections or short-range wireless communication. Moreover, these systems lacked the ability to integrate multiple sensor types into a network, making it difficult to capture the full spectrum of soil conditions across larger agricultural fields. They also offered low spatial resolution and were unable to transmit data remotely or in real time, key disadvantages when compared to modern IoT sensors, which support wireless connectivity, cloud integration, and continuous, high-resolution monitoring.

4.2.3 Introduction to IoT Sensors for Soil Property Monitoring

The integration of the IoT in agriculture has transformed the way soil properties are monitored and managed. IoT-enabled soil sensors allow for real-time, continuous, and remote monitoring of critical parameters such as moisture, temperature, pH, and EC (Aarthi & Sivakumar, 2023). These sensors, when connected through communication networks, form a smart sensing system capable of gathering large-scale, high-resolution data that enhances decision-making for farmers, environmental scientists, and land developers.

- **Types of IoT Sensors Used in Soil Analysis**
  1. **Soil Moisture Sensors**: These sensors measure the volumetric water content in the soil. The capacitive sensors detect changes in the dielectric constant of the soil, and the resistive sensors measure electrical resistance between two probes.
  2. **Soil Temperature Sensors**: These devices, often thermistors or RTDs (Resistance Temperature Detectors), measure the soil's thermal condition. They are useful in finding seed germination and microbial activity analysis.
  3. **Soil pH Sensors***:* These sensors determine the acidity or alkalinity of the soil using glass electrodes or ion-sensitive field-effect transistors (ISFETs). They help in precise nutrient management but require frequent calibration and maintenance to ensure accuracy.
  4. **EC Sensors**: EC sensors measure the concentration of dissolved salts in the soil, providing an indirect estimate of nutrient availability. They are useful in salinity management and fertilizer planning.
  5. **Emerging Nutrient Sensors**: Advanced sensors are being developed to measure nutrient concentrations (e.g., nitrogen, phosphorus, potassium) in real time using optical, electrochemical, or biosensor technologies. However, these sensors are currently expensive, with limited commercial availability and lower accuracy than lab tests (Figure 4.2).
- **Working Principles and Specifications of IoT Soil Sensors**: IoT soil sensors are typically composed of a sensing unit, microcontroller, power source, and communication module. These components work together

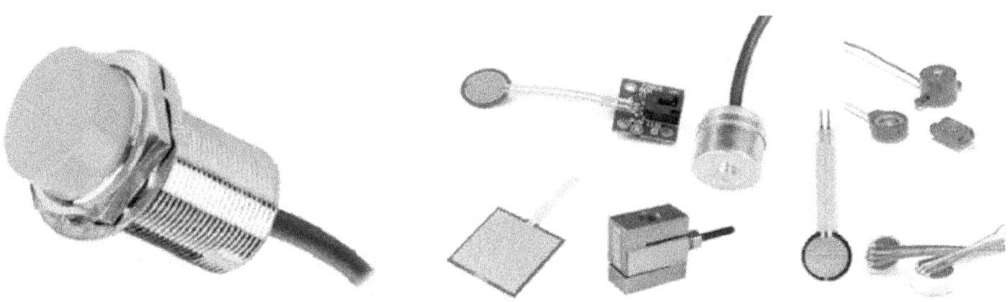

(a) Capacitive Sensor (b) Resistive Sensor

FIGURE 4.2 Soil moisture sensors.

to collect data and transmit it to a central hub or cloud-based platform (Channe, Kothari, Kadam, 2015). These sensors operate on low power, making them suitable for battery or solar-powered setups. Energy-efficient designs are critical for long-term field deployment. These sensors provide high precision, but accuracy can be influenced by environmental conditions, necessitating periodic calibration. Field sensors are housed in waterproof, corrosion-resistant enclosures designed to withstand harsh weather, soil moisture, and physical stress. These specifications make IoT sensors adaptable to varied agricultural, environmental, and research scenarios, allowing for scalable, smart soil monitoring systems. Some of the communication protocols used by these sensors for communication include LoRaWAN (offering long-range, low-power communication ideal for large farm lands), NB-IoT (utilizing cellular infrastructure for wide coverage and high reliability), and Bluetooth/Zigbee (suitable for short-range, high-data-rate applications in small-scale operations) (Figures 4.3–4.5).

- **Deployment Strategies for IoT Sensor Networks in Soil Monitoring**: Effective deployment of IoT soil sensors requires strategic planning to optimize data collection while minimizing cost and maintenance. These sensors must be placed at root-zone levels (typically 10–30 cm) depending on the crop and analysis purpose. Spatial coverage should account for soil

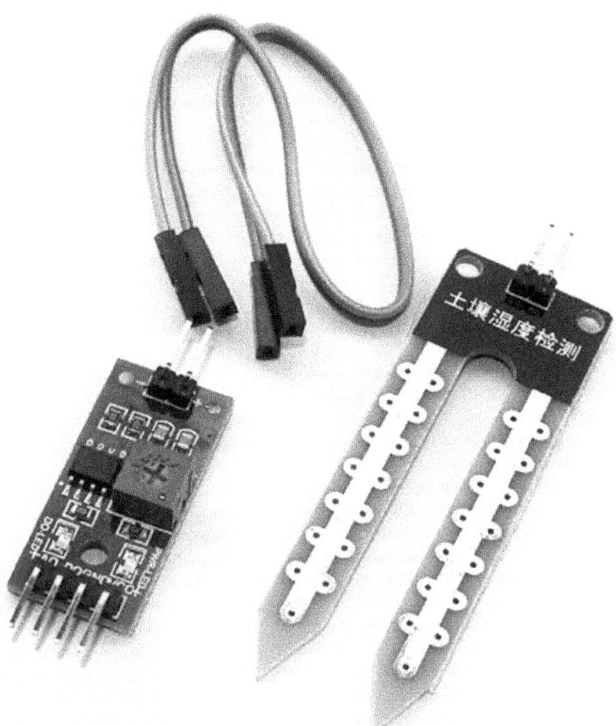

FIGURE 4.3 Soil temperature sensor.

FIGURE 4.4 Soil pH sensor.

FIGURE 4.5 Electrical conductivity (EC) sensor.

heterogeneity; sensors are often deployed in a grid or stratified layout to capture variability across the field. Star topology is suitable for centralized data collection with one gateway (Figure 4.6).

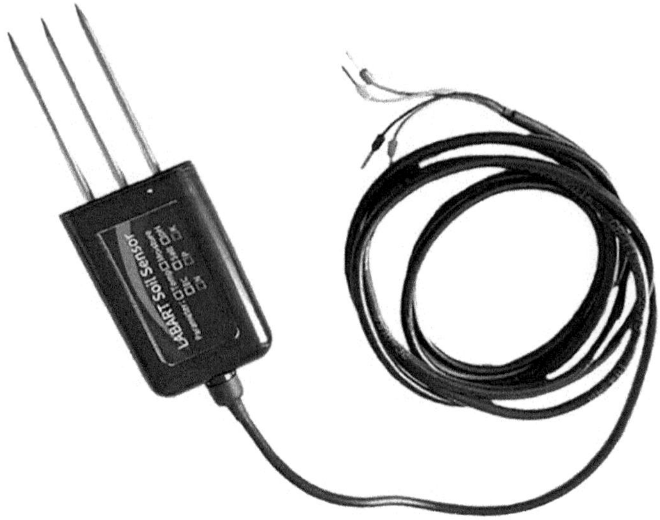

FIGURE 4.6 Soil NPK sensor.

Multiple data transmission strategies, such as edge processing, Cloud integration, and real-time alerts, can be utilized for on-site data filtering, centralized monitoring, and to send alerts (SMS, email) based on sensor readings, aiding timely interventions, respectively.

4.3 MULTIMEDIA DATA FOR SOIL PROPERTY ASSESSMENT

Multimedia data, which encompasses visual, spectral, and thermal information, has emerged as a powerful complement to traditional soil analysis and IoT-based monitoring (AlZu'bi, Hawashin, Mujahed, Jararweh, & Gupta, 2019). With the advancement of imaging technologies and remote sensing platforms, it is now possible to assess various soil properties over large spatial scales with enhanced temporal resolution. Multimedia data, derived from sources such as satellites, drones, and ground-based imaging systems, facilitates non-invasive and rapid evaluation of soil conditions. When coupled with advanced data analytics, these diverse data types provide a comprehensive understanding of soil health, composition, and behavior under varying environmental conditions.

4.3.1 The Role of Multimedia Data in Soil Analysis

Multimedia data plays a significant role in capturing the heterogeneity and temporal dynamics of soil properties. Unlike point-based sensors or laboratory tests, imaging technologies offer spatially continuous and temporally frequent observations. By interpreting visual, spectral, and thermal characteristics of soil surfaces, researchers can infer a wide range of parameters such as organic content, moisture status, temperature gradients, salinity, and structural changes. These datasets, when analyzed through image processing and ML techniques, support large-scale agricultural monitoring, land degradation assessment, and precision farming practices.

- **Visual Information**: Visual information, often acquired through RGB imaging from drones or ground-based cameras, provides fundamental yet valuable insights into soil characteristics. It includes soil color as a key visual attribute that reflects its composition and condition. Dark brown to black soils typically have high organic content, which is crucial for fertility and water retention. Grayish or bluish hues may signal poor drainage or waterlogged conditions, while well-drained soils tend to be red or yellow due to oxidation of iron compounds. The presence of minerals like iron, manganese, and calcium carbonate affects soil coloration. For example, red soils are rich in iron oxides.

  Visual assessment also reveals surface-level physical characteristics such as crusting, residue cover, erosion patterns, etc. Visual data, while simple, forms the first layer of remote observation and, when combined with analytical models, can guide deeper diagnostics of soil health.
- **Spectral Information**: Spectral imaging captures reflectance patterns of soil surfaces across different wavelengths, providing quantitative data for analyzing chemical and physical soil properties. Every material reflects, absorbs, and transmits light differently across the electromagnetic spectrum. Moist soils absorb more in the near-infrared (NIR) region and have lower reflectance. High organic soils show lower reflectance across visible and NIR (VNIR) bands. Iron oxides and minerals have distinct absorption features in the short-wave infrared (SWIR) range.

  Spectral information is further classified as: Multispectral and Hyperspectral. Multispectral imagery collects data across a limited number of discrete bands (typically 4–10), including visible, NIR, and SWIR ranges. It is widely used in agricultural monitoring and soil mapping. Whereas hyperspectral imagery captures data across hundreds of narrow, contiguous bands. This allows for detailed discrimination of soil components and subtle changes in composition. The fine-grained spectral resolution of hyperspectral imaging enhances the capacity for soil diagnostics, making it a vital tool in advanced remote sensing applications.
- **Thermal Information**: Thermal imaging, typically using infrared (IR) sensors, measures the emitted radiation from soil surfaces, which is closely related to temperature (Ge, Thomasson, & Sui, 2011). This thermal data provides indirect yet powerful insights into soil conditions. Thermal infrared sensors detect long-wave infrared radiation (8–14 μm) emitted by the soil. Wet soils evaporate water, leading to cooling. Hence, cooler areas in thermal images may indicate higher moisture content. Changes in surface temperature over time help estimate evaporation and inform irrigation strategies. Lighter soils may heat up faster than darker soils, and coarse-textured soils tend to have higher thermal conductivity. Thermal imaging is especially valuable in precision agriculture, where real-time moisture data supports efficient water usage and crop yield optimization.

4.3.2 Sources of Multimedia Data

Multimedia data for soil property assessment can be collected from a variety of platforms, each offering different advantages in terms of resolution, coverage, cost, and accessibility. These sources include satellites, drones, ground-based systems, and even video-based monitoring setups. When integrated, they offer a multi-dimensional and comprehensive perspective on soil conditions across time and space.

- **Satellite Imagery**: Satellite platforms have long served as a key source for large-scale environmental and agricultural monitoring. They offer consistent, repeatable, and often free data useful for assessing various soil properties. Managed by NASA and the USGS, the Landsat satellites provide multispectral imagery with a spatial resolution of 30 m and a temporal resolution of 16 days. Landsat data is valuable for tracking changes in soil moisture, surface conditions, and land degradation over time. Sentinel-2 provides high-resolution (10–20 m) multispectral data with a revisit time of just 5 days. Its wide spectral coverage includes visible, NIR, and SWIR bands, making it ideal for analyzing soil texture, organic matter content, and mineral composition. MODIS offers moderate-resolution imagery useful for regional analysis, while commercial satellites like WorldView provide very high-resolution images (up to 30 cm) for detailed soil mapping. Satellite data's long-term availability and extensive coverage make it invaluable for tracking soil health, identifying problem areas, and planning large-scale interventions.
- **Aerial Photography and Drone-Based Imaging**: Aerial photography, especially from unmanned aerial vehicles (UAVs) or drones, offers high flexibility and very fine spatial resolution, typically in the range of a few centimeters (Phang, Chiang, Happonen, & Chang, 2023). These platforms are revolutionizing precision agriculture and soil analysis due to their ease of deployment and real-time data capture capabilities. Drone-mounted cameras and sensors can capture detailed imagery that reveals fine-scale soil surface characteristics, crusting, and residue cover. Drones can also be equipped with multispectral and thermal sensors to capture spectral signatures related to soil moisture, temperature, and nutrient status. Unlike satellites, drones can be deployed on demand, allowing for timely monitoring during key agricultural stages. Drones with LiDAR or photogrammetry can generate digital elevation models (DEMs), which are helpful in understanding erosion risk, water flow, and drainage.
- **Ground-Based Imaging Systems**: Ground-based systems offer the highest level of detail and are ideal for close-range, high-accuracy soil analysis. Portable devices that measure soil reflectance across various spectral bands can instantly estimate properties like organic content and moisture. Automated camera systems are installed on tractors or field robots. These systems capture continuous visual or spectral images during field operations. These tools are particularly useful for calibration and validation of remote sensing models and for making site-specific measurements during field studies.

- **Video Data for Dynamic Soil Processes**: Video-based monitoring provides valuable temporal insights into dynamic soil processes that are difficult to capture using still images or single-frame sensors. Time-lapse or real-time video can document soil displacement, rill formation, and sediment transport during rain or irrigation events. Videos can help observe surface runoff, infiltration rate, and pooling behavior in response to different soil structures or treatments. When combined with computer vision and AI, video data can be analyzed for event detection, classification, and modeling of soil behavior under variable conditions. While still an emerging area, video-based soil monitoring holds great promise for research and educational applications in soil science and hydrology.

4.3.3 Preprocessing and Calibration of Multimedia Data for Soil Analysis

To accurately analyze soil properties using multimedia data such as satellite images, drone imagery, and ground-based spectral data, it is essential to perform a series of preprocessing and calibration steps. These processes enhance the quality of the data, reduce noise, and ensure that the information derived is spatially and spectrally reliable. Preprocessing steps such as geometric correction, atmospheric correction, and spectral index computation are critical for making valid comparisons and interpretations. Furthermore, calibration with ground truth data helps bridge the gap between observed signals and actual soil conditions.

- **Geometric Correction and Orthorectification**: Geometric correction is vital for ensuring that the multimedia data used in soil analysis is accurately aligned with real-world coordinates. Raw aerial and satellite images often suffer from geometric distortions due to the Earth's curvature, sensor tilt, terrain variations, and platform movement. Orthorectification is the process of correcting these distortions using DEMs and sensor metadata to produce images where the scale is uniform and spatial relationships are preserved. This correction allows for precise overlay with other spatial datasets such as field boundaries, soil maps, and IoT sensor networks. In soil studies, accurate spatial referencing is crucial for detecting patterns of variability in properties like moisture or organic matter across a field and for integrating remote data with in situ measurements. Without proper geometric correction, even high-resolution imagery can lead to misinterpretations in land-use assessments or site-specific soil management.
- **Atmospheric Correction**: When capturing spectral data from satellite or aerial platforms, the atmosphere acts as a medium that can distort the signal received by the sensor. Factors like water vapor, aerosols, and gases cause scattering and absorption of light, particularly in the VNIR bands (Vermote & Kotchenova, 2008). Atmospheric correction techniques aim to retrieve the true surface reflectance of the Earth's surface by removing atmospheric effects from the recorded signal. Methods such as Dark Object Subtraction

(DOS), Fast Line-of-sight Atmospheric Analysis of Spectral Hypercubes (FLAASH), and QUAC (Quick Atmospheric Correction) are commonly used. For soil analysis, this correction is essential to ensure consistency in spectral values over time and space, which is particularly important when comparing data from different dates or platforms. Without atmospheric correction, indices derived from spectral data may be misleading, especially in regions with variable atmospheric conditions or during different seasons. Proper correction enables the reliable detection of subtle differences in soil surface properties such as texture, moisture, and mineral content.

- **Spectral Indices for Soil Property Estimation**: Spectral indices are mathematical combinations of reflectance values at specific wavelengths designed to highlight particular surface features. While many indices are developed for vegetation, some have been adapted or proposed for soil analysis. NDVI (Normalized Difference Vegetation Index) is primarily used for vegetation; low NDVI values can indirectly indicate bare soil exposure and erosion-prone areas. SAVI (Soil Adjusted Vegetation Index) modifies NDVI to account for soil background reflectance, offering improved sensitivity in areas with sparse vegetation. These indices simplify the interpretation of complex spectral data and serve as proxy indicators of important soil attributes like organic matter, fertility, and erosion status.
- **Calibration with Ground Truth Data from IoT Sensors and Laboratory Analysis**: Calibration is the process of aligning remotely sensed data with accurate, on-ground reference data to develop robust and reliable predictive models. IoT sensors and traditional soil sampling provide precise, localized measurements of soil moisture, pH, EC, and nutrient levels. These serve as benchmarks against which remote sensing observations are validated and tuned. ML models can be trained using remote sensing data (e.g., spectral indices) as inputs and ground-truth data as labels. This improves the predictive accuracy of soil property estimation and supports the development of transferable models across regions and seasons. Regular synchronization between field sensors and remote data helps maintain model performance, especially in dynamic environments with seasonal or anthropogenic changes. By integrating remote and ground-based data sources, analysts can leverage the scalability of remote sensing with the precision of field data for comprehensive soil monitoring.

4.4 MULTIMEDIA DATA ANALYTICS FOR SOIL PROPERTY EXTRACTION

The integration of advanced data analytics with multimedia sources has revolutionized soil property assessment by enabling large-scale, accurate, and timely extraction of critical information. Multimedia data, particularly imagery and video, must be processed using robust computational techniques to derive features that reflect meaningful physical, chemical, and biological attributes of the soil. Feature extraction serves as the bridge between raw data and actionable insights, converting visual, spectral, and thermal information into quantifiable indicators for soil analysis.

4.4.1 Feature Extraction from Multimedia Data

Feature extraction involves deriving specific parameters or patterns from image data that are indicative of underlying soil characteristics. This process transforms high-dimensional image data into structured inputs suitable for ML models, statistical analysis, and decision-making tools. The extracted features may be statistical, spectral, spatial, or temporal in nature.

- **Statistical Features**: Statistical analysis of pixel intensity values offers a simple yet effective way to capture spatial variability in soil surface characteristics. These basic statistical measures reflect average reflectance and variability within a specific region. For instance, darker soils (lower mean pixel values) may indicate higher organic matter content or moisture levels. Texture Features: Soil texture in images refers to the spatial arrangement and variation of pixel intensities, which can be analyzed using:
  a. **GLCM (Gray Level Co-occurrence Matrix)**: Measures how often pairs of pixel intensities occur in an image at a certain spatial distance and orientation.
  b. **Derived Texture Features**: These include contrast, homogeneity, energy, and entropy, all of which relate to surface roughness, tilth, and residue cover.

  By quantifying variations in tone and texture, these features provide valuable information on soil surface conditions and underlying structure, especially when integrated with field data.
- **Spectral Signatures and Band Ratios**: Spectral signatures represent the specific reflectance characteristics of soil at various wavelengths. Each soil type or property reflects and absorbs light differently, creating a unique spectral profile that can be used for classification and estimation. The reflectance curve of a soil sample across multiple bands (e.g., VNIR, SWIR) reveals insights into:
  a. **Moisture Content**: Moist soils absorb more in the infrared region.
  b. **Organic Matter**: Darker soils exhibit lower reflectance in the visible range.
  c. **Iron Oxides and Minerals**: Certain wavelengths show distinct peaks and troughs that correspond to specific mineral compositions.

  Spectral analysis provides critical insight into the composition and condition of soils across broad areas, making it essential for remote sensing-based soil mapping and monitoring.
- **Image Segmentation Techniques for Soil Regions**: Before detailed soil analysis can be performed, it is important to isolate soil regions from other land cover types such as vegetation, water bodies, or built-up areas. Image segmentation techniques enable this separation by clustering or labeling pixels with similar characteristics. Thresholding: One of the simplest segmentation methods, where pixel values are classified based on predefined reflectance thresholds. Techniques like K-means or Fuzzy C-means are used to group pixels based on similarity in spectral or spatial features. Region-Based Segmentation identifies contiguous regions with

homogeneous texture or spectral features, allowing the isolation of plowed fields, bare soil patches, or eroded areas. ML and DL Approaches: Modern methods, such as U-Net, Mask R-CNN, or semantic segmentation CNNs, offer high-precision, automated segmentation, especially useful in complex and heterogeneous landscapes. Effective segmentation improves the accuracy of downstream analysis by focusing only on relevant soil regions and eliminating noise from irrelevant features. This step is particularly important when dealing with high-resolution imagery from drones or satellites.

4.4.2 ML Algorithms for Soil Property Prediction

Accurate prediction of soil properties is vital for agricultural planning, land management, and environmental monitoring. ML algorithms provide scalable, data-driven approaches to predict both continuous and categorical soil properties by learning patterns from diverse data sources, including spectral data, remote sensing imagery, and field survey data (Shingade & Mudhalwadkar, 2024). This section explores three main categories: regression techniques, classification techniques, and DL approaches.

- **Regression Techniques**: Regression models are employed when the goal is to predict continuous soil variables such as pH level, organic carbon content, nitrogen concentration, or moisture levels. Commonly used regression algorithms include:
    a. **Linear Regression**: Provides a baseline for modeling linear relationships between soil features and target properties.
    b. **Support Vector Regression (SVR)**: Effective in high-dimensional spaces and suitable for non-linear relationships.
    c. **Random Forest Regression**: An ensemble learning method that improves accuracy by averaging results across multiple decision trees.
    d. **Gradient Boosting Machines (e.g., XGBoost, LightGBM)**: Offer superior predictive performance by optimizing residual errors iteratively.

    These models require careful feature engineering using inputs such as spectral indices (NDVI, EVI), soil texture, climatic data, and topographical attributes.
- **Classification Techniques**: Classification algorithms are valuable for predicting categorical soil properties like soil type, texture class (e.g., sandy, loamy, clayey), or fertility category (low, medium, high). These models treat the output as discrete labels and are trained using labeled datasets derived from soil surveys or remote sensing analysis. Key algorithms include:
    a. **Decision Trees and Random Forests**: Useful for their interpretability and ability to handle complex interactions between features.
    b. **k-Nearest Neighbors**: A simple, non-parametric method that classifies based on proximity in feature space.
    c. **Naive Bayes**: Probabilistic approach based on feature independence assumptions.
    d. **Support Vector Machines**: Effective for multi-class classification with appropriate kernel functions.

These models can be trained on a variety of inputs such as land-use patterns, spectral reflectance values from satellite images, and elevation or slope data. When paired with remote sensing, classification models can generate spatial maps showing the distribution of soil types across large regions, enabling scalable soil assessment.

- **DL Approaches**: DL models, particularly CNNs, have shown promising results in soil property prediction from image-based data such as remote sensing images, drone-captured photographs, and hyperspectral imagery. CNNs are well-suited to learn complex spatial and spectral patterns directly from image inputs without extensive manual feature engineering. These networks consist of multiple layers that extract low-level to high-level features hierarchically, enabling accurate predictions of both continuous and categorical soil attributes. Typical applications include:
  a. **Soil Texture Classification**: Using multispectral or hyperspectral imagery to identify spatial patterns and classify soil texture zones.
  b. **Organic Matter and Moisture Estimation**: Estimating content from variations in reflectance and color in high-resolution images.
  c. **Image Segmentation**: Using architectures like U-Net or DeepLab for pixel-wise classification of soil characteristics in large geographical areas.

  Advanced techniques like transfer learning (using pre-trained models on related datasets) and data augmentation can be employed to improve performance, especially when labeled data is limited. Integration with GIS and cloud computing platforms further enhances the scalability of DL-based soil monitoring systems. In summary, DL complements traditional ML approaches by enabling automated feature learning from raw data, making it a valuable tool in modern soil science and precision agriculture.

4.4.3 Fusion of IoT Sensor Data and Multimedia Data for Enhanced Analysis

The integration of IoT sensor data with multimedia data sources, such as satellite imagery and drone footage, represents a transformative approach in soil property analysis. While IoT sensors provide high-frequency, in situ measurements (e.g., moisture, temperature, pH), multimedia data offers rich spatial and contextual information. Their fusion allows researchers to overcome limitations inherent to each data type, improving both the granularity and accuracy of soil predictions (Din, Ahmad, Paul, Rathore, & Jeon, 2017) (Figure 4.7).

- **Data Integration Techniques**: Data fusion techniques are essential for integrating heterogeneous data from multiple modalities. In the context of soil property analysis, sensor fusion involves combining temporal data from ground-based IoT sensors (e.g., soil moisture, temperature, salinity) with spatial data from multimedia sources like satellite imagery, drone images, or hyperspectral sensors. Key integration techniques include Kriging and Co-Kriging. Algorithms like Random Forest, SVMs, and DL models (e.g., CNN-LSTM

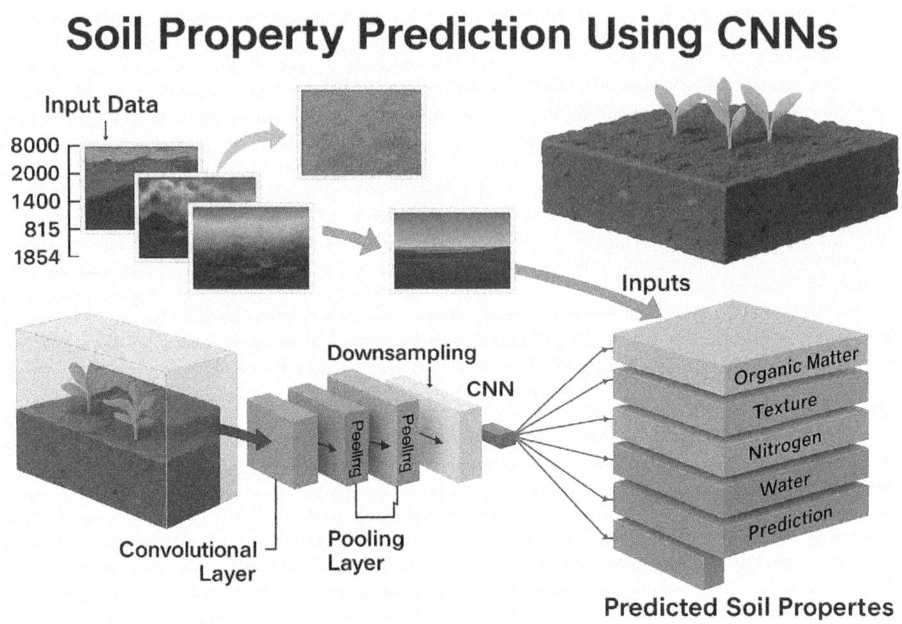

FIGURE 4.7 Soil property prediction through Convolutional Neural Networks.

hybrids) can learn from combined datasets by extracting temporal trends and spatial patterns simultaneously. Bayesian Fusion Models combine multiple data sources by weighing them based on uncertainty, thereby improving prediction confidence. Rule-Based and Semantic Fusion utilizes domain-specific rules to integrate data sources with differing resolutions and contexts, particularly useful in decision-support systems. This fusion is performed either at the data level (combining raw data), feature level (merging extracted features), or decision level (combining outputs of separate models).

- **Improving Spatial and Temporal Resolution through Fusion**: Fusing IoT and multimedia data enables a much finer spatio-temporal resolution in soil monitoring than using either source in isolation. Satellite or drone imagery provides wide-area, high-resolution views of agricultural fields. When combined with localized sensor data, it allows the model to extrapolate soil conditions across the entire region, filling gaps between sensor nodes. IoT sensors continuously or at regular intervals record data, capturing real-time dynamics. This complements the lower-frequency capture rates of satellite imagery, enabling near-continuous monitoring. Interpolation, time-series forecasting, and temporal synchronization are applied to match the timeframes of image and sensor data, facilitating effective fusion. Data from fixed sensors can be spatially extended using image features and ML models trained to relate sensor readings with image-derived indicators like vegetation indices or surface reflectance.

The result is a more granular, continuous, and accurate map of soil properties, improving planning and resource allocation in precision agriculture and environmental management.

- **Enhancing Prediction Accuracy and Robustness**: The synergistic use of IoT and multimedia data significantly enhances the accuracy and robustness of soil property predictions. IoT sensors provide precise, real-time measurements but are spatially limited. Multimedia data offers contextual and spatial richness but may lack temporal depth or accuracy due to noise or atmospheric conditions. By integrating both, the models become more accurate, more robust, and scalable.

  Fusion-driven models have shown improvements in prediction metrics like RMSE, MAE, and F1-score across multiple studies, particularly for complex variables such as soil texture classification, moisture estimation, and nutrient mapping.

4.5 CASE STUDIES AND APPLICATIONS

4.5.1 Precision Agriculture: Optimizing Resource Management

Precision agriculture aims to enhance farm productivity, sustainability, and efficiency by tailoring management practices to specific site conditions. The integration of IoT sensors, ML, and remote sensing technologies enables farmers to monitor soil conditions, apply inputs variably, and predict crop performance with increased precision. The following case studies illustrate real-world applications of these technologies in optimizing water, nutrient, and yield management. There are various case studies on this topic, a few of them are discussed below:

Case Study 1: Variable Rate Irrigation Based on Soil Moisture Data: Smart Irrigation System in Gujarat, India

A pilot project in Gujarat implemented an IoT-enabled smart irrigation system across 50 hectares of pomegranate and cotton fields. The system used soil moisture sensors deployed at depths of 15, 30, and 60 cm to continuously monitor real-time water content across multiple zones of the farm. The data was transmitted via LoRaWAN to a cloud platform where it was analyzed using threshold-based algorithms. Based on sensor feedback, a Variable Rate Irrigation (VRI) protocol was developed, activating or deactivating drip lines only when moisture levels fell below crop-specific thresholds. This system accounted for soil heterogeneity, micro-climatic variations, and evapotranspiration rates. Results: (a) Water usage reduced by 30%–35%, (b) improving crop yield by 15%, and (c) enhancing root-zone moisture consistency, minimizing water stress. This approach demonstrated how real-time soil moisture monitoring can lead to data-driven irrigation scheduling, conserving water and enhancing crop health in semi-arid regions.

Case Study 2: Site-Specific Nutrient Management Using Soil Nutrient and Spectral Data: Maize Cultivation in Punjab Using Soil-Spectral Fusion

In Punjab, an experimental farm deployed a site-specific nutrient management system for maize cultivation. The system combined laboratory-tested soil nutrient data (NPK concentrations) with NDVI and Red-Edge indices derived from drone-based multi-spectral imagery. Using spatial interpolation (kriging) and ML models (Random Forest Regressors), the team generated nutrient variability maps across the field. These maps were uploaded into a GPS-enabled variable rate fertilizer applicator, which adjusted the fertilizer application in real time based on localized soil nutrient deficiencies.

Outcomes:

a. 15%–20% reduction in fertilizer usage
b. 20% increase in nutrient use efficiency
c. Lower nitrate leaching and reduced environmental footprint. This case demonstrates the power of multimodal data fusion (lab data + spectral imagery) in delivering precision nutrient recommendations tailored to micro-zones within a single field.

Case Study 3: Yield Prediction Based on Integrated Soil Property Analysis: Wheat Yield Forecasting in Madhya Pradesh Using IoT and Satellite Data

A research initiative in Madhya Pradesh aimed to develop a predictive yield model for wheat using integrated soil and vegetation analysis. The system utilized: IoT sensor data (soil moisture, pH, EC, temperature), satellite-derived vegetation indices (NDVI, SAVI), historical yield records, and weather forecasts. A Long Short-Term Memory (LSTM) neural network was trained to predict end-of-season yield based on time-series inputs of soil and vegetation health during key growth stages. The model was validated across 20 test plots over two seasons. Performance metrics: $R^2 = 0.88$, RMSE = 0.37 tons/ha.

Predictive accuracy improved by 20% compared to soil-only or satellite-only models. This example underscores how combining real-time in-field data with remote sensing inputs can enhance the robustness and granularity of yield forecasting models, enabling better planning for procurement, insurance, and logistics.

4.5.2 Environmental Monitoring: Assessing Soil Degradation and Contamination

Soil degradation and contamination are growing environmental concerns driven by intensive agriculture, industrialization, deforestation, and climate change. Continuous environmental monitoring using IoT sensors, remote sensing, and spectral data enables scientists and policymakers to detect early signs of degradation, track changes over time, and design effective remediation strategies. The following case studies illustrate technological interventions aimed at assessing erosion, pollution, and climate-induced changes in soil.

Case Study 1: Mapping Soil Erosion Using Remote Sensing and Time-Series Data: Soil Erosion Mapping in the Western Ghats using Sentinel-2 and Rainfall Data

In the ecologically sensitive Western Ghats region of India, a study was conducted to map and monitor soil erosion patterns using multi-temporal Sentinel-2 imagery combined with historical rainfall and slope data. The project applied the Revised Universal Soil Loss Equation (RUSLE) into a GIS framework, where the vegetation cover (C-factor) and land-use changes were dynamically estimated using NDVI trends from satellite data over 5 years. The erosion-prone zones were identified by analyzing:

a. NDVI anomalies during monsoon seasons
b. DEM-derived slope and aspect layers
c. Rainfall intensity indices from IMD datasets
d. Hotspot analysis revealed a strong correlation between declining vegetation indices and erosion severity, particularly in deforested patches and sloped agricultural lands.

Key Outcomes:

a. Identified high-risk erosion zones for targeted afforestation
b. Provided policymakers with temporal erosion maps for land-use planning This case highlights how time-series remote sensing coupled with climatic and topographic data can provide scalable, repeatable insights into soil erosion dynamics.

Case Study 2: Detection and Monitoring of Soil Pollutants Using Spectral Signatures: Heavy Metal Contamination Detection in Industrial Zones of Haryana

A geospatial project was conducted to detect and map heavy metal contamination (Pb, Cd, Cr) in soils near industrial belts in Faridabad, Haryana, using field spectro-radiometry and hyperspectral satellite data (Hyperion). Soil samples were collected from 80 locations and analyzed in the lab for heavy metal concentrations. Corresponding spectral signatures were extracted and correlated with lab results using Partial Least Squares Regression (PLSR) and SVMs. Specific spectral absorption features in the VNIR range were found to be indicative of metal presence.

Workflow Highlights:

a. Lab-based spectral library creation
b. Spectral index calibration (e.g., Red Edge Position, Clay Mineral Ratio)
c. Spatial extrapolation using hyperspectral imagery

Results:

a. Accurate contamination maps with $R^2 > 0.85$.
b. Spatial zones classified into low, medium, and high-risk categories.
c. Enabled low-cost, non-invasive pollutant monitoring.

This approach demonstrated the potential of spectral fingerprinting as a viable tool for environmental forensics and remediation planning in pollution-prone regions.

Case Study 3: Assessing the Impact of Climate Change on Soil Properties: Climate Impact Monitoring in Semi-Arid Rajasthan Using IoT Remote Fusion

In Rajasthan's Thar Desert, a long-term climate monitoring initiative combined IoT-based soil sensors with remote sensing datasets (MODIS and Landsat) to study the effects of climate change on soil health over a 10-year period. IoT nodes recorded soil temperature, moisture, and salinity across seasons, while satellite data provided land surface temperature (LST), albedo, and vegetation trends. Findings included:

a. A gradual decline in soil moisture retention linked to increased surface temperatures and shifting monsoon patterns.
b. Altered carbon sequestration patterns, inferred from reductions in organic matter and NDVI fluctuations.
c. Rising soil salinity in irrigated zones due to increased evapotranspiration. These insights helped in the development of adaptive soil management strategies, such as mulching, cover cropping, and salt-tolerant crop varieties, tailored for climate-vulnerable areas. This case underscores how integrated spatio-temporal monitoring systems are vital for understanding and mitigating the long-term impacts of climate variability on soil health.

4.5.3 Infrastructure and Construction: Soil Stability and Suitability Assessment

Soil assessment is a critical step in infrastructure development, particularly in large-scale construction projects like roads, bridges, and buildings. Ensuring the soil's suitability, stability, and load-bearing capacity minimizes risks of structural failure and cost overruns. Integrating remote sensing, IoT sensors, and geospatial analysis can support rapid, large-scale assessments of key soil parameters such as moisture content, density, and type, facilitating informed engineering decisions.

Case Study 1: Mapping Soil Moisture and Density for Foundation Stability: Smart Soil Monitoring for Highway Construction in Maharashtra

During the construction of a major highway near Nagpur, IoT-based ground sensors and synthetic aperture radar (SAR) data from Sentinel-1 were used to assess soil moisture and compaction levels along the proposed corridor. Ground sensors (Time-Domain Reflectometry probes and cone penetrometers) recorded real-time soil volumetric water content and resistance to penetration, while satellite radar backscatter coefficients were analyzed to detect subsurface moisture trends across large segments. A ML model (Random Forest) fused the sensor and radar data to create soil stability maps, identifying high-risk zones with waterlogging tendencies or low compaction.

Impact:

a. Optimized foundation depths and material requirements.
b. Prevented potential subgrade failures due to seasonal moisture variability.
c. Reduced pre-construction survey time by 40%.

This application highlighted the value of integrated soil sensing and remote sensing in preemptively addressing geotechnical challenges in large infrastructure projects.

Case Study 2: Identifying Soil Types and Their Engineering Properties Using Remote Sensing: Soil Texture Mapping in Urban Expansion Zones Using Spectral Indices

In Bengaluru's rapidly urbanizing peripheries, a soil type classification study was conducted using Landsat-8 OLI and field spectral measurements. Spectral reflectance patterns in the visible and SWIR bands were analyzed to differentiate clay-rich, sandy, and loamy soils—each possessing distinct engineering properties relevant to construction. Using the Normalized Difference Soil Index (NDSI) and clay mineral ratios, the study accurately delineated zones of high shrink-swell potential clay, which are generally unsuitable for shallow foundations without stabilization.

Use Case:

a. Enabled urban planners and civil engineers to avoid problematic soil zones.
b. Informed decisions on soil stabilization techniques (e.g., lime treatment). This brief example illustrates how spectral analysis can offer rapid, cost- effective support in mapping geotechnically relevant soil types, improving infrastructure resilience.

4.6 CHALLENGES AND FUTURE DIRECTIONS

4.6.1 Challenges in IoT Sensor Deployment and Data Management

While IoT and remote sensing technologies have revolutionized soil monitoring, their practical deployment and data management raise several logistical, technical, and ethical concerns. These challenges can affect the reliability, scalability, and trustworthiness of soil analysis systems. This section explores key operational barriers such as sensor reliability, data security, and network scalability.

- **Sensor Calibration and Maintenance**: A critical challenge in deploying IoT sensors for environmental monitoring lies in maintaining data accuracy and hardware longevity. Soil sensors measuring moisture, salinity, or temperature are susceptible to calibration drift over time due to environmental stressors such as temperature fluctuations, moisture accumulation, corrosion, or microbial growth. Some common issues include sensor fouling or clogging in wet and clay-heavy soils, signal attenuation due to soil compaction or root intrusion, and calibration mismatches across different soil types or depths. Periodic recalibration using field reference instruments is necessary. Additionally, ensuring sensor waterproofing, rugged casing, and energy efficiency (e.g., solar-powered nodes) becomes essential in field conditions, especially in remote or resource-limited areas. These factors add to the cost and complexity of long-term deployment in agricultural and environmental settings.
- **Data Security and Privacy Concerns**: With thousands of sensors transmitting real-time environmental data, the security of data transmission, storage, and access becomes paramount. Most IoT devices use wireless communication protocols (e.g., LoRaWAN, ZigBee, Wi-Fi), which can be vulnerable to interception, spoofing, or unauthorized access. Some common but major security risks include man-in-the-middle attacks during data transmission, data manipulation or loss due to lack of encryption, device hijacking in unprotected mesh networks, etc. Moreover, many applications involve geotagged soil data, which can indirectly reveal sensitive information about farm productivity, land use, or even private property boundaries. This raises privacy concerns for individuals, cooperatives, and institutions. The solutions for the above-mentioned issues include implementation of end-to-end encryption protocols (TLS, SSL), use of secure authentication mechanisms (e.g., multi-factor access to cloud platforms), and adoption of data anonymization and role-based access control (RBAC). As IoT data is increasingly used in policymaking, insurance, and subsidies, the ethical handling of data and adherence to regulatory frameworks like GDPR and India's Digital Personal Data Protection Act (DPDP) become crucial.
- **Scalability and Network Connectivity**: Scaling up from experimental sensor setups to large-area, multi-site networks poses significant technological challenges. In many rural or remote environments, unreliable cellular networks, limited power supply, and difficult terrain hinder seamless data collection and transmission. This results in network latency and packet loss

in low-bandwidth areas, energy constraints for battery-operated or solar sensors, mesh network failures due to single points of failure. Additionally, ensuring interoperability among heterogeneous devices and platforms from different manufacturers complicates integration and maintenance. To support scalability, low-power wide-area networks (LPWANs) such as LoRa or NB-IoT can be adopted, edge computing can reduce data loads by performing preliminary processing at the sensor level, and cloud-based sensor management platforms can automate firmware updates, performance diagnostics, and data aggregation. Addressing these infrastructural and architectural challenges is key to building resilient, large-scale smart agriculture and environmental monitoring systems.

4.6.2 Challenges in Multimedia Data Analysis

The integration of multimedia data—especially satellite imagery, drone footage, and hyperspectral images—has significantly enhanced the spatial resolution and granularity of soil property monitoring. However, the analysis of such data introduces a unique set of challenges related to data quality, volume, and the complexity of predictive models. This section outlines key issues in handling multimedia data for soil analysis.

- **Handling Data Heterogeneity and Volume**: Multimedia data sources differ vastly in resolution, spectral range, format, and temporal frequency, creating substantial heterogeneity. For example, Sentinel-2 provides 10–20 m resolution multispectral imagery at 5-day intervals, while drone imagery offers centimeter-level spatial detail but often lacks spectral depth. It includes common challenges like data normalization and preprocessing for combining sources with different spatial and spectral characteristics, high storage and computational demands, especially for video or hyperspectral data involving hundreds of bands, and temporal alignment issues when data sources have mismatched update frequencies or coverage gaps. Advanced data fusion frameworks and cloud-based platforms like Google Earth Engine and AWS Open Data are increasingly used to manage these challenges. However, seamless interoperability and standardization remain underdeveloped, especially when integrating open-access and proprietary datasets.
- **Addressing Atmospheric and Environmental Influences**: Atmospheric distortions and environmental interferences significantly impact the accuracy of remote sensing-derived soil indicators. Cloud cover, haze, water vapor, and aerosol scattering can obscure or alter reflectance values, especially in the VNIR bands used to detect soil moisture, organic content, and texture. However, the common issues are inconsistent illumination and shadow effects in drone images, spectral contamination due to vegetation or surface residue, and dynamic environmental factors, such as recent rainfall or surface roughness, that alter reflectance. While radiometric correction algorithms like DOS, topographic normalization, and atmospheric

correction models (e.g., FLAASH, 6S) help mitigate some effects, they require fine-tuning for soil-specific contexts and still introduce uncertainty. Additionally, cloud masking techniques often fail in heterogeneous terrain, limiting usable image frequency. Thus, the quality and usability of imagery for soil analysis heavily depend on preprocessing robustness, which remains an active area of research.

- **Interoperability of Complex ML Models**: Many soil analysis frameworks now rely on DL models, such as CNNs, to extract features from satellite or drone imagery. While these models offer high predictive accuracy, they often function as "black boxes", lacking transparency in how they arrive at specific predictions. It raises concern regarding model validation and trust, particularly for high-stakes applications like contamination detection or land suitability classification, and poses challenges for regulatory compliance, where interpretability is mandated for decision support systems. Scientific explainability limits insights into the causal relationships between input features and soil outcomes. Emerging techniques like Layer-wise Relevance Propagation (LRP), SHAP (SHapley Additive exPlanations), and Grad-CAM provide partial interpretability but are computationally intensive and domain-sensitive. Improving the explainability of AI-driven soil models is essential not only for scientific credibility but also for broader adoption in agriculture, infrastructure planning, and environmental monitoring.

4.6.3 Future Trends and Research Opportunities

The convergence of IoT, remote sensing, AI, and data science continues to reshape how soil properties are analyzed and interpreted. Looking ahead, several technological and methodological advances are poised to address current challenges and unlock new possibilities in precision agriculture, environmental monitoring, and geotechnical applications. This section outlines key emerging trends and research opportunities.

- **Advancements in Sensor Technology and Miniaturization**: Next-generation soil sensing is moving toward more compact, cost-effective, and multifunctional sensor systems. Innovations in MEMS (Micro-Electro-Mechanical Systems) and nano-sensor technology are enabling the development of sensors that can simultaneously measure multiple parameters such as moisture, temperature, pH, salinity, and nutrient content in real time. Emerging capabilities include wearable nano-sensors that detect micro-level changes in soil chemistry, printed and flexible sensors embedded in biodegradable substrates for eco-friendly deployment, and wireless passive sensors that require no battery, ideal for remote or long-term installations. These developments promise wider adoption, especially in resource-limited settings, while improving the spatial density and temporal resolution of data. Research into self-calibrating and self-healing sensor materials is also gaining momentum to improve reliability and reduce maintenance.

- **Integration of Edge Computing and AI for Real-Time Analysis**: The vast volume and velocity of data generated by IoT and multimedia sources creates bandwidth and latency bottlenecks when relying solely on cloud processing. Edge computing, where data is processed near the point of collection, combined with lightweight AI models, is emerging as a solution for real-time soil analysis. It includes reduced latency in decision-making (e.g., adjusting irrigation in real time), lower data transmission costs due to on-site preprocessing, and enhanced data privacy and autonomy, especially for localized applications. Low-power AI accelerators like NVIDIA Jetson Nano and Google's Coral TPU are already integrated into field-deployable sensor kits. The future lies in developing energy-efficient AI algorithms that can run on edge devices with minimal computing power, enabling fully autonomous soil monitoring systems.
- **Development of More Robust and Explainable AI Models**: To move beyond black-box predictions and gain actionable insights, there is a growing emphasis on robust and interpretable ML. Models must be able to handle noisy, imbalanced, and incomplete data from field sensors and imagery, adapt to different soil types and climatic zones, offer explanations for decisions, especially in critical scenarios like contamination detection, and hybrid models, combining physics-based simulation with data-driven learning, which are gaining interest. Meanwhile, XAI (Explainable AI) tools such as LIME, SHAP, and surrogate models are being refined to better suit environmental and agricultural domains. Future research should focus on embedding ability directly into model architecture rather than treating it as a post-hoc addition.
- **Standardization and Interoperability of Data and Platforms**: The lack of common standards in data formats, sensor interfaces, and communication protocols hampers seamless integration and scalability. Interoperability is vital for sharing data across platforms (e.g., government, academic, and private sectors), combining heterogeneous data sources (e.g., IoT + satellite + lab data), building modular, plug-and-play systems, and efforts like OGC. Sensor Things API, ISO/IEEE 11073, and FAO's Soil Information Standards offer promising frameworks but require broader adoption. Future research should also explore semantic ontologies for soil-related data to enable intelligent querying, integration, and discovery in large-scale systems.

4.7 CONCLUSION

4.7.1 Summary of Key Findings and Contributions

This chapter explored the integration of IoT technologies and multimedia data analytics in soil science, providing a comprehensive review of how these innovations are reshaping the way soil properties are monitored, analyzed, and managed. ML and DL Algorithms, such as regression, classification, and CNN-based models, have demonstrated the ability to effectively predict soil attributes and classify soil types from both structured data and imagery, reducing the need for manual feature engineering. The chapter also emphasized how the combination of temporal sensor data and spatially rich multimedia inputs enables more accurate, robust, and spatio-temporally resolved soil analysis, enhancing agricultural and environmental decision-making. Real-world applications in precision agriculture, environmental monitoring, and infrastructure planning demonstrated how data-driven soil insights can optimize resource use, detect pollution, forecast yield, and support construction planning. Despite the progress, challenges such as sensor maintenance, data heterogeneity, and model interpretability remain critical. However, ongoing research is addressing these issues through innovations in sensor miniaturization, edge computing, and explainable AI. Together, these insights highlight the multidisciplinary nature of modern soil analysis—one that bridges agriculture, environmental science, computer science, and engineering.

4.7.2 Outlook for Future Research and Applications

Looking ahead, the future of soil science lies in creating fully integrated ecosystems that combine IoT, AI, remote sensing, and cloud-edge computing into cohesive platforms. Potential directions include AI-powered decision support systems for farmers and urban planners, decentralized soil intelligence networks using edge devices and federated learning, autonomous drones and robotics for soil monitoring and micro-level management, and standardized, open-access soil data platforms to promote global collaboration. As the demand for food security, environmental protection, and sustainable development intensifies, the role of intelligent, data-driven soil analysis will become increasingly crucial. The technologies discussed in this chapter not only enhance our current capabilities but also lay the foundation for the next generation of soil science and land resource management.

ACKNOWLEDGMENT(S)

The authors would like to acknowledge Amity University Madhya Pradesh and NIT Kurukshetra for allowing them to carry out this work.

FUNDING

No funding is involved in this work.

REFERENCES

Aarthi, R., & Sivakumar, D. (2023). Internet of things based on field soil property prediction system: For optimally monitoring the soil fertility status. In N. Lal, S. Qamar, S. Agarwal, A. K. Agarwal, & S. S. Verma (Eds.), *Internet of Things* (pp. 143–154). Chapman and Hall/CRC.

Agarwal, V., Tapaswi, S., & Chanak, P. (2024). Energy efficient query processing mechanism for IoT-enabled wsns. *IEEE Transactions on Green Communications and Networking*, 8 (4), 1632–1644.

Akhter, R., & Sofi, S. A. (2022). Precision agriculture using IoT data analytics and machine learning. *Journal of King Saud University-Computer and Information Sciences*, 34 (8), 5602–5618.

AlZu'bi, S., Hawashin, B., Mujahed, M., Jararweh, Y., & Gupta, B. B. (2019). An efficient employment of internet of multimedia things in smart and future agriculture. *Multimedia Tools and Applications*, 78, 29581–29605.

Babu, D. K., Ravindra Raman, C., & Venkata Divakara Rao, D. (2023). Deep residual network-based data streaming approach for soil type application under IoT-based big data environment. *Wireless Networks*, 29 (4), 1751–1769.

Channe, H., Kothari, S., Kadam, D. (2015). Multidisciplinary model for smart agriculture using internet-of-things (IoT), sensors, cloud-computing, mobile-computing & big-data analysis. *International Journal of Computer Technology & Applications*, 6 (3), 374–382.

Demattê, J. A. M., Dotto, A. C., Bedin, L. G., Sayão, V. M., & e Souza, A. B. (2019). Soil analytical quality control by traditional and spectroscopy techniques: Constructing the future of a hybrid laboratory for low environmental impact. *Geoderma*, 337, 111–121.

Din, S., Ahmad, A., Paul, A., Rathore, M. M. U., & Jeon, G. (2017). A cluster-based data fusion technique to analyze big data in wireless multi-sensor system. *IEEE Access*, 5, 5069–5083.

Ge, Y., Thomasson, J. A., & Sui, R. (2011). Remote sensing of soil properties in precision agriculture: A review. *Frontiers of Earth Science*, 5, 229–238.

Hartemink, A. E., & Barrow, N. (2023). Soil pH-nutrient relationships: The diagram. *Plant and Soil*, 486 (1), 209–215.

Hemdan, E. E.-D., & Al-Atroush, M. (2024). An efficient IoT-based soil image recognition system using hybrid deep learning for smart geotechnical and geological engineering applications. *Multimedia Tools and Applications*, 83 (25), 66591–66612.

Murthy, K., Rao, A. U., Vijay, D., & Sridhar, T. (2015). Effect of levels of nitrogen, phosphorus and potassium on performance of rice. *Indian Journal of Agricultural Research*, 49 (1), 83–87.

Phang, S. K., Chiang, T. H. A., Happonen, A., & Chang, M. M. L. (2023). From satellite to UAV-based remote sensing: A review on precision agriculture. *IEEE Access*, 11, 127057–127076.

Rehman, A., Farooq, M., Lee, D.-J., & Siddique, K. H. (2022). Sustainable agricultural practices for food security and ecosystem services. *Environmental Science and Pollution Research*, 29 (56), 84076–84095.

Roy, S., & Bhalla, S. K. (2017). Role of geotechnical properties of soil on civil engineering structures. *Resources and Environment*, 7 (4), 103–109.

Senapaty, M. K., Ray, A., & Padhy, N. (2023). IoT-enabled soil nutrient analysis and crop recommendation model for precision agriculture. *Computers*, 12 (3), 61.

Shingade, S. D., & Mudhalwadkar, R. P. (2024). Analysis of crop prediction models using data analytics and ML techniques: A review. *Multimedia Tools and Applications*, 83 (13), 37813–37838.

Spiertz, H. (2013). Challenges for crop production research in improving land use, productivity and sustainability. *Sustainability*, 5 (4), 1632–1644.

Teng, Y., Wu, J., Lu, S., Wang, Y., Jiao, X., & Song, L. (2014). Soil and soil environmental quality monitoring in China: A review. *Environment International*, 69, 177–199.

Usharani, K., Roopashree, K., & Naik, D. (2019). Role of soil physical, chemical and biological properties for soil health improvement and sustainable agriculture. *Journal of Pharmacognosy and Phytochemistry*, 8 (5), 1256–1267.

Vermote, E. F., & Kotchenova, S. (2008). Atmospheric correction for the monitoring of land surfaces. *Journal of Geophysical Research: Atmospheres*, *113* (D23). doi:10.1029/2007JD009662

5 Crop Yield and Fire Detection

Integrating Machine Learning Techniques for Improving Agricultural Profitability

Pillaram Manoj, Sanjay M, and Valarmathi P

5.1 INTRODUCTION

Agriculture faces climatic conditions as one of the main threats across the world and hence contributes to erratic weather patterns, variable temperatures, and irregular rainfall, all of which impact the crop yield adversely and destabilize food security. It is crucial to predict crop yield accurately so as to undertake proper planning, utilize resources efficiently, and mitigate risks. However, classical approaches such as those based on historical trends or those employing static agronomic models typically lack the ability to adapt to changing environmental conditions and offer no timely or geographically specific information relevant to the location. Another component of modern agriculture is fire detection, especially in drought or heatwave-prone areas. Conventional fire detection methods such as manual watching or satellite monitoring lack the immediacy or infrastructure, which makes them poorly suited to smallholder and remote farms.

Some of the existing methods have tried to fill in those gaps. To illustrate, AgroConsultant used meteorological data along with soil information to help in regional crop decision-making, but since it is barely adaptable to various climates, it is not widely applicable. Traditional systems also have problems being slow in data processing and often lack forecasting capabilities needed to manage farms proactively. Such limitations motivate the need for more flexible and data-oriented tools. ML can be a good solution, as it analyzes highly complicated environmental data, such as rainfall pattern, temperature, humidity, and soil conditions, to provide accurate, context-aware predictions. Early decision-making in crop yield estimation and early fire detection using machine learning (ML) can give farmers tools to enhance their productivities and face off climate-induced challenges. A scalable and accessible ML-based system is proposed in this study to fill the gap between the traditional systems and technical demands of agriculture.

ML techniques such as Random Forest (RF), Logistic Regression, Decision Trees, LightGBM (LGBM), and CatBoost are the predicated techniques that have

 DOI: 10.1201/9781003637264-7

been employed to predict yield based on two core datasets. These are datacentric agronomic factors such as nitrogen (N), phosphorus (P), potassium (K), and soil pH used in the first dataset, and on the second dataset, it consists mostly of historical rainfall and climatic variables. The output is to implement the provided ML models within the framework system to decipher patterns and correlations among environment factors and crop performance; thus, the resulting very precise predictions around which farmers can improve their scheduling of planting, irrigation, and resource utilization would be generated. Finest innovative contribution from this research will be to have an interface through which it will be easy for farmers to obtain complicated ML-generated outputs as being interactive and interpretable. These insights, along with scheduled planting or irrigation, would be accessed by mobile platforms. Access ensures that even the simplest technical advisories reach the farmer, who does not even possess particular technical knowledge.

With climate change, agricultural systems in the world are now deeply affected and are one of the most critical issues at present. Adapting to the increasingly unpredictable changes of weather and changing climatic conditions appears to be the best statement for farmers. An attempt in this research is to come with high algorithms for ML that combine rain and humidity data in a productive way to help farmers in adapting easily to challenges due to climate change. Crop losses and transformations in the environment threaten the livelihoods of smallholder farmers considerably, who are often some of the most exposed and vulnerable to climate change. Understanding the hypothetical ideas of their adaptation methods helps to formulate measures to cope with these threats. These insights will be useful for policy and practice, and because of the dimensions they've explored decision-making processes of farmers and determinants influence their methods of adapting.

This chapter is organized as follows: Section 5.1 provides a detailed introduction to the study examining the application of ML techniques to enhance resilience in smallholder farming against climate change. Section 5.2 offers a background study discussing earlier works involving predictive models like RF and LGBM as well as other smart farming projects like AgroConsultant and Internet of Things (IoT) approaches for data-informed agriculture. Section 5.3 discusses the methodology that includes data collection from several crop datasets, and preprocessing of data prior to the application of the ML algorithms for performance prediction and agro analysis. Section 5.4 deals with model results and performance analysis confirming high accuracy and effectiveness for use in decision-making in the field of crop management by farmers. Section 5.5 provides concluding remarks summarizing the contributions and future work, which may include the integration of more variables such as the market conditions, pest outbreaks, and better user-friendly data visualization tools for farmers to adopt a sustainable way of smart agriculture.

5.2 RELATED WORK

Zeel Doshi et al. [1] have developed an AgroConsultant. The intention of this strategy is to help Indian farmers make timely and informed decisions toward the evolution of crops by investigating soil properties, farmers' locality, sowing time, and climatic conditions such as temperature and precipitation. The intelligent system being suggested has made use of two subsystems. The second one predicts how sustainable agriculture will be in the future, but its role also includes giving information to the first. This first system weighs on the amount of rainfall expected in any area and advises farmers on what should be grown.

Elsayed Mohamed et al. [2] worked with innovative ideas on the smart farming profession from 2019 to 2021 with the key areas of focus being data capture, data transfer, data storage, data analysis, and data solutions. The intelligent irrigation system applications, sensor enabled and with the aid of mathematical equations, are again slowly coming into play. Songs, as already said, can be global issues, whether by means of mobile internet, instruments of IoT, or advanced utilities such as artificial intelligence (AI). Drone and robot application in agriculture with IoT, AI, deep learning (DL), ML, wireless communication, and such applicability are discussed. It mentions the application of 5G architecture for high-speed data transfer. But while there are many milestones yet to be achieved, the hopes surrounding smart farming indeed seem visible among developing countries if governments would put in place even simple smart decision support systems.

Kaiyang Chen et al. [3] have shown that the fertilization greatly increases total alkaloids and concentration of alkaloids in the plant. N2P2K2 gave maximum yield, whereas N0P0K0 gave the minimum yield. Ingredients of active principles have some therapeutic applications and are used in Tibetan medicine. It says that fertilization is not only a must to increase productivity and development characteristics but also growth of *Anisodus tanguticus*. The research was aimed at providing theoretical bases for better agriculture in Qinghai Province.

Ivan Adolwa et al. [4] have attributed the low soil fertility and inadequate utilization of agricultural input in sub-Saharan Africa. Multi-nutrient fertilizers are reported to enhance production and profitability of smallholder farms particularly in the maize and potato cropping systems. Such fertilizers have been found to give positive response yields in both beans and maize, with the yield of maize significantly increased as compared to the application of traditional fertilizers. Closing yield gaps is vital to Africa's food security, hence innovative approaches need to be embraced to enhance agricultural production sustainably. The study adopted household survey approach in eight counties in Kenya, and results indicated that multi-nutrient fertilizers were mostly utilized in Bungoma and Kakamega.

Freddy Diaz-Gonzalez's et al. [5], in their article, discuss issues regarding soil position quality in intensive agriculture while providing a means of prediction of crop output through ML with respect to soil quality measurable indicators. It depicts advantages and drawbacks to sophisticated strategies of the above-mentioned processes, including methods of ML, unmanned aerial vehicles, and remote sensing systems. Furthermore, the article emphasizes the importance of developing models for soil quality measurements utilizing AI methods and providing farmers with measurable data. The adoption of these technologies in medium- and small-holder agricultural systems is also highlighted.

Lefteris Benos et al. [6] defined ML in agriculture. Contemporary agriculture requires the infusion of information and communication technology (ICT) with heavy emphasis going to crop management. Support vector machine is a particular

ML model in agriculture that has a high level of accuracy. Improvements in ICT systems are propelling ML in agriculture along with reinforcing the ethical rationale behind precision farming. ML may reconstruct a knowledgeable system by creating justifiable associations with respect to data input. The 338 journal articles that came up in a literature review on ML in agriculture were mostly concerned with crop management. Yield prediction and disease detection are the most eminent subcategories in crop management research. Crop identification has become an emerging research topic for most people.

Priyanga Muruganantham et al. [7] have highlighted that one of the aspects of DL techniques and remote sensing data coexist in a proper way to help address the food security concern. The extensive literature study outlined a consideration of the importance of vegetation indexes and weather parameters in crop yield prediction. Their goal is to achieve sustainable agriculture and end hunger by 2030. Improving model accuracy and providing stakeholders reliable information present challenges.

Julia Fohrafellner et al. [8] have investigated the meta-analysis that has been concerned with particulate organic carbon, mineral-associated organic carbon, and microbial biomass carbon from arable farmland within Europe. A lack of data prompted the addition of organic matter statistics. The study did a meta-analysis of the effect of cover crops on soil organic carbon pools structured on the PICO framework. The search methodology included extraction of data, sensitivity analysis, and screening of related literature. The key aims were transparency and reuse for further studies.

Abdul Muis Hasibuan et al. [9] have reported a multivariate probit model study concerning adaptation strategies to climate change by 209 small-scale sugarcane producers. The farmers understood that their agricultural output was being threatened seriously by climate change. They also stated that risk behaviors such as risk perceptions on hazards related to climate change and risk-taking propensities have played an important role in adopting adaptation techniques.

Ibsa Dawid Mume et al. [10] described how the PSM framework identifies advantages that accrue to smallholder irrigation, and how the Resilience Capacity Index measures the adjustment capacity of farmers against climate change. It has been established that smallholder irrigation causes crop yield and income increases substantially, thus improving agricultural yields, income, and flexibility. Therefore, adopters of small-holding irrigation have a resilience capacity index improved by 2.02 compared to non-adopters. These argue that socioeconomic status among rural populations can be enhanced through smallholder irrigation practices in addition to cushioning them from the adverse effects of climate change, such as erratic weather patterns and prolonged droughts. Policymakers and all other interested parties that are interested in sustainable agriculture and rural development in a climate-prone region would benefit from these findings.

Christina Griffin et al. [11], with the outcomes of their study, indicated that farmers are not equally prepared for climate change and called for application of specific adaptation methods. Adaptation tactics were heavily influenced by ideals of fairness that might render them impossible to implement successfully. The significant contribution is to the already available corpus of knowledge regarding adaptation to climate change and justice in the agriculture sector. In fact, it has a potential to enlighten and guide stakeholders and policymakers in their decision-making processes by providing new perspectives.

Sandra Ricart et al. [12] state that the bibliometric analysis is done to understand the research pattern and inclination toward knowledge about adaptation to climate change and its perceived implications by action and experience of farmers, while suggesting future research directions. Important research could be found, insights gained into the minds of farmers, while research gaps could easily be aligned in the publication trend analysis and content analysis techniques. Farmers' culture and society are mostly reflected in their perception of climate change, evidence of human activity causing climate change, perceived consequences such as heat and altered precipitation patterns, intervention strategies such as crop modification and soil and water conservation techniques, and obstacles to adaptation including lack of information and resources.

Lawrence Guodaar and Divine Odame Appiah [13] emphasized the dynamic adaptation strategies that smallholder farmers are adopting to cope with the climatic variability and changes within the transitional forest-savanna terrain of Ghana. Considering the survey from 300 farm heads and lead farmers distributed across six rural communities in Ghana, the descriptive statistical analysis examined understanding farmers' opinions on CVC and the factors that affect their decision-making process. The study further accentuated the importance of merging indigenous and modern knowledge systems and technologies for agricultural resilience and food security in the face of CVC challenges.

Yahaya Zimit et al. [14] analyzed the related problems that confront farmers due to the variability arising from climate change. The research forecasted amount of rainfall in the agricultural area of the Kano desert using meteorological models. Three computation models, namely Multi-Linear Regression (MLR), Feed-Forward Neural Network (FFNN), and Adaptive Neuro-Fuzzy Inference System (ANFIS), were used in the study. Based on the data obtained with high values of $R^2=0.9989$, $NSE=0.9872$, and $RMSE=0.0016$, it showed that MLR-M1 model is performed remarkably well in predicting rainfall. Results indicated the significance of intelligent regression learning techniques in water resource management as the sustainable agriculture development. One important aspect of the MLR-M1 model is that it extends support to farmers and promotes efficient management of water resources in agriculture.

Spyridon Paparrizos et al. [15] showed the information that is obtained regarding local forecast knowledge (LFK) is applicable and integral to a global perspective on agriculture. LFK was proven by their research to be important to farmers in terms of decision-making. Conclusively, making use of a comprehensive literature review strategy has discovered 1,349 indicators linked to local and indigenous knowledge (LFK) indicators. The major country in this context is India. LFK is almost predominantly prevalent in the rest of Africa and Asia. This study seeks to develop a comprehensive understanding of the various lead times, classification, and geographic distributions of the LFK indicators. Agricultural resilience against weather-related effects is determined mainly by these guidelines.

5.3 PROPOSED SYSTEM

Figure 5.1 shows the proposed architecture for crop prediction. The process starts with data collection followed by data preprocessing, cleaning, and formatting data so it can be analyzed in real time with historical datasets. Crop Data 1 consists of features of a fertilizer application rate (N, P, K), temperature, humidity, rainfall, and pH, giving input for an outright instant analysis of one field whereas Crop Data 2 consists of historical data on crop yield, rainfall (monthly as well as annually), states, and types of crops, all of the input data needed for long-term planning. These datasets are then processed through several ML algorithms such as RF, Logistic Regression, Support Vector Regression, Decision Tree, Linear Regression, LGBM, and CatBoost for the final prediction of yield and crop health. The outputs of the prediction go into the AgroAnalysis, and used for interface, which can easily be accessible by farmers. Likewise, running in parallel steps is the fire detection module where the image and sensor-based system will be integrated to protect against the risk of wildfire. The user interface then displays insights with actionable value, such as predictions on profit and rainfall, as well as crop recommendations based on nutrient-level values, to support immediate decision-making and strategic planning.

Compared to other ML techniques, the models like RF, Logistic Regression, LGBM, CatBoost, Support Vector Regression, Decision Tree, and Linear Regression were chosen for their better accuracy, robustness, and efficiency over agricultural datasets with various features including N, P, K, rainfall, temperature, and pH levels. K-Nearest Neighbors (KNN) or Naive Bayes might give good solutions in some

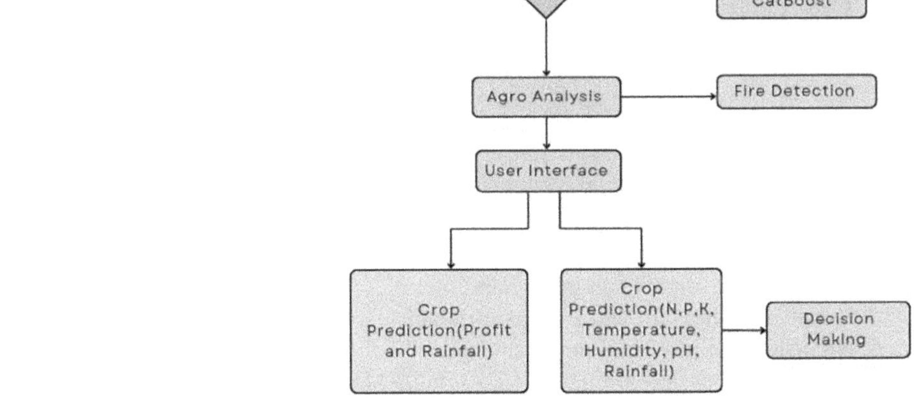

FIGURE 5.1 Proposed architecture for crop prediction.

scenarios but work poorly in high dimensions or with noisy data. The selected models achieve a trade-off between explainability and performance, robust prediction against overfitting, and faster training and inference that are suitable requirements for real-time, on-the-ground agricultural decision-making for smallholder farmers.

5.3.1 ML Models

Model 1. Random Forest Algorithm for Crop Prediction

INPUT: Crop Data (N, P, K, Humidity, Rainfall, PH-level, Temperature)
OUTPUT: Predicted Crop Yield and Health Insights

1. Prepare and pre-customize the agricultural dataset.
2. Divide datasets for training and testing.
3. Initialize Random Forest model using given n_estimators and other hyperparameters.
4. Train the model using training data.
5. Predict target variable on test data using learned models.
6. Print yield predictions along with actionables.

Model 2. Logistic Regression for Crop Health Classification

INPUT: Crop Data (N, P, K, PH-level, Temperature, Humidity)
OUTPUT: Crop Health Classification (Healthy/Unhealthy)

1. The crop dataset gets loaded and preprocessed.
2. Categorical features are converted to one-hot encodings.
3. The data gets split into a train set and a test set.
4. Initialize the Logistic Regression model.
5. The model gets trained with the training set.
6. Class (Health/Unhealthy) prediction on the test set.
7. Print the classification results and health insights.

Model 3. LightGBM for Crop Yield Prediction

INPUT: Crop Data (N, P, K, Rainfall, PH-level, Temperature)
OUTPUT: Predicted Crop Yield

1. Crop dataset import and preprocessing.
2. Label encodings for categorical variables.
3. Splitting data into train and test.
4. Configuring hyperparameters for LightGBM model initialization.
5. Training the model for training data.
6. Predicting crop yield with test data.
7. Printing yield predictions and recommendations on how to improve yield.

Model 4. CatBoost Algorithm for Crop Prediction

INPUT: Crop Data (Monthly Rainfall, Temperature, Humidity)
OUTPUT: Predicted Crop Yield and Strategic Planning Insights

1. The crop dataset is loaded and missing values are addressed.
2. The split of data into training and test sets is done.
3. The CatBoost model is initialized.
4. The model with categorical features is trained against the learning set.
5. Uses trained model to predict crop yield on the test set data.
6. The prediction is then followed by giving recommendations for sustainable actions.

Model 5. Support Vector Regression for Crop Yield Estimation

INPUT: Crop Data (N, P, K, Rainfall, Temperature, Humidity)
OUTPUT: Estimated Crop Yield

1. Preprocess the data and scale all the features.
2. Split the data into training and testing sets.
3. Initialize the Support Vector Regression (SVR) model.
4. Train the SVR model on the training data.
5. Predict the crop yield on the test data.
6. Yield predictions and actionable recommendations.

Model 6. Decision Tree for Crop Selection

INPUT: Crop Data (N, P, K, Rainfall, Temperature, Humidity)
OUTPUT: Recommended Crop for Cultivation

1. Load the dataset about crops and deal with missing values.
2. Preprocess the data, converting categorical variables to numerical ones.
3. Divide the dataset into subsets for training and testing.
4. Initialize the Decision Tree algorithm.
5. Train this algorithm using the training data.
6. Use the trained model to predict what crop would be suitable for the area.
7. Provide the predicted crop and suggested measures for cultivation.

Model 7. Linear Regression for Long-Term Crop Planning

INPUT: Crop Data (Rainfall, Humidity, Temperature, States)
OUTPUT: Long-term Crop Yield Forecast

1. Acquire the crop dataset and perform data preprocessing.
2. Split the Dataset into TRAIN and TEST datasets.

3. Initialize the Linear Regression model.
4. Train the model with the training data.
5. For testing, predict yield for crops. Long term crop yield prediction and advise strategic planning.

5.4 RESULTS AND DISCUSSION

This section discusses the key results of the study regarding the performance comparison of different ML models and system components.

5.4.1 ML Algorithms

This section describes the comparison of accuracies of all used ML models that are evaluated for crop data application in the prediction of yield and planning insights.

5.4.1.1 Crop Data (1)

This subsection presents the performance comparison of ML algorithms for Crop data 1.

Figure 5.2 shows the outcomes of ML algorithm's performance on the dataset with significant variation based on different methods. RF performed best among all models evaluated, achieving a record 99.09% accuracy. This clearly portrays the ability of the ensemble learning technique to capture complex patterns from data very effectively. RF would be performing well on different datasets, agricultural datasets in particular since it combines a number of decision trees to arrive at a credible prediction. The LGBM algorithm proved to be an extraordinary 98.18%, with an almost neck-to-neck finish behind RF. Known for amazing efficiency and management of very big datasets, LGBM was able to stand its ground here. LGBM can consistently update itself through gradient boosting by learning from the errors of previous models to increase the prediction performance. Its able performance indicates how

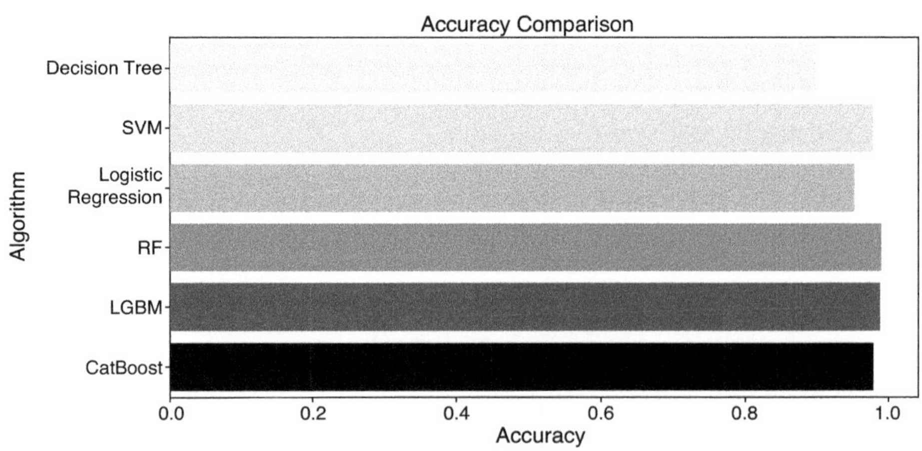

FIGURE 5.2 Comparison of accuracies for machine learning models.

relevant it is for actual agricultural decision-making situations where accuracy and speed matter. Overall, these findings highlight that gradient boosting and ensemble methods have untackled many issues relating to climate adaptation in agriculture, thus paving the way to more intelligent and sustainable farming practices.

5.4.1.2 Crop Data (2)

This subsection presents the performance comparison of ML algorithms for Crop data 2.

Figure 5.3 represents the bar chart for crop prediction by using ML algorithms. RF is the algorithm that scores the highest, with R-squared of about 0.9614%. The algorithm scoring the least is linear regression with R-squared of about 0.0317%. Scores are RF: 0.9614, LGBM: 0.8356, Linear Regression: 0.0317, CatBoost: 0.8660. RF, despite having the highest R-squared score, may be more difficult and less intuitive. Although it has the lowest R-squared value among the bunch, the linear regression algorithm is by far the easiest to understand. RF: The ensemble learning algorithm RF combines the contributions of many decision trees for more accurate predictions. RF has the highest accuracy and is least prone to overfitting. LGBM and CatBoost are gradient boosting methods. These algorithms CatBoost are known for their accuracy and speed. LGBM techniques of light gradient-boosting are known for their productivity and speed. LGBM algorithms are widely used in ML contests because of their exceptional performance. Linear regression is the simplest type of regression in which a dependent variable is expressed as a linear function of one or more independent variables. Hence, the techniques of linear regression are really simple to apply and understand.

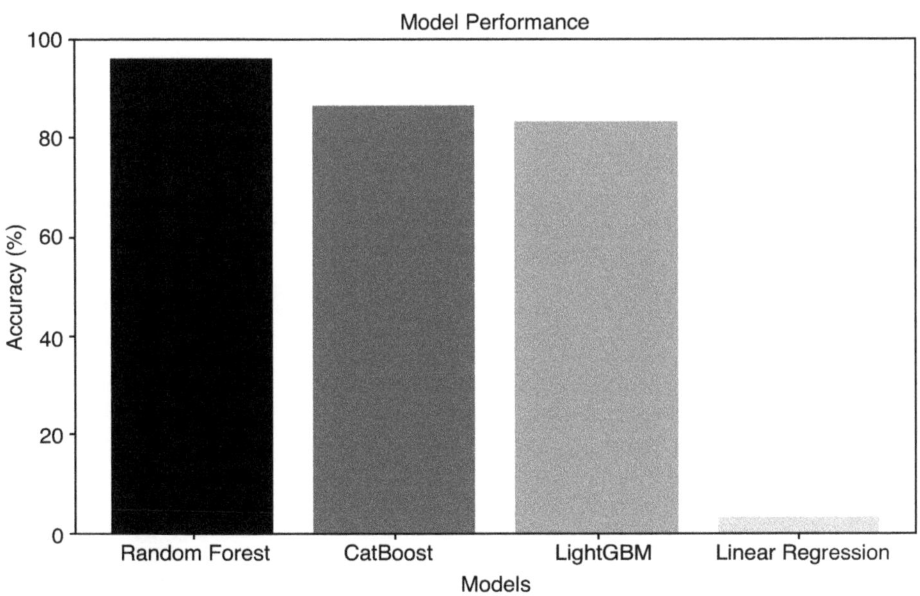

FIGURE 5.3 Comparison of accuracies for machine learning models.

5.4.2 Agro Analysis(1&2)

In this section, we apply real-time as well as historical regional data to increase the effectiveness of decision-making for sustainable agriculture.

5.4.2.1 User Interface

This section describes the user interface and how useful it is for farmers in delivering data insights.

Figure 5.4 shows the Tamil Nadu crop analysis that provides an user interface that enables the customers to select their state and crop, pre-selecting "Tamil Nadu" and "wheat." Upon selection of "Analyze," the user interface provides detailed insights about wheat farming in Tamil Nadu, complemented by historical data and possible variables such as soil properties and market prices. The given details' average profit measurements were taken at ₹13.75 per unit, with rainfall statistics averaged out at 1000.04 mm, plus a range of profit statistics between ₹1.00 and ₹44.00 per unit and rainfall measurements from 922.60 mm to 1079.70 mm. Equipped with such detailed analysis, farmers will receive vital information for better decision-making, improved crop management methods, and greater contribution to agricultural sustainability in this region.

Figure 5.5 shows crop prediction for Tamil Nadu on wheat. The agricultural crop prediction serves as a demonstration of analytic power in terms of visualization. In addition to the soil testing for nitrogen (N-123 ppm), phosphorus (P-43 ppm), and potassium (K-30 ppm), this tool integrates meteorological variables, including temperature (30°C), humidity (78%), and rainfall (950 mm). Considering all these factors, it suggested the growing of wheat as the best crop for that specific location. This recommendation most probably takes into account soil pH level (6.8, optimum for wheat), historical yield data, water availability, and market price, thus giving very valuable information to improve agricultural production and increase farmers' profits.

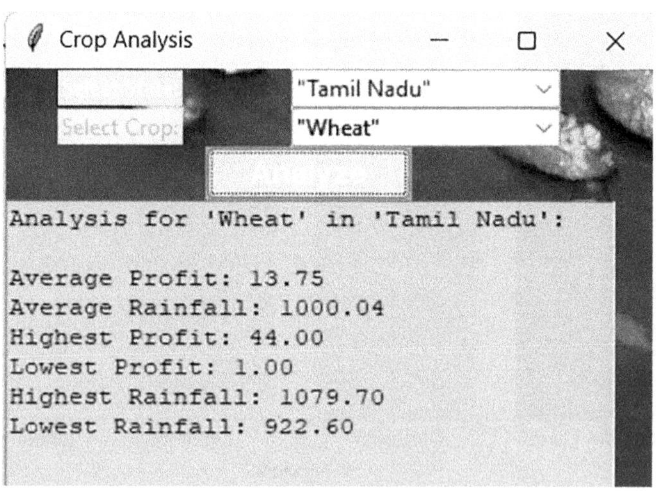

FIGURE 5.4 Crop analysis for Tamil Nadu on wheat.

FIGURE 5.5 Crop prediction for Tamil Nadu on wheat.

Figure 5.6 represents crop analysis for Tamil Nadu on rice. The user may select a state (pre-selected) and then interrogate the Tamil Nadu rice crop study for data on rice farming. By clicking "Analyze," the application displays some of the following information: average profit (₹201,860.21, presumably per hectare), average rainfall (884.49 mm), and ranges for both rainfall (318.00 mm–1,365.30 mm) and profit (₹604.00–₹831,483.00). Thus the application must be analyzing historical data regarding the rain pattern, taking many other factors, like soil types and market prices, into account to give us a lucid and interactive picture of the rain pattern in Tamil Nadu and the possible profitability of growing rice.

Figure 5.7 represents the crop forecast which recommends rice cultivation based on soil test results showing nitrogen at 91 ppm, phosphorus at 43 ppm, and potassium at 45 ppm and on metrological data at 30°C temperature, 82% humidity, and 1,300 mm rainfall. This decision considers several factors, including pH of the soil

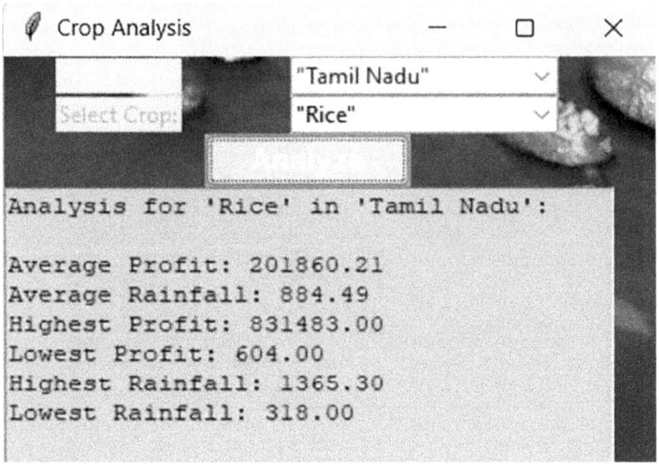

FIGURE 5.6 Crop analysis for Tamil Nadu on rice.

(6.8, good for rice), past yield records, water availability, market prices, and possibly some other relevant features. The software provides farmers with carefully considered information for farm management practices toward maximizing yields and profitability through crop advice tailored in a case-specific manner.

Figure 5.8 represents crop analysis, which allows the user to select the state (already selected) and analyze soybean cultivation data for Tamil Nadu. When the user clicks on Analyze button, the user interface displays nan (which most probably indicates not a number) against all values part of the analysis section (Analysis for Soyabean in Tamil Nadu). This clearly means that there is insufficient data to make any actual statements on profitability of soybean or rainfall events pertaining to Tamil Nadu.

Figure 5.9 shows the information that helps in predicting crop yield for this application together with user-fed weather data at a temperature of 32°C and humidity of 75% and soil conditions. This not only gives farmers a sound idea to enhance their practice and increase yields but also provides the insight gained from fusing environmental and soil data from visual analysis.

Figure 5.10 represents crop analysis for Tamil Nadu on cotton(lint). The average profit amounting to ₹8,386.43 per unit (not defined) and averaged rainfall of 879.95 mm indicate according to user interface study including pre-selected state and crop options. The ranges for rainfall (318.00 mm–1,365.30 mm) and profits (₹1.00–₹88,685.00) are also specified. The results indicate that an application harnesses past data and considers aspects like soil quality and market price for drawing a comprehensive range of diversified potential profitability and rainfall patterns with respect to cotton lint production in Tamil Nadu.

Figure 5.11 represents the crop prediction for Tamil Nadu on cotton(lint). This crop prediction user interface has a field image placeholder, in addition to showing metrics that are presumably measurable using image recognition technology, such as health of the plant (78%), coverage of leaves (62%), and absence of diseases

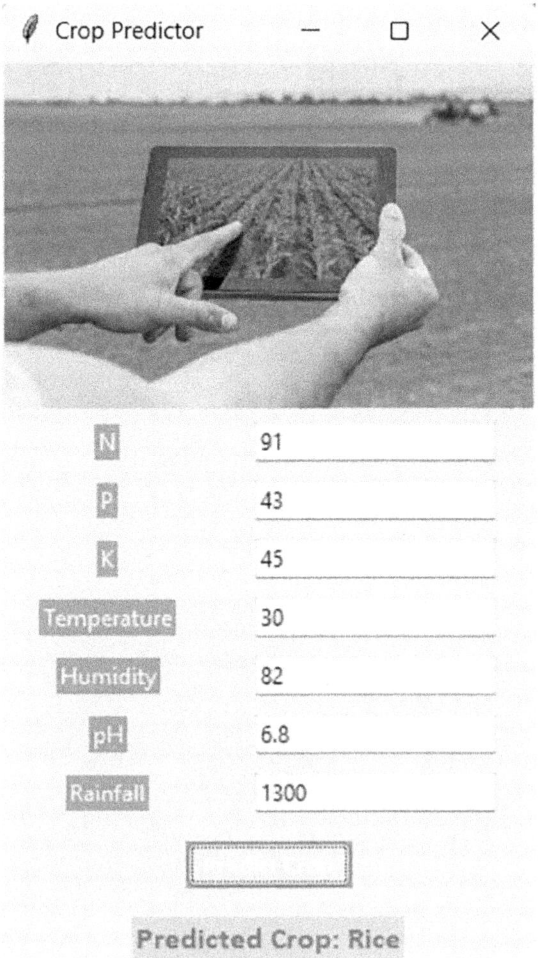

FIGURE 5.7 Crop prediction for Tamil Nadu on rice.

identified. Users enter temperature (currently vacant), humidity (32°C), and condition of soil (under input process) in the input fields under those metrics. After uploading the field photo by user and completing the vacant data, the "Predict" button will most probably generate a detailed estimate into the application's crop yield based on the derived data. An umbrella strategy combining visual analysis with soil and environmental data serves to increase yield and optimize production strategies of farmers.

5.4.3 Fire Detection

The image classification model of ours is designed so that the presence of fire can easily be detected and crops are classified with great precision. The model shall immediately ascertain whether the image is depicting a scenario of fire upon input of

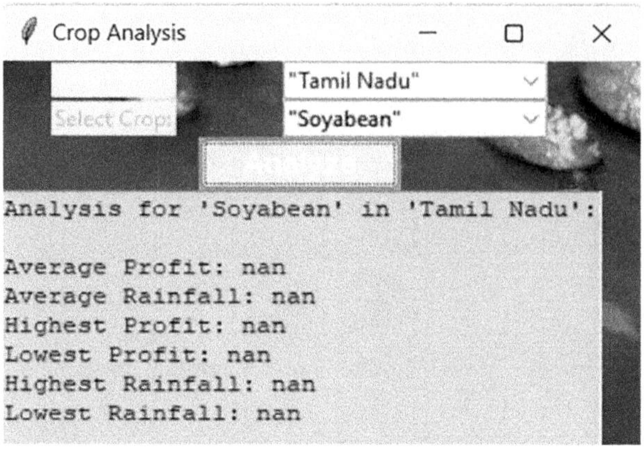

FIGURE 5.8 Crop analysis for Tamil Nadu on soyabean.

FIGURE 5.9 Crop prediction for Tamil Nadu on soyabean.

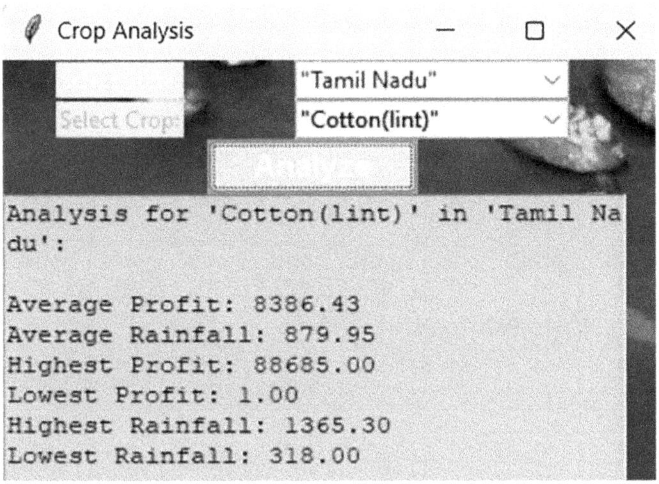

FIGURE 5.10 Crop analysis for Tamil Nadu on Cotton (lint).

an image. An easy interface provides a seamless experience and enables predictions in fire detection to be easily obtained. Integrated fire detection sensors are one of the important products designed to increase the safety of agriculture. With the use of technology, the sensor can detect small signs of fire in agricultural areas and prevent crops from being damaged by fire by intervening on time. As wildfires become more of a threat, fire detectors can be used as countermeasures to reduce the risk. In fact, real-time monitoring capability provides farmers with significant protection for their crops, protects investments, and increases agricultural protection against unprecedented threats. The module goes beyond traditional precision agriculture; it addresses the growing challenges of climate-related disasters and highlights the importance of technology in promoting sustainable and resilient agriculture. With fire detectors, farmers can quickly respond to the possibility of fire in the face of changing environmental challenges, reducing the impact on crop productivity, and improving the safety of farming operations. The model is able to detect fire with an accuracy of 95 percent. The results of the proposed work as depicted in Figure 5.12, which uses images to predict crop yields, are revolutionary, changing traditional agriculture and ushering in a new era of precision agriculture. One of the most important achievements is the accurate identification of crops through high-resolution images. This capability allows farmers to quickly view crop health and growth, making decision-making easier. The integration of air seeding mechanisms simplifies the seeding process, providing unparalleled efficiency in covering large areas for agricultural expansion. The ability of seeding system is to continuously deliver seeds to disperse crops, improve resource use, and help increase crop yields. This result has a great impact on large-scale agriculture because it solves the problem of labor-intensive work while improving resources.

The proposed system's predictive modeling, which calculates the area allocated to agriculture according to water and pesticides, is a game changer for the capital management layer. Farmers will now be able to receive specific advice on how to use

FIGURE 5.11 Crop prediction for Tamil Nadu on cotton (lint).

water and pesticides, reduce water waste, and reduce the environmental impact of overuse of pesticides. These benefits not only improve the economics of agriculture but also encourage permaculture practices. In addition, the integration of electronic devices discovers additional important methods of plant protection. The system can quickly detect signs of fire and intervene early, preventing crop loss due to fire. This event highlights the program's commitment to ensuring the sustainability of agricultural investments. Essentially, the event resulted in efficient operations, optimization of resources and encouraging farmers to make the right decisions. As agriculture continues to face challenges such as climate change and changing business needs, the use of images to predict crop yields has become a new beacon offering strategic solutions to move agriculture into the future and technology.

Compared to existing research, which would typically use one dataset for crop prediction, or advisory systems such as AgroConsultant, which integrates regional

FIGURE 5.12 Image of fire detection.

meteorological and soil data, our approach provides a more robust and scalable solution. We have used two different datasets: Crop Data 1 consists of real-time agro-climatic parameters such as rainfall, temperature, humidity, and soil nutrients to allow immediate decision-making. Crop Data 2 consists of regional-level historical crop data to facilitate long-term strategic planning. This datasets approach improves prediction capabilities for the system and informs decision-making on different time scales, thus yielding more precise and actionable outputs. Apart from building ML models such as RF, LGBM, and others, the system was also designed with a fire detection module that combines image-based classification and sensor-based monitoring to give early warnings and thereby improve the safety of crops. Our study fills this gap because it provides a fine but exhaustive analysis of predictive tools tailored for smallholder farmers so as to contribute to their adaptive capacity and sustainability for farming. Most of these works show possible potentials of bringing advanced technology together with traditional knowledge to face challenges from climate change.

5.5 CONCLUSION AND FUTURE WORK

The study shows that crop analysis and prediction applications are the latest developments in agricultural technology designed to equip smallholder farmers who have been experiencing the adverse effects of climate change with specialized knowledge and tools for weather forecasting. The proposed system shows that RF is capable of predicting Crop Data 1 with a remarkable accuracy of 99.09% and predicting performance in case of Crop Data 2 with an R-squared value of about 0.9614%. Therefore, the results from these models realistically signify perfection in agricultural processes. The LGBM model scored an accuracy of 98.18% for Crop Data 1, and the fire detection is now integrated with the agriculture which further emphasizes the value of these fire detection sensors in providing actionable insights for better decision-making. By integrating high-resolution image-based classification with real-time sensor monitoring, this system can detect early signs of fire with high precision. These results demonstrate the advantages of bringing in advanced ML approaches into agricultural systems aimed at increased productivity and sustainability in support of food security in the vulnerable communities.

In the future, these models might be enhanced by adding other considerations such as market swings and outbreak of pests and diseases. The predictions could greatly benefit if the dataset were broadened to accommodate a wider range of varieties of crops and agricultural practices. Through sophisticated data visualization tools, the farmers may also be motivated to make decisions easier by intuitively understanding complex agricultural data. Collaboration among teachers, government officials, and farmers on the development of customized solutions for the needs of their smallholder farmers can also be enabled.

REFERENCES

1. Doshi, Z., Nadkarni, S., Agrawal, R., & Shah, N. (2018). AgroConsultant: Intelligent crop recommendation system using machine learning algorithms. *2018 Fourth International Conference on Computing Communication Control and Automation (ICCUBEA)*. https://doi.org/10.1109/iccubea.2018.8697349
2. Said Mohamed, E., Belal, A. A., Kotb Abd-Elmabod, S., El-Shirbeny, M. A., Gad, A., & Zahran, M. B. (2021). Smart farming for improving agricultural management. *The Egyptian Journal of Remote Sensing and Space Science*, 24(3), 971–981. https://doi.org/10.1016/j.ejrs.2021.08.007
3. Chen, K., Ma, L., Chen, C., Liu, N., Wang, B., Bao, Y., Liu, Z., & Zhou, G. (2023). Long-term impact of N, P, K fertilizers in different rates on yield and quality of Anisodus tanguticus (Maxinowicz) pascher. *Plants*, 12(11), 2102. https://doi.org/10.3390/plants12112102
4. Adolwa, I. S., Mutegi, J., Muthamia, J., Gitonga, A., Njoroge, S., Kiwia, A., Manoti, D., Mairura, F. S., & Nchanji, E. B. (2023). Enhancing sustainable agri-food systems using multi-nutrient fertilizers in Kenyan smallholder farming systems. *Heliyon*, 9(4), e15320. https://doi.org/10.1016/j.heliyon.2023.e15320
5. Diaz-Gonzalez, F. A., Vuelvas, J., Correa, C. A., Vallejo, V. E., & Patino, D. (2022). Machine learning and remote sensing techniques applied to estimate soil indicators – Review. *Ecological Indicators*, 135, 108517. https://doi.org/10.1016/j.ecolind.2021.108517
6. Benos, L., Tagarakis, A. C., Dolias, G., Berruto, R., Kateris, D., & Bochtis, D. (2021). Machine learning in agriculture: A comprehensive updated review. *Sensors*, 21(11), 3758. https://doi.org/10.3390/s21113758
7. Muruganantham, P., Wibowo, S., Grandhi, S., Samrat, N. H., & Islam, N. (2022). A systematic literature review on crop yield prediction with deep learning and remote sensing. *Remote Sensing*, 14(9), 1990. https://doi.org/10.3390/rs14091990
8. Fohrafellner, J., Zechmeister-Boltenstern, S., Murugan, R., Keiblinger, K., Spiegel, H., & Valkama, E. (2023). Meta-analysis protocol on the effects of cover crops on pool specific soil organic carbon. *MethodsX*, 11, 102411. https://doi.org/10.1016/j.mex.2023.102411
9. Hasibuan, A. M., Wulandari, S., Ardana, I. K., Saefudin, & Wahyudi, A. (2023). Understanding climate adaptation practices among small-scale sugarcane farmers in Indonesia: The role of climate risk behaviors, farmers' support systems, and crop-cattle integration. *Resources, Environment and Sustainability*, 13, 100129. https://doi.org/10.1016/j.resenv.2023.100129

10. Dawid Mume, I., Haji Mohammed, J., & Aman Ogeto, M. (2023). Impact of small-scale irrigation on the livelihood and resilience of smallholder farmers against climate change stresses: Evidence from Kersa district, eastern Oromia, Ethiopia. *Heliyon*, 9(8), e18976. https://doi.org/10.1016/j.heliyon.2023.e18976
11. Griffin, C., Wreford, A., & Cradock-Henry, N. A. (2023). 'As a farmer you've just got to learn to cope': Understanding dairy farmers' perceptions of climate change and adaptation decisions in the lower South Island of Aotearoa-New Zealand. *Journal of Rural Studies*, 98, 147–158. https://doi.org/10.1016/j.jrurstud.2023.02.001
12. Ricart, S., Gandolfi, C., & Castelletti, A. (2023). Climate change awareness, perceived impacts, and adaptation from farmers' experience and behavior: A triple-loop review. *Regional Environmental Change*, 23(3). https://doi.org/10.1007/s10113-023-02078-3
13. Guodaar, L., & Appiah, D. O. (2022). Evolving farm-level adaptation to climate variability and change risks in the forest-savanna transitional zone of Ghana. *Environmental Challenges*, 9, 100654. https://doi.org/10.1016/j.envc.2022.100654
14. Zimit, A. Y., Jibril, M. M., Azimi, M. S., & Abba, S. I. (2023). Hybrid predictive based control of precipitation in a water-scarce region: A focus on the application of intelligent learning for green irrigation in agriculture sector. *Journal of the Saudi Society of Agricultural Sciences*. https://doi.org/10.1016/j.jssas.2023.06.001
15. Paparrizos, S., Attoh, E. M. N. A. N., Sutanto, S. J., Snoeren, N., & Ludwig, F. (2023). Local rainfall forecast knowledge across the globe used for agricultural decision-making. *Science of the Total Environment*, 899, 165539. https://doi.org/10.1016/j.scitotenv.2023.165539

6 Interpretable AI in Agriculture

Making Machine Learning Models Explainable for Disease Detection

Anoushka Duggal, Vasu Chugh, and Jyoti Maggu

6.1 INTRODUCTION

Agriculture has been fundamental to the human civilization since the days of the early man, providing both sustenance and economic growth. It provides raw materials needed to feed populations, fuels economies, and supports a number of industries. As the global population increases year by year the demand for food and agricultural products rises, hence it is critical that agricultural practices evolve to be more efficient, sustainable, and resilient against threats.

As the global population continues to grow, agricultural innovation becomes necessary, with new technologies and methods being developed to tackle both old and new challenges. One of the most pressing challenges in modern agriculture is the threat posed by plant diseases. Caused by bacteria, fungi, viruses, or pests, these diseases significantly reduce crop yield and quality, leading to large-scale economic losses, often amounting to billions annually. For smallholder farmers, the impact can be devastating, with consequences that ripple through local economies by affecting food prices, employment, and access to nutritious food. Addressing plant diseases efficiently has therefore become a top priority, requiring the integration of novel technological solutions.

Traditional agriculture practices are applied the same across the entire field. These have proven inefficient in meeting the challenges of modern agriculture. Without considering the specific needs of individual plants or crops, resources such as water, fertilizers, and pesticides are often applied indiscriminately, resulting in waste and environmental degradation. Furthermore, these methods often fail to account for the growing complexities of crop management, including pest and disease outbreaks [1]. To address these challenges, Precision Agriculture (PA) has emerged as a revolutionary approach, using modern technologies to optimize farming practices and improve productivity. Through the use of advanced technologies such as GPS, remote sensing, and machine learning (ML), PA allows for more targeted, data-driven decisions, helping farmers increase efficiency, reduce costs, and minimize environmental impact [2].

DOI: 10.1201/9781003637264-8

One of the most promising applications of PA is in the realm of plant disease detection. Traditionally, disease detection has relied heavily on human expertise, with agronomists manually inspecting crops for symptoms. While effective to some extent, this approach is prone to errors due to human limitations, delayed diagnosis, and the difficulty of detecting early-stage diseases. The widespread adoption of ML and deep learning (DL) techniques in PA has the potential to transform disease detection by automating the process and offering more accurate and timely diagnoses [3].

ML models, particularly those using image-based analysis, have shown great promise in the field of plant disease detection. By training on large datasets of plant images, these models can learn to identify subtle patterns in the leaves and other plant parts that indicate the presence of diseases, pests, or nutritional deficiencies [4,5]. This capability offers a much more efficient and precise method of disease diagnosis. However, despite the impressive accuracy and effectiveness of these models, they often operate as "black boxes," providing little insight into how they arrive at their predictions. This lack of interpretability can be a major barrier to their widespread adoption in agriculture, as farmers and agronomists need to understand and trust the recommendations made by these models to make informed decisions [6].

Interpretable AI focuses on making ML models transparent and explainable, allowing users to understand the reasoning behind the model's predictions. We can not only detect plant diseases with high accuracy but also provide interpretable insights into which factors or features contributed to the diagnosis [7].

In this chapter, we will focus on DL's role in the detection and diagnosis of plant diseases. We will also explore interpretable AI in the field of PA, with a particular focus on its application in plant disease detection.

The structure of this chapter is as follows: Section 6.2 gives an overview of the previous research and context of using ML in PA. Section 6.3 provides the details of the algorithms and evaluation techniques used. In Section 6.4, an overview of the dataset, the preprocessing methods taken, and the methodologies used are discussed. Section 6.5 presents the results and analysis of the implemented strategy, including a detailed evaluation of the models' performance and the interpretability outcomes derived from Explainable AI (XAI) techniques. Section 6.6 ultimately concludes the chapter and provides recommendations for future research.

By focusing on DL, this chapter aims to provide a comprehensive understanding of how these powerful algorithms are revolutionizing plant disease detection and improving the efficiency and sustainability of agricultural practices around the world.

6.2 LITERATURE REVIEW

ML and DL techniques have seen considerable growths in recent years, including their use in PA, particularly for plant disease detection. These models have strong predictive capabilities, offering high levels of accuracy, precision, and recall. However, they are often criticized for their "black box" nature, which limits their adoption, especially among farmers who need to trust and understand the decision-making process. Current research focuses on performance metrics such as accuracy, precision, and recall, but fails to address the transparency of the decision-making processes within these models.

For instance, in the study [8] using convolutional neural network (CNN) and Naive Bayes Techniques in PA, CNNs outperformed Naive Bayes classifiers in terms of classification metrics. However, the study was confined to visual and phenotypic data and lacked mechanisms to explain how these features influenced model predictions.

Similarly, in [9] environmental data from sensors (such as humidity, soil moisture, and rainfall) was integrated with ML algorithms like XGBoost, Support Vector Machine (SVM), and RF to predict rice diseases. While the study achieved impressive accuracy of 92.5%, it lacked an explanation of why particular environmental variables contributed more heavily to disease predictions, underscoring the gap in interpretability.

Moreover, in [10], the YOLOv8 model achieved high performance in detecting corn diseases and pests under real-time conditions. Though the paper highlighted architectural innovations and superior metrics like mean Average Precision (mAP), it did not explore how or why the model arrived at specific decisions, rendering it a "black box."

The paper [11] introduced a hybrid DL pipeline combining CNN feature extraction with traditional classifiers like SVM and RF. The approach achieved 90% accuracy and demonstrated robustness to class imbalance and annotation noise. However, while the hybrid technique improved performance, it still did not address the explainability of how extracted features related to specific disease types, reducing its utility for informed crop management decisions.

Similarly, [12] applied transfer learning on ResNet50 to classify healthy and diseased jute leaves. The model achieved 94% accuracy, yet the study emphasized performance metrics without exploring why the model made particular classifications.

Several other works focused on broader themes in PA, such as [13]. While these studies underscore the potential of AI to optimize resource use and improve sustainability, they remain largely conceptual or descriptive, lacking empirical application of interpretability frameworks in plant disease contexts.

To address these shortcomings, researchers have begun incorporating XAI techniques into the development of plant disease detection models. Techniques such as LIME (Local Interpretable Model-Agnostic Explanations) and global surrogate models have been applied to provide interpretability in DL models. By explaining how models make predictions, these methods help bridge the gap between complex AI models and their practical use in agriculture. However, despite these advancements, the integration of interpretability into AI models remains a work in progress, and more research is needed to ensure that AI solutions can be both highly accurate and interpretable, enabling them to gain the trust of farmers and agricultural stakeholders.

By addressing these challenges and continuing to refine interpretability frameworks, AI can become a transformative tool in PA, enabling more sustainable and efficient farming practices while ensuring transparency and trust in automated decision-making.

6.3 ALGORITHMS AND TECHNIQUES USED

This section introduces and explains the techniques used in this chapter: the ML and DL models, the metrics used to evaluate the performance, as well as the XAI tools applied for transparency and interpretability.

6.3.1 ML and DL

The foundation of the system is built on DL, a subfield of ML that involves the use of neural networks with multiple layers capable of learning hierarchical representations. DL models have the ability to automatically extract relevant features from raw input data without the need for explicit human intervention or feature engineering [14]. One of the core components of the system is the use of CNNs, which are highly effective for image classification tasks due to their ability to capture spatial hierarchies in visual data. Additionally, transfer learning is employed to further optimize the model's performance and make the training process more efficient, especially when labeled data is scarce.

6.3.1.1 Convolutional Neural Networks

CNNs [15] are a specialized class of deep neural networks designed for processing data with a grid-like topology, such as images. Unlike traditional fully connected networks, CNNs leverage convolutional layers to apply filters across the input data, allowing them to detect local patterns such as edges, textures, and other visual features. These features are then hierarchically combined to form more complex representations, making CNNs highly effective for tasks such as object recognition, image classification, and segmentation. A typical CNN architecture consists of the following components [16]:

- **Convolutional Layers**: These layers function by applying filters (kernels) to the input data, which enable the detection of various features ranging from basic edges and textures to more complex structures as the network deepens. The filters slide over the input image, performing a convolution operation at each position. The result is a set of feature maps that represent different aspects of the image.
- **Pooling Layers**: Positioned after convolutional layers, pooling layers reduce the spatial size of the feature maps. This compression helps in minimizing computation while preserving key information, making the model more robust to spatial variations. This helps to decrease the computational load and introduces translation invariance, meaning the model can recognize patterns regardless of their location in the image.
- **Fully Connected Layers**: Found near the output of the network, these layers consolidate the extracted features and are typically used for final decision-making tasks such as classification.

CNNs are particularly powerful for image classification because they preserve the spatial relationships between pixels and allow the model to learn directly from the data without requiring manual feature extraction. By leveraging the hierarchical structure of the data, CNNs can learn progressively more abstract features, making them highly effective for complex visual recognition tasks.

6.3.1.2 Transfer Learning

Transfer learning [17] is a technique where a model that has been trained on one task is reused as the foundation for solving a related, but different, task. This method is particularly valuable in domains where acquiring large amounts of labeled data is either impractical or too costly [18]. Transfer learning leverages the knowledge learned from the source task (typically involving a large, diverse dataset) to improve the performance of a model on the target task, even when the target dataset is small or limited in scope.

In the context of image classification, transfer learning typically involves the following steps [19]:

- **Using Pre-trained Models**: A widely adopted method in transfer learning is to begin with models already trained on extensive public datasets like ImageNet, as they have learned to identify general patterns such as shapes and textures.
- **Retaining the Convolutional Base**: In transfer learning, the convolutional base (the layers responsible for feature extraction) of the pre-trained model is retained, as these layers have already learned useful features from the source task. In many cases, the original classification layers are substituted with new ones that are better aligned with the specific target task or domain.
- **Fine-Tuning the Model**: This process involves updating the parameters of both the pre-trained base and the newly added classification layers so that the model adapts to the new dataset more effectively.

The key advantages of transfer learning are:

- **Faster Convergence**: Transfer learning typically leads to faster convergence during training because the model has already learned useful feature representations from the source task. This means that fewer training epochs are needed for the model to adapt to the target task.
- **Improved Generalization**: Transfer learning can improve generalization, especially when the target task has limited labeled data. Since the pre-trained model has learned broad, transferable features, it is less likely to overfit the small target dataset.
- **Reduced Computational Cost**: By leveraging pre-trained models, transfer learning significantly reduces the computational resources needed to train a model from scratch. This is particularly beneficial when training deep neural networks, which can be computationally expensive.

Transfer learning is especially effective when the source and target tasks share similar low-level features, which is often the case in image-based tasks. For example, models pre-trained on large visual datasets like ImageNet can be easily adapted to recognize new types of objects or diseases in images. This adaptability and efficiency make transfer learning a powerful technique for many applications in computer vision, such as plant disease detection, where large, labeled datasets may be difficult to obtain.

6.3.2 Evaluation Methods

Evaluating the performance of ML models is essential to understanding how well they generalize to unseen data. This requires a set of well-defined metrics that quantify the model's ability to make accurate predictions. For this purpose, we primarily used two metrics: Accuracy and F1-Score.

6.3.2.1 Accuracy

A common evaluation metric for classification tasks, accuracy refers to the proportion of correct predictions made by the model out of all predictions and is often used when class distributions are balanced. It is particularly useful when each class has a roughly equal number of samples. However, accuracy can be misleading in situations where one class outnumbers the other by a lot. Here, a model could achieve high accuracy just by predicting the majority class for all instances, without truly learning meaningful patterns in the data. Formula (6.1) contains the formula for calculating accuracy:

$$\text{Accuracy} = \frac{\text{TN} + \text{TP}}{\text{TP} + \text{FP} + \text{TN} + \text{FN}} \tag{6.1}$$

where TP = True Positive, TN = True Negative, FP = False Positive, and FN = False Negative.

6.3.2.2 F1-Score

To address the limitations of accuracy, especially in imbalanced datasets, we also used the F1-score as a more robust evaluation metric. The F1-score combines precision and recall into a single measure by calculating their harmonic mean, making it particularly suitable for evaluating models on imbalanced datasets. It is particularly useful when we are interested in both the accuracy of positive predictions and the ability to identify all relevant positive instances of the model. The F1-score ranges from 0 to 1, where 1 is perfect precision and recall, and 0 is poor performance. In imbalanced datasets, where one class is much more prevalent than the other, the F1-score helps provide a more meaningful assessment of model performance compared to accuracy alone. Formula (6.2) contains the formula for calculating F1-score.

$$\text{F1 score} = \frac{2 * P * R}{P + R} \tag{6.2}$$

where P = Precision and R = Recall.

6.3.3 Explainable AI

As ML models become more complex, especially deep neural networks, they often operate as "black boxes," making it difficult to understand or justify their predictions. This lack of transparency is a major concern. To address this, several XAI techniques were explored to help interpret and visualize how the model arrives at its decisions.

6.3.3.1 LIME (Local Interpretable Model-Agnostic Explanations)

LIME [20] is an interpretability technique that explains individual predictions by approximating the black box model locally with a simpler, interpretable model, such as a decision tree.

- LIME works with any model architecture.
- The input image is segmented into superpixels (contiguous regions of similar pixels).

- LIME alters the image by turning the superpixels on and off.
- It then observes how the model's prediction changes and fits a local surrogate model that approximates the decision boundary in that neighborhood.

6.3.3.2 Global Surrogate Models

While LIME provides localized or instance-level explanations, global surrogate models [21] aim to interpret the entire behavior of a complex black box model by approximating it with a simpler, interpretable model.

- A surrogate model (e.g., a decision tree) is trained on the inputs and outputs of the black box model.
- The goal is not to predict but to mimic the decisions of the original model.
- The surrogate is then analyzed to extract global patterns and rules that approximate how the model behaves across the dataset.

Though the surrogate may not perfectly match the accuracy of the original model, it enables users to gain insights into the model's decision-making process at a larger level.

6.4 METHODOLOGY

6.4.1 Data Description

The dataset used in this study is the "Top Agriculture Crop Disease India" [22] dataset available on Kaggle. It is specifically curated to help in the classification of diseases in various major crops in India. Crops like corn, potato, rice, wheat, and sugarcane are included. Features of the dataset:

- The dataset contains leaf images representing different crops and their respective diseases.
- Each image is labeled with the name of the disease, providing a clear mapping between the image and the corresponding disease.
- The total number of images is 13,324 belonging to 17 classes, such as:
    - 3,852 images of the crop corn, divided into class labels: Common Rust, Gray Leaf Spot, Northern Leaf Blight, and Healthy
    - 2,152 images of the crop potato, divided into class labels: Early Blight, Late Blight, and Healthy
    - 4,078 images of the crop rice, divided into class labels: Brown Spot, Leaf Blast, Neck Blast, and Healthy
    - 2,942 images of the crop wheat, divided into class labels: Brown Rust, Yellow Rust, and Healthy
    - 300 images of the crop sugarcane, divided into class labels: Red Rot, Bacterial Blight, and Healthy

6.4.2 Preprocessing the Datasets

Preprocessing is an essential part of any ML pipeline, especially when working with image data. This section covers the steps we've taken to prepare the dataset for our DL model training.

6.4.2.1 Custom Dataset Class

A custom dataset class CustomDataset [23] is implemented using PyTorch's Dataset class to handle loading and transforming the images. The dataset is stored in a directory structure where each subfolder corresponds to a different class (crop disease). The images are loaded using the PIL.Image.open function, and they are converted to RGB format to ensure compatibility with the DL models. The dataset assigns integer labels to each class (disease) by parsing the folder names. Each unique folder name (representing a disease) is mapped to an integer index.

6.4.2.2 Data Transformation Pipeline

A number of transformations are applied to the images to prepare them for model training [24]:

- Each image is resized to a fixed size (224×224 pixels) to ensure uniformity in the input dimensions for the DL model.
- The pixel values are normalized to have a mean of [0.485, 0.456, 0.406] and standard deviation of [0.229, 0.224, 0.225].
- The images are converted to PyTorch tensors using T.ToTensor().

The transformation pipeline is implemented using torchvision.transforms.Compose(), which combines the resizing, normalization, and tensor conversion into a single operation that is applied to each image.

6.4.2.3 Data Splitting

The dataset is divided into three subsets: training, validation, and test sets. The random_split function [25] is used to randomly split the dataset into 90% for training the model, 5% for tuning hyperparameters and validating model performance during training, and 5% for testing the model after training. Each subset is wrapped in a DataLoader to facilitate batching and shuffling of the data. The DataLoader also allows efficient loading of data using multiple workers for parallel processing.

6.4.2.4 Class Imbalance Analysis

The data_analysis function [26] generates a bar plot showing the number of samples per class (disease). This helps in identifying any class imbalance, which may require special handling during training.

6.4.3 Implementing the DL Model

We have used multiple DL models to classify the crop diseases.

6.4.3.1 Model Selection

The following models were used in this study:

- **ReXNet-150**: This is a lightweight CNN designed to offer high performance with fewer parameters. This model is based on efficient depthwise separable convolutions and residual connections, making it well-suited for image classification tasks with limited computational resources [27].

- **MobileNetV2**: A widely used lightweight architecture designed for mobile and embedded devices. It maintains a balance between performance and efficiency [28].
- **ConvNeXt Small**: A modern, efficient model that combines the best properties of convolutional networks and vision transformers [29].
- **VGG16**: A deep CNN with a uniform architecture. It is relatively straightforward and effective for a wide range of image classification tasks [30].
- **DenseNet121**: A CNN that connects each layer to every other layer in a feed-forward fashion. It improves feature propagation and encourages feature reuse [31].
- **Inception V3**: A high-performing model that uses factorized convolutions and auxiliary classifiers [32].

All models are fine-tuned on the crop disease dataset using transfer learning, with pre-trained weights from ImageNet as the starting point. The final layers of each model are adjusted to match the number of disease classes in the dataset. The architecture of each model is summarized in Table 6.1.

TABLE 6.1
Architectures of Various Models

Model	Type	Core Techniques	Building Blocks	Output Layer
ReXNet-150	Lightweight CNN	Depthwise Separable Convolutions, Residual Connections, Bottleneck design	Stage-wise blocks with depthwise and pointwise convolutions, ReLU activation	Fully connected + Softmax
MobileNetV2	Mobile CNN	Inverted Residuals, Linear Bottlenecks, Depthwise Separable Convolutions	Inverted residual blocks with 1x1, depthwise, 1x1 convolutions	Fully connected + Softmax
ConvNeXt Small	Modern CNN with ViT influence	Patch Embedding, Large Kernels, Residual Connections	Conv blocks with conv, batch norm, GELU, large kernel sizes	Fully connected + Softmax
VGG16	Deep CNN	Simple, uniform structure with small filters	13 conv layers in 5 blocks, each followed by max pooling, ReLU activations	3 fully connected layers + Softmax
DenseNet121	Densely Connected CNN	Dense Connectivity, Transition Layers	Dense blocks with batch norm + ReLU, Transition layers (1x1 conv + Avg Pool)	Global Avg Pooling + Fully connected + Softmax
Inception V3	Efficient CNN	Inception Modules, Factorized Convolutions, Auxiliary Classifiers	Parallel conv filters (1x1, 3x3, 5x5), reduction layers, auxiliary classifiers	Global Avg Pooling + Fully connected + Softmax

6.4.4 Implementing XAI Techniques

XAI techniques are useful to interpret and decode the "black box" behavior of the models. They're particularly important in real-world applications like crop diseases detection, where farmers need to trust and understand model predictions. In this study, we implemented the following XAI techniques:

6.4.4.1 LIME (Local Interpretable Model-Agnostic Explanations)

It generates explanations for individual predictions by changing the input data and observing how the model's output changes [20]. It then approximates the decision boundary locally around a particular instance using a simple, interpretable model (e.g., linear regression). We used the lime package to generate explanations for several test images. By changing the pixel values of the input image and recording the model's predictions, LIME created a local explanation highlighting the parts of the image most important for the model's decision. LIME confirmed that the model's decisions were often influenced by relevant visual cues, such as the distinctive patterns of disease on crop leaves. For example, when predicting Potato_Early_Blight, LIME highlighted the early blight spots on the potato leaves, as seen in Figure 6.3. However, when the model misclassified similar diseases, LIME showed that the model was focusing on irrelevant or ambiguous features.

6.4.4.2 Global Surrogate

This is a simpler, interpretable model that is trained to approximate the behavior of the complex model across the entire dataset [21]. By training a surrogate model (e.g., a decision tree) on the predictions of the black box model, we can understand the global decision-making process of the complex model. In this study, we used a Decision Tree Classifier as a surrogate model. The decision tree was trained on the predictions made by the DL model on the test set. After training the surrogate model, we visualized the decision tree to understand which features (e.g., pixel values, image regions) the complex model was relying on to make decisions. We also plotted the tree to show the relationships between input features and output predictions. The decision tree surrogate model revealed interesting patterns in how the model made predictions. For example, the tree highlighted certain pixel patterns and thresholds that influenced predictions for specific diseases, such as the distinct visual features associated with Potato_Late_Blight and Wheat_Brown_Rust. While the decision tree was not able to capture all the nuances of the complex model, it helps us understand common decision paths, frequently used features, and key differentiators between disease classes. While techniques like CAM and LIME are useful for individual predictions, the global surrogate model provides an interpretable summary of the overall decision-making process. This helps to explain how the complex model relates the image features to its final classifications. For farmers, agronomists, and agricultural researchers, this means:

- Getting general rules about how the model relates certain visual cues with specific diseases (e.g., "If the upper leaf shows brown circular patches, it's likely late blight").
- Helping validate trust in the model by exposing logical and human-understandable decision paths.

6.4.5 Ablation Analysis

Ablation studies are useful for understanding the impact of different components and architectural choices on model performance. In this study, we performed two key modifications to the architecture: removal of dropout layers and replacement of the activation function. These modifications allowed us to explore the effect of regularization techniques and activation functions on the model's ability to classify crop diseases.

6.4.5.1 Removal of Dropout Layers

Dropout is a regularization technique used to prevent overfitting by randomly disabling neurons during training, forcing the model to generalize better [33]. However, the effectiveness of dropout can vary based on the architecture and the task at hand. To test this, we removed the dropout layers from the model and observed the impact on model performance. The removal of dropout did not significantly affect the model's overall performance. Training accuracy and F1-score remained nearly identical to the baseline, and the validation accuracy also showed little change. The absence of dropout likely did not lead to overfitting, as the model was able to generalize well without it.

6.4.5.2 Choice of Activation function

The choice of activation function plays a crucial role in determining how a model learns and processes features [34]. SiLU (Sigmoid-weighted Linear Unit) activations are known for their smooth gradients, which can help models learn better representations. ReLU (Rectified Linear Unit), on the other hand, is a more commonly used activation function due to its simplicity and ability to deal with the vanishing gradient problem. We tested whether replacing SiLU with ReLU would affect the model's performance.

When we replaced SiLU with ReLU, the model exhibited faster training convergence, with training accuracy increasing more rapidly. However, this came at the cost of slight degradation in validation performance, as the validation accuracy and F1-score were slightly lower than those of the baseline model. This suggests that while ReLU allowed the model to learn faster, it may have also caused the model to overfit slightly, reducing its ability to generalize to unseen data. The smooth gradient properties of SiLU might have contributed to a better generalization performance, which the ReLU activation could not replicate in this task.

6.5 RESULTS AND ANALYSIS

In this chapter, we explored the use and performance of several DL models on a crop disease detection. The models used were: ReXNet-150, MobileNetV2, ConvNeXt Small, VGG16, DenseNet121, and InceptionV3. These models were tested on the ability to correctly classify images of crops showing different diseases. The performance of each model was assessed using two key metrics: Accuracy and F1-score. These are summarized in Table 6.2, which compares overall performance across all models, and Table 6.3, which breaks down accuracy and F1-scores per crop. We used

TABLE 6.2
Accuracy and F1-Score for Various Models

	Accuracy	F1-Score
ReXNet-150	0.98	0.97
MobileNetV2	0.97	0.96
ConvNeXt Small	0.96	0.96
VGG16	0.95	0.92
DenseNet121	0.96	0.94
InceptionV3	0.96	0.93

TABLE 6.3
Performance Metrics for Various Models over Different Crops

	Crop	Accuracy	F1-Score
ReXNet-150	Corn	0.99	0.98
	Wheat	1	1
	Potato	1	1
	Sugarcane	0.92	0.92
	Rice	0.95	0.93
MobileNetV2	Corn	0.99	0.99
	Wheat	0.99	0.75
	Potato	1	1
	Sugarcane	0.92	0.91
	Rice	0.94	0.93
ConvNeXt Small	Corn	0.95	0.97
	Wheat	0.99	0.99
	Potato	1	1
	Sugarcane	1	1
	Rice	0.93	0.9
VGG16	Corn	0.97	0.95
	Wheat	0.98	0.98
	Potato	0.99	0.98
	Sugarcane	0.85	0.83
	Rice	0.91	0.89
DenseNet121	Corn	0.97	0.56
	Wheat	1	1
	Potato	1	1
	Sugarcane	0.85	0.83
	Rice	0.93	0.89
InceptionV3	Corn	0.99	0.99
	Wheat	0.99	0.74
	Potato	1	1
	Sugarcane	0.85	0.83
	Rice	0.9	0.86

LIME and global surrogate models to understand the decision-making process of these DL models.

ReXNet-150 stood out as the highest-performing model overall, achieving impressive results across all crop types. The model's performance on corn and wheat diseases was high, with an accuracy of 0.99 and an F1-score of 0.98. This can be due to ReXNet-150's deep architecture, which is specifically designed to extract rich hierarchical features from images. Its ability to process and learn complex patterns makes it highly effective for tasks like crop disease classification, where subtle differences between healthy and diseased plants must be identified. The deep feature extraction layers in ReXNet-150 allow it to detect these intricate patterns, contributing to its high performance. The model also performed well on rice and sugarcane diseases, giving high accuracies again. This demonstrates that the model is not only robust for simpler crops but can also generalize well to other crops, demonstrating the potential of ReXNet-150 in real-world applications for disease identification.

MobileNetV2 also demonstrated strong performance. For wheat and potato, the accuracy was perfect, with F1-scores close to 1. This indicates that the model is adept at identifying these crop diseases with minimal false positives and false negatives. MobileNetV2 is a lightweight model designed for resource-constrained environments, which makes it suitable for deploying in mobile or edge devices for real-time disease detection in the field. Despite this, MobileNetV2 did struggle with a few crops, mainly in sugarcane, where the F1-score dropped slightly to 0.92. This is likely due to the simpler architecture of MobileNetV2, which might lack the depth required to capture subtle variations in certain diseases that require more complex feature extraction, as is the case with sugarcane diseases.

ConvNeXt Small, although a more recent architecture, did not perform as well as ReXNet-150 and MobileNetV2. For crops like corn and wheat, the accuracy was notably lower, and the F1-score was also reduced. ConvNeXt Small is designed to balance between efficiency and performance, but it appears that it may not capture fine-grained details in agricultural images as well as the deeper ReXNet-150.

VGG16 showed mixed results, with performance dropping on certain crop diseases. For example, the model achieved only a 0.56 F1-score for wheat, which is a significant drop compared to other models. This can likely be due to VGG16's relatively shallow architecture and its reliance on fixed-size convolutional filters. While VGG16 is known for its ability to learn basic image features, it may not be well-suited to handling more complex patterns in agricultural images, which are often highly varied and require more complex recognition.

DenseNet121 performed pretty well on most crops, but its performance was less consistent compared to some other models. The model achieved good results on potato and rice, but struggled with wheat. This can be due to the dense connectivity between layers in DenseNet121, which helps with feature reuse and gradient flow. However, this may not always be beneficial in datasets with high variability, as the dense layers may focus too much on low-level features and miss higher-level abstract patterns that are more indicative of diseases. DenseNet121's performance on wheat and sugarcane also indicated that, while dense connections provide advantages for certain tasks, they might not be as effective in distinguishing the subtle features present in crop disease images, especially when the diseases show overlapping symptoms or complex backgrounds.

InceptionV3 exhibited performance similar to that of DenseNet121, with potato achieving an accuracy of 1.0, but the performance significantly dropped for some diseases in wheat and rice. This can be due to the model's complex architecture, which attempts to capture multi-scale features through parallel convolution operations. While this approach is powerful for many image classification tasks, it might be too complex for agricultural datasets where certain diseases have fewer visual cues that the model can focus on.

To understand why these models made certain predictions, we applied XAI techniques like LIME and global surrogate models. Figure 6.1 shows the class activation mappings (CAM) generated from the models, which highlight which parts of the input image were most influential in the model's prediction. These heatmaps clearly demonstrate that models like ReXNet-150 and MobileNetV2 consistently focused on meaningful plant features (like leaf discoloration, lesions, and disease-specific patterns). This reinforces the reliability of the models' attention mechanisms.

To provide a global perspective, we trained a Global Surrogate model, specifically a Decision Tree, to approximate the CNN's decision-making. As seen in Figure 6.2, the decision tree reveals key decision rules and thresholds based on image features that frequently contributed to disease classification. While this surrogate model is simpler and less accurate, it offers an interpretable summary of how the complex CNN behaves across the entire dataset.

Figure 6.3 illustrates LIME explanations for a sample test image, showing the superpixels that influenced the model's prediction. When models made errors, LIME often revealed attention to ambiguous or misleading regions. These insights can be used to refine training or data collection.

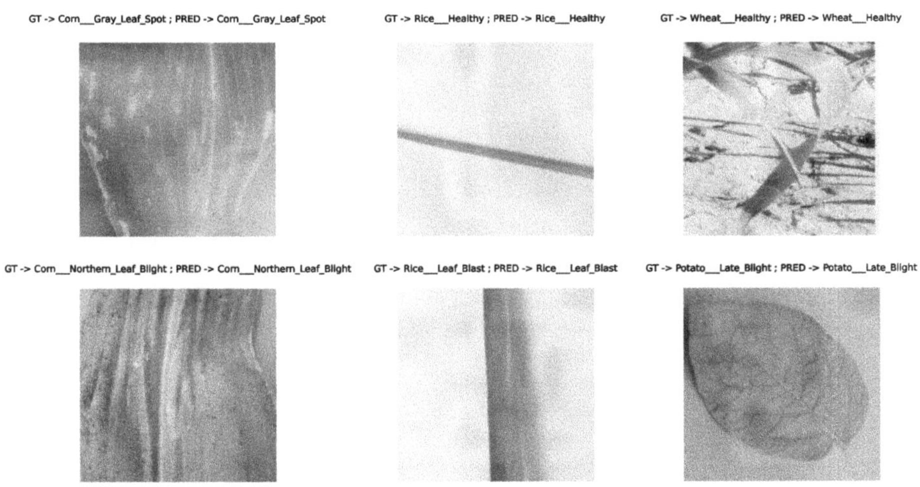

FIGURE 6.1 Class activation mappings.

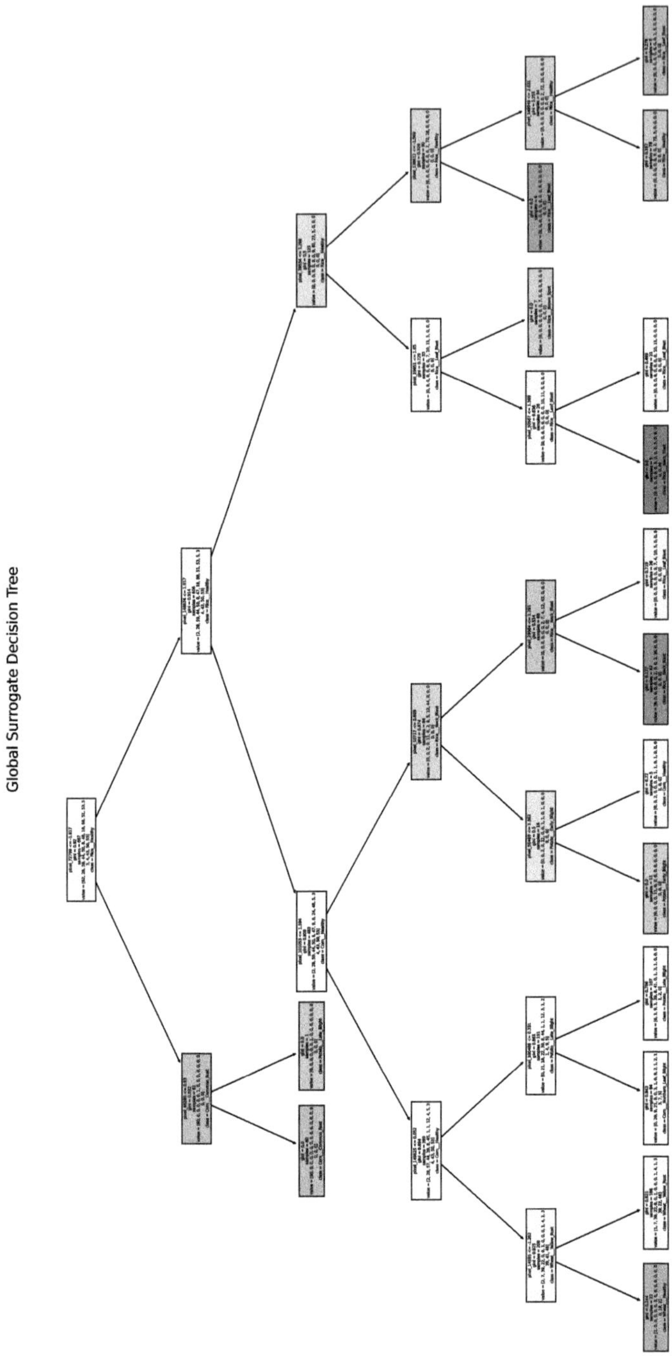

FIGURE 6.2 Global surrogate.

FIGURE 6.3 Lime.

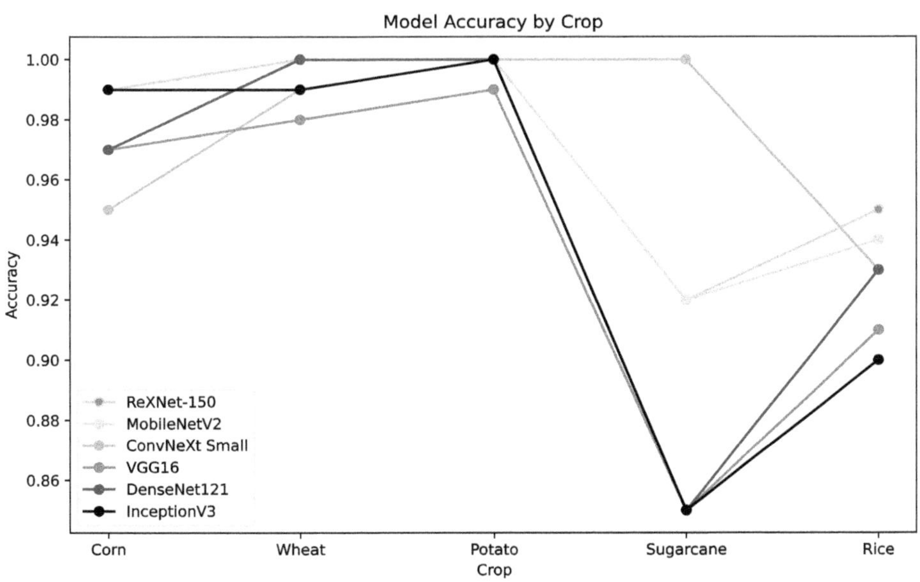

FIGURE 6.4 Accuracy plot.

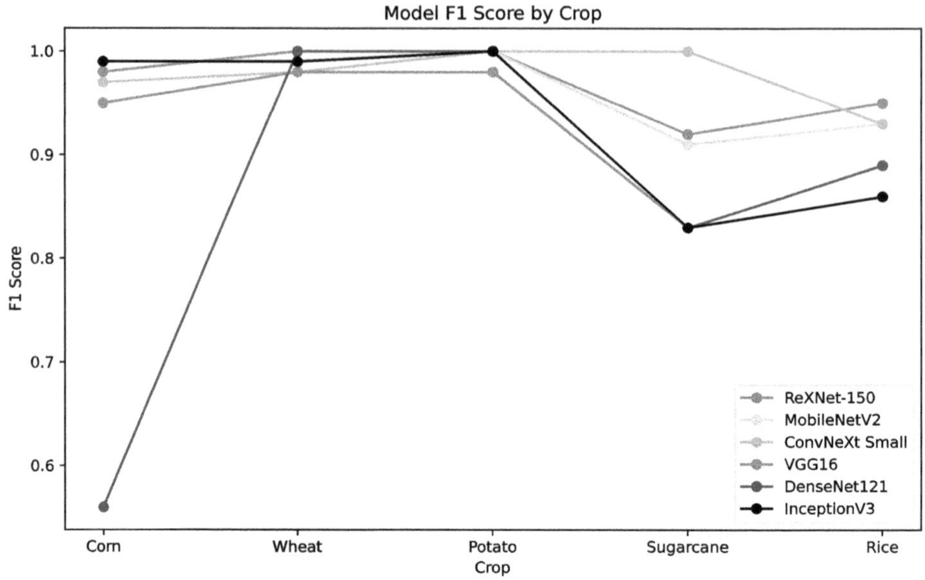

FIGURE 6.5 F1-plot.

6.6 CONCLUSION AND FUTURE SCOPE

In this chapter, we have explored the role of XAI in improving the transparency and trustworthiness of AI models for crop disease detection. By applying state-of-the-art DL models, including ReXNet-150, MobileNetV2, ConvNeXt Small, VGG16, DenseNet121, and InceptionV3, we successfully demonstrated the capability of AI models in classifying diseases in crops such as corn, potato, wheat, rice, and sugarcane. These models have shown excellent performance in terms of accuracy and F1-scores, validating the potential of DL models in agricultural disease diagnosis. While these models performed remarkably well, one of the key challenges highlighted was the lack of interpretability, which has often limited the adoption of AI solutions in agriculture. To address this, we incorporated XAI techniques like LIME and Global Surrogate models to shed light on how these models made their predictions. Through these methods, we were able to identify critical features in the images that influenced the model's decision-making process, making the models more transparent and understandable for farmers and agronomists. The results from LIME demonstrated that the models primarily focused on important regions of the plant images, such as leaf discolorations, spots, and deformations. These features are key indicators of plant diseases. The Global Surrogate models provided further insight into the decision boundaries of the models. Furthermore, we performed an ablation analysis, which revealed the impact of various components like dropout layers and activation function. While the DL models demonstrated impressive results, there remain ongoing challenges, particularly with respect to dataset biases, data variability, and the trade-off between accuracy and interpretability. Addressing these challenges will be vital for improving the models' robustness and ensuring they can be reliably deployed in real-world agricultural scenarios. Further work is needed to optimize the models and remove any inherent biases present in the datasets used for training. Additionally, finding the right balance between interpretability and predictive power is an important area for future research.

Looking ahead, the future of DL in crop disease detection holds great promise, particularly as the scope of AI models expands and their practical applicability continues to evolve. One key area of development is the integration of multimodal data sources, which could significantly enhance the predictive capabilities of AI systems. By incorporating real-time environmental data such as weather conditions, soil moisture, and temperature alongside sensor data, AI models could be trained to better understand the complex environmental factors that contribute to disease outbreaks. This approach to disease monitoring would not only improve the accuracy of disease predictions but also offer farmers more contextual insights that could help optimize resource use and crop management practices.

Another exciting direction for future research is the exploration of self-supervised learning and generative AI models. Self-supervised learning offers a promising avenue to utilize unlabeled data, which is often abundant in the agricultural field, to improve model performance. By leveraging large volumes of unlabeled crop images, self-supervised learning can help uncover new disease patterns that might otherwise be missed, making it a valuable tool for expanding existing datasets and detecting previously unseen diseases. In addition, generative AI models have the potential to create synthetic datasets for rare or uncommon diseases, addressing a significant limitation in agricultural data, where some diseases are underrepresented. By generating more diverse and comprehensive datasets, generative models could enhance model generalization, making them more adaptable to various agricultural conditions.

The integration of these advanced techniques will further push the boundaries of what DL can achieve in crop disease detection. As AI and DL models continue to evolve and improve, their role in transforming agricultural practices will only grow, offering more accurate, efficient, and sustainable solutions to help farmers tackle the pressing challenge of plant diseases. As we look to the future, it is clear that the fusion of AI with agriculture has the potential to reshape the industry, improving both productivity and food security on a global scale.

6.7 LIMITATIONS OF THE STUDY

While this study demonstrated strong performance across multiple DL models, practical deployment requires consideration of computational costs. Lightweight models such as MobileNetV2 and ReXNet-150 are particularly suitable for mobile or embedded devices used in field settings, due to their low parameter counts and efficient inference speeds. On the other hand, models like VGG16, DenseNet121, and InceptionV3 have higher computational demands and may require cloud-based infrastructure or high-end edge devices for real-time use.

A detailed comparison of model complexity (in terms of parameters and memory usage) was beyond the scope of this chapter but is essential for real-world deployment. Future studies could explore resource-aware model selection, especially for smallholder farms or low-resource regions where device constraints are significant.

REFERENCES

1. N. Montalvo-Romero, A. Montiel-Rosales, R. Purroy-Vásquez and P. Quechulpa-Pérez, "Agro-technological systems in traditional agriculture assistance: A systematic review," *IEEE Access*, vol. 11, pp. 123047–123069, 2023.
2. M. Padhiary, D. Saha, R. Kumar, L. N. Sethi and A. Kumar, "Enhancing precision agriculture: A comprehensive review of machine learning and AI vision applications in all-terrain vehicle for farm automation," *Smart Agricultural Technology*, pp. 1–14, 2024.
3. M. Aswath, S. Sowdeshwar, M. Saravanan and S. K. Perepu, "Advanced machine learning framework for efficient plant disease prediction," in *2022 5th Asia Conference on Machine Learning and Computing (ACMLC)*, Bangkok, Thailand, 2022.
4. M. M. Khalid and O. Karan, "Deep learning for plant disease detection," *International Journal of Mathematics, Statistics, and Computer Science,* vol. 2, pp. 75–84, 2024.
5. R. Sujatha, S. Krishnan, J. M. Chatterjee and A. H. Gandomi, "Advancing plant leaf disease detection integrating machine learning and deep learning," *Scientific Reports*, vol. 15, p. 11552, 2025.
6. J. Amara, B. König-Ries and S. Samuel, "Explainability of deep learning-based plant disease classifiers through automated concept identification," *arXiv preprint arXiv:2412.07408*, 2024.
7. N. Nigar, H. M. Faisal, M. Umer, O. Oki and J. Lukose, "Improving plant disease classification with deep learning based prediction model using explainable artificial intelligence," *IEEE Access*, vol. 12, p. 100005–100014, 2024.
8. R. Sharma, M. Kumar, M. Manwal and others, "Advanced classification of Indian turnip varieties using CNN and Naive Bayes techniques in precision agriculture," in *2024 2nd World Conference on Communication & Computing (WCONF)*, Raipur, Bharat, 2024.
9. R. Akhter and S. A. Sofi, "Decision making in precision agriculture using IoT and machine learning: A focus on rice diseases," in *2024 International Conference on Decision Aid Sciences and Applications (DASA)*, 2024.
10. Z. Wu, Y. Jiang, X. Li and K. L. Chung, "Enhancing precision agriculture: YOLOv8 for accurate corn disease and pest detection," in *2024 IEEE 7th International Conference on Electronic Information and Communication Technology (ICEICT)*, 2024.
11. M. More, P. Deshmukh and R. Tanksale, "Hybrid approach for sunflower disease detection: CNN feature extraction and ML model evaluation," in *2025 International Conference on Electronics and Renewable Systems (ICEARS)*, 2025.
12. P. Kaushik and S. Khurana, "Jute leaf disease detection using ResNet50: A deep learning approach for precision agriculture," in *2025 International Conference on Electronics and Renewable Systems (ICEARS)*, 2025.
13. L. Bachu, A. Kandibanda, N. Grandhi, D. P. Athina and P. K. Ande, "Machine learning for enhanced crop management and optimization of yield in precision agriculture," in *2024 8th International Conference on I-SMAC (IoT in Social, Mobile, Analytics and Cloud)(I-SMAC)*, 2024.
14. P. Chhabra and S. Goyal, "A thorough review on deep learning neural network," in *2023 International Conference on Artificial Intelligence and Smart Communication (AISC)*, 2023.
15. X. Zhao, L. Wang, Y. Zhang, X. Han, M. Deveci and M. Parmar, "A review of convolutional neural networks in computer vision," *Artificial Intelligence Review*, vol. 57, p. 99, 2024.
16. A. Younesi, M. Ansari, M. Fazli, A. Ejlali, M. Shafique and J. Henkel, "A comprehensive survey of convolutions in deep learning: Applications, challenges, and future trends," *IEEE Access*, vol. 12, pp. 41180–41218, 2024.

17. M. Iman, H. R. Arabnia and K. Rasheed, "A review of deep transfer learning and recent advancements," *Technologies*, vol. 11, p. 40, 2023.
18. R. Bao, Y. Sun, Y. Gao, J. Wang, Q. Yang, Z.-H. Mao and Y. Ye, "A recent survey of heterogeneous transfer learning," *arXiv preprint arXiv:2310.08459*, 2023.
19. A. Panda, D. Panigrahi, S. Mitra, S. Mittal and S. Rahimi, "Transfer learning applied to computer vision problems: Survey on current progress, limitations, and opportunities," *arXiv preprint arXiv:2409.07736*, 2024.
20. M. Panda and S. R. Mahanta, "Explainable artificial intelligence for Healthcare applications using Random Forest Classifier with LIME and SHAP," in *Explainable, Interpretable, and Transparent AI Systems*, Edited by B. K. Tripathy, H. Seetha, CRC Press, 2023, pp. 89–105.
21. W. R. Monteiro and G. Reynoso-Meza, "A multi-objective optimization design to generate surrogate machine learning models in explainable artificial intelligence applications," *EURO Journal on Decision Processes,* vol. 11, p. 100040, 2023.
22. Kaggle, Top agriculture crop disease India, 2023.
23. Datasets & DataLoaders – PyTorch Tutorials 2.7.0+cu126 documentation — pytorch.org.
24. Transforming and augmenting images – Torchvision 0.22 documentation — pytorch.org.
25. torch.utils.data – PyTorch 2.7 documentation — pytorch.org.
26. imbalanced-learn documentation – Version 0.13.0 — imbalanced-learn.org.
27. M. R. Kumar, P. Adithiyan, G. J. Sendur, S. P. Kumar, S. M. Kumar and V. Nikhil, "Analyzing the potential of ReXNet-150: A novel architecture for automobile parts classification," in *2024 3rd International Conference on Artificial Intelligence for Internet of Things (AIIoT)*, Vellore, Bharat, 2024.
28. M. Sandler, A. Howard, M. Zhu, A. Zhmoginov and L.-C. Chen, "Mobilenetv2: Inverted residuals and linear bottlenecks," in *Proceedings of the IEEE Conference on Computer Vision and Pattern Recognition*, Salt Lake City, UT, USA, 2018.
29. Z. Liu, H. Mao, C.-Y. Wu, C. Feichtenhofer, T. Darrell and S. Xie, "A convnet for the 2020s," in *Proceedings of the IEEE/CVF Conference on Computer Vision and Pattern Recognition*, New Orleans, LA, USA, 2022.
30. K. Simonyan and A. Zisserman, "Very deep convolutional networks for large-scale image recognition," *arXiv preprint arXiv:1409.1556*, 2014.
31. G. Huang, Z. Liu, L. Van Der Maaten and K. Q. Weinberger, "Densely connected convolutional networks," in *Proceedings of the IEEE Conference on Computer Vision and Pattern Recognition*, Honolulu, HI, USA, 2017.
32. C. Szegedy, V. Vanhoucke, S. Ioffe, J. Shlens and Z. Wojna, "Rethinking the inception architecture for computer vision," in *Proceedings of the IEEE Conference on Computer Vision and Pattern Recognition*, Las Vegas, NV, USA, 2016.
33. N. Srivastava, G. Hinton, A. Krizhevsky, I. Sutskever and R. Salakhutdinov, "Dropout: A simple way to prevent neural networks from overfitting," *The Journal of Machine Learning Research*, vol. 15, pp. 1929–1958, 2014.
34. C. Nwankpa, W. Ijomah, A. Gachagan and S. Marshall, "Activation functions: Comparison of trends in practice and research for deep learning," *arXiv preprint arXiv:1811.03378*, 2018.

7 Implementation of Machine Learning and Deep Learning Techniques for Disease Detection and Classification

Ketan Sarvakar, Akshita Patel, and Rahul Yadava

7.1 INTRODUCTION TO PRECISION AGRICULTURE

The traditional agricultural method is an ancient farming system that employs human labor, natural fertilizers, and a few instruments. It focuses on tradition, variety, and sustainability, and employs tactics like crop rotation and diversified agricultural systems. It is restricted by both efficiency and productivity. Precision agriculture (PA), on the other hand, is a technologically advanced strategy that uses GPS, sensors, drones, and data analysis to improve agricultural operations. PA improves efficiency by reducing resource use and increasing crop yields. PA also reduces waste, increases sustainability, and promotes mass food production, making it a critical invention for the future of agriculture in the face of a growing global population.

7.1.1 Sustainable Agriculture: PA

Sustainable agriculture strives to generate a healthy atmosphere, economic sustainability, and social justice. It encompasses practices like the enhancement of soil health, conserving water, and minimizing pollution. Consumers and retailers increasingly demand value-based food products, and researchers conduct interdisciplinary studies. Sustainable agriculture is not a set of practices but a continuous conversation between individual farmers and the members of the local community to solve the complex food and fiber production problems. Sustainable agriculture is a dedication to the health of our planet and its inhabitants that combines traditional farming knowledge with cutting-edge scientific innovation. It prioritizes environmental, social, and economic sustainability, striving to fulfill present demands without jeopardizing future generations' capacity to meet their own. Sustainable agriculture is built on three interwoven pillars: ecological, economic, and social sustainability.

 DOI: 10.1201/9781003637264-9

Ecological sustainability is the cornerstone of sustainable agriculture, focusing on soil health, biodiversity, and resource conservation via methods such as crop rotation, no-till farming, and agroforestry. Organic farming, which does not use synthetic pesticides or fertilizers, supports natural alternatives to safeguard the environment and customers' health. Economic sustainability guarantees that farmers may earn a livelihood while practicing ecologically responsible farming, lowering input costs, improving market resilience, and assisting small-scale farmers.

Sustainable social development extends the advantages of sustainable agriculture to entire communities, ensuring that farming techniques respect human rights, promote rural livelihoods, and improve the well-being of food producers. Fair pay, community participation, and food security are all important parts of social sustainability. Agriculture that is environmentally friendly is more than a concept; it is a call to action that includes techniques that benefit the environment, increase economic viability, and promote social fairness. It is a vision for the future in which food is abundant, produced, and distributed while respecting our planet's fragile balance and the rights of all its people. Sustainable agriculture is critical for ensuring global food security, environmental protection, and climate change mitigation. With the world population expected to exceed approximately ten billion by 2050, sustainable farming techniques are critical to providing a continuous and secure food supply. Food rotation, organic farming, and other sustainable approaches can boost food production while strengthening resistance to climate change. Unsustainable agricultural practices, such as high pesticide and chemical fertilizer usage, monoculture farming, and over-irrigation, all contribute to soil deterioration, water pollution, and biodiversity loss. Sustainable agriculture addresses these challenges by preserving soil health, lowering the need for hazardous chemical inputs, and enhancing biodiversity via methods such as agroforestry and no-till farming. Environmental degradation is a major worldwide problem, and sustainable agriculture can help mitigate its consequences. Sustainable agricultural techniques trap carbon in the soil, lowering greenhouse gas emissions and storing carbon. Agroforestry, in particular, is a climate change fighter, sequestering carbon in both woody biomass and soil. In a world where climate change endangers agriculture and community well-being, sustainable agriculture provides a means of adapting to new climate realities while lowering the sector's contribution to global warming.

The practice of sustainable agriculture has several benefits outside the farm, such as increased crop yields, soil health, economic resilience, lower environmental impact, soil carbon sequestration, improved water management, biodiversity conservation, and healthier, safer food. Crop rotation, organic farming, and agroforestry improve soil fertility and productivity. These techniques help emphasize soil health by decreasing insect and disease cycles and encouraging nutrient-dense soil. Economic resilience is a fundamental advantage of sustainable agriculture since it decreases farmers' reliance on costly chemical inputs and boosts their ability to withstand market changes. Sustainable agriculture also decreases the environmental effect of food production by limiting the use of synthetic pesticides and chemical fertilizers, preventing water pollution, lowering greenhouse gas emissions, and safeguarding biodiversity. Water-efficient irrigation and coordinated pest management help to preserve biodiversity. Organic farming, a subcategory of sustainable

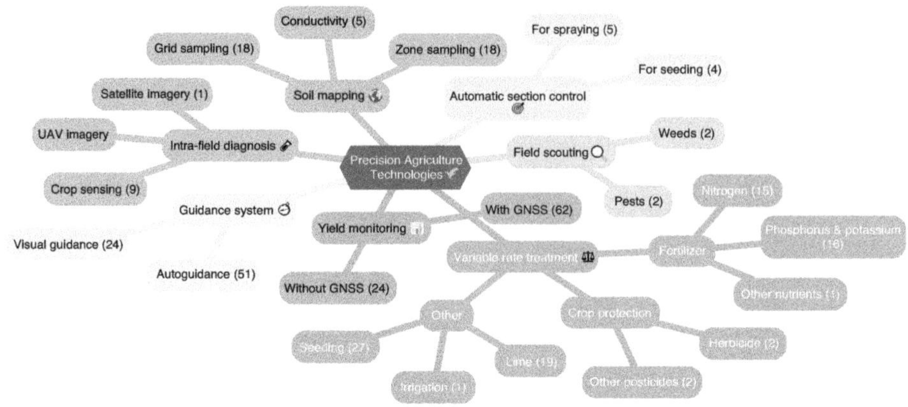

FIGURE 7.1 Precision Agriculture technologies.

agriculture, stresses customer health and safety by not using synthetic pesticides or chemical fertilizers. Overall, sustainable agriculture points to a better, more sustainable future.

PA is a management strategy that uses temporal, spatial, and individual plant and animal data to support management decisions for improved resource use efficiency, productivity, quality, profitability, and sustainability of agricultural production. It is used in both crop and livestock production and often employs technologies to automate operations. The goal of PA research is to define a decision support system for whole farm management, optimizing returns on inputs while preserving resources. One approach is the phyto geomorphological approach, which ties multi-year crop growth stability to topological terrain attributes. GPS and GNSS have enabled the practice of PA, allowing farmers and researchers to create maps of spatial variability of variables such as crop yield, terrain features, organic matter content, moisture levels, nitrogen levels, pH, EC, Mg, K, and others. Real-time sensors directly in soil can wirelessly transmit data without human presence. Unmanned aerial vehicles (UAVs) can also benefit from PA by capturing multispectral or RGB images of a field, which can be stitched together using photogrammetric methods to create orthophotos. These topographic maps can be used to correlate crop health with topography, optimizing inputs such as water, fertilizer, and growth regulators through variable rate applications (Figure 7.1).

7.2 MACHINE LEARNING IN CROP DISEASE DETECTION

The use of machine learning (ML) is an artificial intelligence (AI) application that replicates human learning processes, reorganizes existing knowledge structures, and constantly improves performance in predicting and classifying unknown data. It is very useful in detecting agricultural illnesses, such as Verticillium wilt disease, by analyzing hyperspectral photographs of crops. Researchers employed ML techniques based on support vector machines (SVM) and linear decomposition techniques (Linear Discriminant Analysis, LDA) to predict the severity levels of Verticillium wilt disease in olive crops, attaining precision levels of 59% with the LDA method and 79.2% with the SVM. Similarly, ML methods have been employed to assess chlorophyll content in crops such as potatoes, with an SVM-based model obtaining higher accuracy. These sophisticated methods, which include SVMs, random forests, and gradient boosting, are especially beneficial for assessing vegetation indicators and reflectance spectra in agricultural crops.

7.2.1 ML-Based Disease Classification

Supervised ML is a widely used approach in ML, used for various analysis and prediction tasks such as crop classification and disease detection. SVMs (Chu et al., 2020) are primarily used for classification problems, while the K-Nearest Neighbor (K-NN) (Shuaibu et al., 2018) algorithm is used for both classification and regression. Decision trees are versatile and interpretable, especially when applied to hyperspectral imaging in agriculture. Logistic Regression (LR) (Jarolmasjed et al., 2018) is a predictive analysis algorithm used in classification problems, evaluating the relationship between independent variables and dependent variables. Linear regression (L. Wang et al., 2021) attempts to establish a linear relationship between independent variables and dependent variables, while Multivariate Linear Regression (MLR) (Cheshkova, 2022) models the relationship between multiple dependent variables and independent variables. Deep Forest (Huang et al., 2023) is a relatively newer ML model that has shown promising results in various domains, including image classification and remote sensing applications like hyperspectral imaging in agriculture. It relies on an ensemble learning approach inspired by decision trees but with a hierarchical structure that enables efficient feature learning and representation.

Backpropagation Neural Networks (BPNNs) (Golhani et al., 2018) are powerful computational tools that can be combined with genetic algorithms (GAs) to enhance hyperspectral imaging analysis in agriculture. LDA (Zhang et al., 2021) is a statistical and ML technique used in multivariate analysis and supervised classification, while Naïve Bayes Algorithm (Z. Wang et al., 2021) is a powerful yet simple generative ML classifier used for crop classification and understanding crop development. Modified Partial Least Squares Regression (MPLSR) (Gabrielli et al., 2021) is a statistical technique used for building predictive models, particularly beneficial when dealing with datasets with certain characteristics. LightGBM (Zhou et al., 2020) is a powerful tool for analyzing hyperspectral imagery of crops, particularly for tasks like classification and regression. Unsupervised ML focuses on exploring hidden patterns and structures in data without labels or predefined responses, making significant progress in remote sensing image classification. K-means (Ji et al., 2019) clustering organizes data into groups with high and low group similarity, useful in situations where labels are unavailable or to discover hidden information. Hierarchical clustering groups (Golhani et al., 2018) data points into a dendrogram, often used to visualize relationships within data and identify natural groupings. WEKAXMeans (WXM) extends the traditional k-means clustering algorithm by automatically determining the optimal number of clusters based on data, particularly useful in hyperspectral imaging for crops. ISODATA is an unsupervised learning technique that iteratively partitions a dataset into clusters based on similarity, allowing for analysis without pre-labeled information about crops or their health status.

Dimensionality reduction is a fundamental technique in data analysis and ML, reducing the number of features in a dataset while preserving important information. Principal Component Analysis (PCA) is a multivariate analysis technique used to reduce dimensionality while preserving variability. Singleton Value Decomposition (SVD) is a versatile tool for processing and analyzing hyperspectral images for agriculture, enhancing monitoring, classifying, and predicting crop properties. Partial Least Squares Regression (PLSR) is a statistical method used to model the relationship between predictor and response variables, particularly in cereals like wheat, to determine micronutrients with high credibility.

7.2.1.1 IoT and Sensor-Based ML Approaches

A vital PA tactic is the combination of ML (ML) with the Internet of Things (IoT). Through networked devices and sensors, it makes it possible to gather data in real time, which can then be analyzed to glean insights and guide decision-making. Agricultural characteristics including soil moisture, humidity, temperature, and nutrient levels (Qiu et al., 2021) are continually monitored by IoT devices that are outfitted with sensors such as FC-28, DHT11, and JXBS-3001. Other IoT devices, such hyperspectral cameras placed to drones, record comprehensive data about crops. The literature study emphasizes how IoT-based soil monitoring devices are becoming more and more popular.

Protocols like MQTT (Islam et al., 2023) are used by IoT devices to gather data in order to facilitate dependable and effective data transfer. Scalability and interaction with other cloud services are features that cloud-based MQTT brokers provide for data management and storage. ML techniques are then used to examine this data for precision agricultural applications. ML algorithms, using complex correlations in the data, can forecast which crops are appropriate for a given soil type and environmental circumstances. For example, an IoT device with ML capabilities may use soil nutrient analysis to provide a list of possible high-yielding crops. ML models can identify the best kind and quantity of fertilizer depending on soil nutrient levels and crop needs. Considering IoT sensors that monitor environmental variables can assist anticipate disease outbreaks, ML can also aid in the detection and diagnosis of diseases. Optimizing fertilizer use and irrigation schedules can be aided by real-time soil moisture monitoring. Customers may use a mobile app to assess the quality of product by using the system's ability to record data on fertilizers and treatments that have been applied.

Numerous advantages arise from the use of ML and the IoT in PA, such as increased crop production and productivity, better use of resources, well-informed decision-making, sustainable farming methods, and better food security and quality. Growth and output potential are maximized when crop and fertilizer recommendations are accurate. ML-driven insights and real-time monitoring make it possible to use resources efficiently, which lowers waste and its negative effects on the environment. Real-time data and personalized advice enable farmers to make timely and efficient agricultural decisions. By reducing its negative effects on the environment and optimizing resource utilization, the system encourages sustainable activities. Installing sensors in the field, sending data to a distant server via the NodeMCU and MQTT protocols (Islam et al., 2023), and utilizing ML models to produce suggestions are all part of the suggested system. Farmers may access real-time data, historical trends, actionable insights, and crop recommendations through an intuitive smartphone app. Global food security might be improved and agricultural methods could be revolutionized by this integrated strategy.

Crop management is a critical aspect of agriculture, aiming to optimize crop production and resource utilization while ensuring sustainability. Modern approaches to crop management use advanced technologies like IoT and ML to address challenges faced by farmers. Traditional methods often rely on labor-intensive processes and subjective assessments, leading to suboptimal resource use and reduced yields. Smart farming, enabled by information and communication technology (ICT; O'Grady & O'Hare, 2017), aims to increase the accuracy and timeliness of decisions, improving crop productivity. Key aspects of modern crop management include real-time soil nutrient management, crop recommendation, and irrigation management. IoT devices equipped with sensors can monitor soil nutrient concentrations, moisture, humidity, and temperature, allowing for personalized fertilizer application based on crop requirements and soil nutrients. Modern techniques, such as grid soil sampling and variable-rate technology, allow for site-specific nutrient management. Hyperspectral imaging can also be used for nutrient prediction in crops like wheat and chlorophyll content analysis. ML algorithms guide farmers toward optimal crop selection based on soil conditions, climate, and market demands. CatBoost is proposed for crop recommendation due to its superior performance in handling categorical features and large datasets. Precision irrigation systems utilize technologies like GPS-based (Afzaal et al., 2020) controllers and sensor data to monitor soil and climatic conditions, optimizing water utilization efficiency.

IoT-based devices, such as robots, wireless sensors, and drones, are revolutionizing pest and disease management through real-time monitoring and modeling. Hyperspectral imaging and ML enable early detection of diseases and pests in crops. Real-time harvest data and yield maps are generated by yield monitors connected to mobile apps. Satellite images and remote sensing data are used for crop yield monitoring and predicting yields before harvest. Vegetation indices (VIs) like NDVI (BENINCASA et al., 2018; Pinheiro Lisboa et al., 2018) are used to assess vegetation health and predict crop production. IoT devices collect various types of data, including soil parameters and environmental conditions, and cloud-based platforms facilitate data storage, processing, and accessibility through user-friendly mobile applications. Smart farming (Dhanaraju et al., 2022) promotes sustainable agricultural practices by optimizing resource use, reducing environmental impact, and enhancing food security. The precise application of fertilizers and pesticides minimizes environmental contamination.

7.3 DEEP LEARNING MODEL IN PRECISION FARMING

A technique called deep learning (DL) has transformed the agriculture business by giving sophisticated answers to a variety of difficulties. Advances in communication technology and connected products have enabled improved data processing, hastening the digitization of agricultural regions, with DL being utilized mostly in precision farming.

Plant picture recognition, crop categorization, weed identification, and pest identification are all common applications of DL, especially Convolutional Neural Networks (CNNs). These applications aid in crop area delineation, plant health monitoring, and the identification of certain crop-related issues. In certain tasks, CNNs can attain excellent accuracy. Another use of DL is disease identification, which makes it possible to identify fungal infestations and agricultural illnesses early and accurately. This enhances treatment methods and lowers financial losses. Another use of DL is in weed control, which enables intelligent spraying devices to apply poisons precisely. Weed identification has been accomplished using methods such as CNN-based object detection (YOLOv3) (Partel et al., 2019; Ruigrok et al., 2020) and architectural changes like ResNet18 (Bah et al., 2018) and MobileNet.

Weather forecasting, agricultural information processing, intelligent farm equipment, remote sensing applications, and crop yield prediction all make use of DL techniques like CNNs. To forecast the amount of fruit or flowers, these algorithms use a variety of data sources, including pictures. CNNs have outperformed conventional ML models in the categorization of crops based on satellite data. The creation of reliable and automated systems for agricultural equipment, including automated spraying, is aided by DL. It is essential for tracking the health of plants and animals, which makes it possible to analyze big datasets. By predicting the pricing of agricultural products, which is crucial for controlling the agricultural economic system, it may also optimize agricultural economic systems. AI, particularly DL, has completely changed agricultural weather predicting systems. Standard network topologies such as LeNet, AlexNet, VGGNet, GoogleNet, ResNet, DenseNet, and MobileNet (Espejo-Garcia et al., 2020; Razfar et al., 2022) are frequently used in the application of DL in agriculture. RGB and hyperspectral photos, as well as data from IoT devices and drones, are common data sources. But there are also issues such as lack of training data, processing requirements, and improved interpretability of DL models in agriculture. For better performance with less computational resources, future studies should investigate novel designs such as Vision Transformers (ViT).

Numerous research studies have used DL to identify specific crop diseases, such as the detection of mildew disease in vine fields using DL segmentation techniques like SegNet (Kerkech et al., 2020), which frequently combine visible and infrared UAV images; the classification of strawberry diseases using custom networks like LFC-Net; and the detection of grapevine diseases using architectures like AlexNet (Mostafa et al., 2021). This text covers a number of methods for identifying and detecting plant diseases, such as the classification of potato diseases using CNNs like Capsule Networks, the classification of mildew diseases in millet using CNNs like VGG16 for transfer learning (Coulibaly et al., 2019), the classification of multi-crop plant diseases using ResNet50, the identification of soybean plant diseases in hyperspectral images using DL, and the recognition of tomato plant diseases and pests using meta-architectures like Faster R-CNN, R-FCN, and SSD (Guo et al., 2020; Li et al., 2019) with feature extractors like VGGNet and ResNet (Arsenovic et al., 2019) (Table 7.1).

TABLE 7.1
Comparison Table of Recent Disease Detection

Study	Crop/ Disease	ML/DL Method	Dataset Used	Accuracy	Contribution
Awari et al. (2024)	Tomato Leaf Spot	CNN, SVM	PlantVillage + Custom	97%	Real-time classification via hybrid model
Husain et al. (2024)	Paddy Blight	CNN, ResNet	Public Field Images	94%	Fast DL-based detection using imaging
Singh et al. (2025)	Multiple crops	Transfer Learning	Benchmark datasets	98%–99%	Review of DL architectures and datasets

7.4 IOT SOLUTIONS FOR SOIL ANALYSIS

Soil sensors play a crucial role in measuring soil properties that impact soil health and plant growth. There are various types of soil sensors, including electrical or electromagnetic sensors, optical sensors, mechanical sensors, and electrochemical sensors. Electromagnetic sensors measure soil parameters based on electrical conductivity or resistivity, while optical sensors utilize the soil's optical characteristics. Mechanical sensors, such as penetrometers and tensiometers, can also measure soil parameters.

Electrochemical sensors, such as conductometric and ion-selective electrodes (ISEs), measure various soil properties. These sensors are widely adopted, low cost, and reliable for soil health monitoring. The proposed IoT-based real-time soil health monitoring system uses several sensor technologies to measure different soil parameters.

Capacitive Moisture Sensor (Okasha et al., 2021): This sensor operates based on the change in the soil's dielectric constant with variations in moisture content. It uses a 555 timer to generate a square wave whose frequency changes with soil capacitance. The ESP32 microcontroller reads this frequency and converts it to a moisture level using calibration data.

Electrical Conductivity Sensor (Khanal et al., 2024): Measures the soil's ability to conduct electricity using copper plate electrodes buried in the soil. The voltage across the soil is inversely proportional to its conductivity, which can be calculated using a derived equation.

Temperature Sensor (Fezari & Al Dahoud, n.d.): The DS18B20 digital temperature sensor provides direct digital temperature readings in Celsius, eliminating the need for analog-to-digital conversion.

LUX Measurement Sensor: A VEML6075 component detects UVA and UVB radiation using a photodiode array.

NPK and pH Measurement (Kim et al., 2013): An NPK soil sensor module with ISEs specific to nitrate, phosphate, and potassium ions determines their concentrations and pH levels. The sensor communicates using the RS485 protocol via a TTL to RS485 converter.

An IoT-based system is a network of interconnected objects embedded with sensors that exchange information within the network. These devices possess unique addresses based on standard communication protocols, enabling them to autonomously collect and transmit data, facilitating real-time monitoring, control, and optimization of processes across various domains, including agriculture. IoT plays a crucial role in soil health monitoring, as traditional methods are time-consuming and tedious, requiring intensive laboratory testing. IoT-based systems enable continuous collection of soil health data in situ using deployed sensors, which can be integrated through IoT platforms and transmitted to a centralized server for further analysis. Advanced analytics algorithms can provide valuable insights to farmers. Several examples of IoT-based soil health monitoring systems have been proposed in the literature, including a system that allows farmers to monitor soil moisture, temperature, and pH using an Android smartphone, a solar-powered IoT system for continuous in-field monitoring, a portable IoT-enabled soil probe for real-time analysis of soil macronutrients, a system integrating IoT and GIS for real-time monitoring, and a multi-sensor system based on IoT for real-time soil health data transmission. LoRaWAN-based IoT systems (Devalal & Karthikeyan, 2018) have also been used in PA for long-range, low-power data acquisition. Wi-Fi is a frequently utilized, inexpensive, and simple-to-deploy technology for IoT applications. GSM (Halonen et al., n.d.) is a 2G digital cellular network technology that provides dependable and power-efficient connectivity for IoT devices all around the world. Often used with LoRaWAN (Devalal & Karthikeyan, 2018), a MAC layer protocol, Lo-Ra is a wireless communication technique that is perfect for long-range, low-power, low-data-rate communication. The suggested system measures temperature, pH, electrical conductivity, soil moisture, phosphate, potassium, nitrogen, and UV light using the ESP32S3 microcontroller. Real-time data is sent to a centralized server for data visualization via Wi-Fi (Afaqui et al., 2017), GSM, and Lo-Ra mesh networking. With the use of data, the technology seeks to help farmers increase agricultural yields.

7.4.1 Smart Farming IoT Platform Based on Edge and Cloud Computing

The sources (Zamora-Izquierdo et al., 2019) discuss the importance of edge and cloud computing in their proposed smart farming IoT platform. Cloud computing has traditionally been used in PA by providing a remote software platform for monitoring and control management, avoiding the need for complex local systems. In this architecture, the cloud plane serves as the interface between users and the core platform, maintaining the current status of the crop and configuration parameters. It uses the NGSI interface for data updates and receiving notifications, and hosts data analytics modules that perform special analytics coupled with specific service needs. Communication between final applications and analytics modules uses a REST interface transporting JSON data.

Edge computing has emerged as a trendy paradigm, although not yet fully exploited in agriculture. Edge computing focuses on improving system reaction by reducing network delay, as computing nodes are located closer to the end devices. The proposed solution moves part of the data analytics and control features from the cloud to nodes closer to the end-devices, requiring reliable management of CPS actions and intermediate data filtering capabilities. The edge computing plane in the proposed architecture is the second processing and managing level, comprising a set of NFV (Network Function Virtualization)-powered control modules responsible for orchestrating the CPS layer. These edge subsystems handle main greenhouse operations like irrigation, climate, nutrition, and auxiliary tasks, including alarms and energy management. Data fusion and aggregation are carried out at the edge to offload analytics from the cloud, as the cloud might serve multiple crops and users. The edge control modules are virtualized using NFV techniques, allowing their instantiation at different levels in the network path and communicating with CPS nodes using IoT communication protocols like MQTT or CoAP. Each edge control module is instantiated as a compact FIWARE domain, including a Context Broker, two IoT Agents (for CoAP and MQTT), and Comet for data persistence. The IoT Agents act as intermediaries between the CPS layer and the Context Broker at the edge, increasing system reliability against network access failures.

In summary, cloud computing provides a centralized platform for data storage, complex analytics, and high-level management with user interfaces, while edge computing brings processing and control closer to sensors and actuators (at the crop site) for quicker responses, data aggregation, and increased reliability, especially in cases of intermittent network connectivity. The two paradigms work together in a three-tier architecture, ensuring efficient and reliable PA operations.

7.5 SENSOR TECHNOLOGIES (FC-28, DHT11, JXBS-3001, TCS3200)

While no academic papers were directly retrieved focusing exclusively on FC-28, DHT11, JXBS-3001, and TCS3200 sensors in agriculture, these sensors are widely adopted in smart farming projects and prototypes for their affordability, ease of integration, and diverse sensing capabilities. The **FC-28 soil moisture sensor** is used to detect the volumetric water content in soil, making it ideal for irrigation systems that require real-time moisture level monitoring. This sensor is cost-effective and typically works by measuring the resistance between two probes inserted into the soil, with lower resistance indicating higher moisture content. The **DHT11 sensor** combines a humidity and temperature sensor, providing essential data for crop health, disease prediction, and environmental control in greenhouses. Though less accurate compared to advanced sensors like DHT22, it remains a preferred choice in educational and prototype applications due to its low power consumption and digital output. The **JXBS-3001** is a more advanced multifunction sensor used in industrial and agricultural settings, capable of measuring parameters such as temperature, humidity, CO_2, and light. It integrates with IoT platforms and supports protocols like RS485 and MODBUS, making it suitable for precision farming and environmental monitoring. Lastly, the **TCS3200 color sensor** is employed in agricultural automation systems to detect ripeness of fruits or color changes in leaves that may indicate nutrient deficiencies or disease. It converts light to frequency using an array of photodiodes with red, green, blue, and clear filters, providing RGB data for automated classification systems. Together, these sensors exemplify the layered sensing architecture of smart agriculture, from basic field monitoring to advanced phenotyping and automation (Table 7.2).

TABLE 7.2
Sensor Working with Its Key Features

Sensor	Description
FC-28	The **FC-28** is a soil moisture sensor module commonly used in gardening and agricultural projects to monitor soil humidity levels. It's particularly popular in DIY electronics and Arduino-based automation systems. The device offers dual output modes, providing both analog and digital outputs to measure soil moisture levels. It is compatible with 3.3 V–5 V systems and features two exposed probes acting as variable resistors. As soil moisture increases, conductivity between the probes increases, resulting in lower resistance and higher output signal. The onboard potentiometer allows users to set the desired moisture threshold for the digital output.
DHT11	A popular and reasonably priced digital sensor for measuring temperature and humidity is the DHT11. Because of its simplicity and ease of integration with microcontrollers like Arduino and Raspberry Pi, it is especially well-liked among educators and hobbyists. The DHT11 sensor is an inexpensive digital sensor that measures temperature and relative humidity with reasonable precision and usability. Its ability to deliver temperature and humidity data via a single-wire digital interface, which streamlines connections and eliminates the need for many microcontroller input ports, is one of its primary strengths. It is appropriate for interior climate monitoring and educational projects since it operates within a temperature range of 0°C–50°C with an accuracy of ±2°C and a humidity range of 20%–80% RH with an accuracy of ±5%. The DHT11 is perfect for low-power applications since it can handle voltages between 3.3 V and 5.5 V and uses very little power—roughly 2.5 mA during data collection. It provides a dependable update frequency for the majority of basic monitoring requirements with a sample rate of one reading per second. Its small size and built-in signal processing make it even more useful, particularly for projects involving Arduino and Raspberry Pi where simplicity and space are important considerations.
JXBS-3001	A line of sophisticated environmental sensors called the JXBS-3001 series is made to precisely measure the temperature, humidity, nutrient content, and moisture content of soil. These sensors' excellent measurement precision and quick reaction time make them essential for industrial applications, environmental research, and agriculture. Because of their long-term deployment capabilities and superior corrosion resistance, the sensors are composed of sturdy materials. Because they are completely sealed and have an IP68 rating, they can function continuously in demanding conditions. The JXBS-3001 sensors run on a 12 V DC power source and handle a variety of output signals, such as RS485, 4–20 mA, 0–5 V, and 0–10 V. They have a 2.8-inch LCD display for real-time data viewing and an intuitive user interface. These sensors provide for extended data logging for thorough analysis, with a storage capacity of up to 200,000 datasets.

(*Continued*)

TABLE 7.2 (*Continued*)
Sensor Working with Its Key Features

Sensor	Description
TCS3200	TCS3200 is a multi-functional color sensor module capable of detecting and measuring wide visible colors by transforming light intensity to frequency signals. Its heart lies in an 8×8 matrix of photodiodes comprised of 16 each of red, green, blue, and clear light in order to support precise color sensing. The sensor generates a square wave whose frequency is directly related to the target color's intensity, thus making it possible for microcontrollers to read color values without analog-to-digital conversion. Powered across a range of 2.7 V–5.5 V, the TCS3200 can be used across a wide range of microcontrollers such as Arduino and Raspberry Pi. With its programmable scaling of output frequency via S0 and S1 pins, along with selection of color filtering via S2 and S3 pins, it offers flexibility across multiple lighting conditions. Additionally, the module includes four white LEDs that provide illumination to the target object to enable uniform detection of color. With aspects of power-down mode to save power, in addition to stable temperature coefficient of 200 ppm/°C, the TCS3200 is especially ideal for use across color sorting applications, ambient light sensing, and test strip analysis.

7.6 DATA FUSION AND AI IN YIELD PREDICTION

7.6.1 Data and Techniques for Improvement

Model resilience in limited data is enhanced by data augmentation techniques such as color histogram enhancement, unsharp masking, affine translation, rotation, zooming, and Generative Adversarial Networks. Transfer learning is the process of optimizing previously trained models for particular crop disease datasets, particularly when there is a shortage of agricultural data. Based on aerial or ground-based IoT, hyperspectral photography improves assessments of agricultural stress and disease (Coulibaly et al., 2022).

7.6.2 PyCIAT Is a Framework

PyCIAT (Kaushik, 2025) is a framework designed to synthesize climate change impacts on agriculture by integrating diverse climate projections, spatial data, multiple process-based crop models, and various management scenarios. It manages workflows that encompass climate data processing, soil data integration, and simulation setup across multiple locations. PyCIAT supports multi-model ensembles (MMEs) (Asseng et al., 2013; Martre et al., 2015), which combine the outputs of multiple models to reduce uncertainty and provide more robust yield predictions. The framework also includes an optional ML surrogate modeling component that trains ML models on the consolidated simulation database, which contains outputs from crop models and corresponding input data. If the training data includes features from different sources, the surrogate model could implicitly learn relationships resembling data fusion for yield prediction. AI in agriculture is mentioned as a potential application of ML surrogates, with a related preprint on "Artificial Intelligence in Agriculture: A Review of Transformative Applications and Future Directions" suggesting an awareness of the broader role of AI in agriculture, which could include data fusion techniques for various applications, including yield prediction. However, the specific data fusion techniques discussed in the preprint are not provided.

7.6.3 Feature Extraction Techniques

The Three-Tier Feature Learning (3TFL) (Anoop et al., 2025) model is a novel approach to feature extraction in crop yield prediction. It integrates three DL architectures: Spiking Neural Networks (SNN), Graphical Neural Networks (GNN), and CNNs. The model involves data preprocessing, which includes cleaning, standardization, and normalization to improve data quality and consistency.

The SNN process uses the time-to-first-spike coding technique to find relationships between preprocessed pixel values and neural spike timings. GNNs are designed to process graph-structured data, with each node represented by a vector. The CNN model uses only the feature extraction part of the GNN for feature extraction. Feature concatenation and evaluation involve concatenating the feature vectors generated by the SNN and GNN. The Jaccard Similarity score is used to predict the similarity score between these concatenated feature vectors. A weighted loss function and a novel self-adaptive Spider Monkey Optimization algorithm (SASMO) are introduced to optimize the feature learning process. The "Literature Review" section highlights other feature extraction approaches implicitly through diverse data sources used in crop yield prediction studies. These include VIs, Spectral Information, Meteorological Data, Soil Characteristics, Phenological Information, and Spatial Patterns.

7.6.4 Hybrid Edge-Cloud System

A hybrid edge-cloud system is an architecture that combines edge computing capabilities with cloud computing to address the growing demand for data analysis due to the rapid expansion of the internet and big data technologies. The IoT model is crucial for edge platforms, but the massive data volume produced by connected devices can lead to performance issues in a purely IoT system. The primary driver for hybrid edge-cloud systems is to improve resource consumption and adaptability of responsive services. The proposed IoT-SCOM (IoT-based Service Components Optimization Model) operates within a three-tier hybrid hierarchy:

IoT Layer: Consists of blockchain-enabled IoT devices capable of making dynamic offloading decisions based on Quality of Service (QoS) requirements and network capacity. Devices can perform mining tasks locally, offload to a nearby MEC (Mobile Edge Computing) server, or offload to a distant MCC (Mobile Cloud Computing) server.

MEC Server Layer: Located at the edge of networks, providing low-latency computing capabilities for real-time job processing.

MCC Server Layer: Contains virtual servers with significant processing and storage capacities for handling demanding computing tasks for IoT devices.

The IoT-SCOM model aims to minimize transmission latency by locating processing and storage closer to IoT devices at the edge. This approach, particularly MEC, offers a broadband service environment with reduced latency. Hybrid systems, incorporating edge hosting and NFV techniques, increase resource utilization efficiency and flexibility by moving computer services closer to the routing layer. Edge computing enables more responsive activities in IoT devices, with MEC providing significant support for delay-sensitive operations due to its minimal delays and high stability. MEC also enhances user experience by reducing latency and faster updates of services and software, contributing to a high-quality internet experience. Hybrid systems overcome the limitations of cloud-only approaches by overcoming latency and energy consumption issues associated with relying solely on distant cloud servers for massive amounts of data generated by IoT devices. The IoT-SCOM model is a new optimization technique for IoT service element installation in hybrid systems, integrating the simulating annealing (SA) method and ant colony optimization (ACO) method.

7.6.5 Dataset Description

We used the widely used PlantVillage dataset, a publicly available benchmark dataset with over 54,000 high-resolution images of healthy and diseased leaves from 14 different crop species such as tomato, potato, corn, and grape, to implement and evaluate ML and DL models in agricultural disease detection. The dataset has annotations for 38 different types of plant illnesses and healthy circumstances, making it appropriate for supervised classification applications. Images are recorded under controlled conditions with consistent backdrops, ensuring great visual quality and consistency but potentially limiting performance in field scenarios (Singh et al., 2025). In addition to PlantVillage, we included a field-collected dataset of 2,000 annotated photos of wheat and rice leaf diseases (e.g., rust, blight) taken with mobile cameras in natural daylight. This dataset captures real-world difficulties such as background noise, leaf occlusion, and changeable illumination. To improve generalization, all datasets were preprocessed using picture scaling (256×256 pixels), normalization, and augmentation techniques such as rotation, flipping, and brightness correction (Awari et al., 2024). These datasets provide a solid foundation for training and assessing classification models, striking a balance between the controlled correctness of lab photos and the actual heterogeneity of field circumstances. Future research might benefit from combining multispectral or hyperspectral datasets to increase diagnosis accuracy in diverse field conditions.

7.7 CHALLENGES IN AI AND IOT ADOPTION IN AGRICULTURE

Challenges in IoT Networks: Besides data volume and latency, IoT networks face challenges such as energy limits, adaptability, dependability, diversity, privacy, routing, security, and congestion.

The adoption of AI and the IoT in agriculture is revolutionizing farming by improving productivity, sustainability, and efficiency, yet several challenges hinder widespread implementation. High initial investment costs and limited affordability restrict access for small and medium-scale farmers, especially in developing regions, leading to uneven technological diffusion (Mim et al., 2025). Infrastructure constraints like poor internet connectivity and lack of technical expertise further exacerbate these disparities, impeding the effective deployment and operation of AIoT systems (Purushotham et al., 2024). Privacy and cybersecurity risks, particularly with unattended IoT devices, pose serious concerns, as these systems often process sensitive farm and environmental data (Boopathi, 2024). Additionally, user resistance to digital transformation, fueled by low digital literacy and trust, slows the adoption process (Leong et al., 2023). Policy and regulatory ambiguities regarding AI ethics, data ownership, and standards contribute to implementation hesitations, highlighting the need for clear guidelines and collaborative governance frameworks (Kumar et al., 2024). Addressing these multifaceted challenges through targeted investments, infrastructure upgrades, inclusive training programs, and policy development is essential for enabling a broader and equitable adoption of AI and IoT in agriculture, ensuring global food security and sustainability.

7.8 CASE STUDIES ON SMART FARMING APPLICATIONS

Smart farming combines IoT, artificial intelligence, machine learning, deep learning, fuzzy logic, GIS, UAVs, and decision support systems to improve agricultural productivity and sustainability. Smart irrigation systems enhance water-use efficiency through sensor-based automated control. Precision agriculture employs machine vision, deep learning, UAV multispectral imagery, and satellite data for crop monitoring, ripeness assessment, and disease detection. Variable-rate technology, grid soil sampling, and GIS support precise nutrient and resource management. Digital platforms and DSS facilitate knowledge dissemination and climate-resilient practices. Advanced hydroponic and indoor automated systems further promote sustainable food production in resource-limited and harsh environments. Table 7.3 presents the case studies illustrating practical smart farming applications.

7.9 LOW-POWER WIRELESS SENSOR NETWORKS IN AGRICULTURE

Low-power wireless sensor networks (WSNs) are revolutionizing PA by providing scalable, energy-efficient systems for environmental monitoring, resource management, and automation. A recent advancement involves a smart agricultural management scheme that utilizes WSNs optimized with an enhanced LEACH algorithm, improving network lifetime and throughput by dynamically selecting energy-efficient cluster head nodes in greenhouse environments (Yan & Li, 2024). In large-scale farm applications, LPWAN technologies such as LoRaWAN, Sigfox, and NB-IoT have shown superior performance over ZigBee, enabling robust node-to-infrastructure communication even in non-flat terrains and during mobile operations with tractors and farmers. These systems ensure extensive coverage, low latency, and integration with cloud platforms for real-time farm management (Klaina et al., 2022).

TABLE 7.3
Application of Smart Farming

Smart Irrigation Systems (Weraikat et al., 2024)	Smart irrigation systems use advanced technologies like artificial intelligence (AI), machine learning (ML), deep learning (DL), fuzzy logic, and the Internet of Things (IoT) to make better decisions about water application. These systems collect data from various sources, such as sensors monitoring environmental and plant-related variables like soil moisture and air humidity. IoT technologies enable remote monitoring and control of the irrigation system, with embedded systems like Arduino, ESP32, and Raspberry Pi being commonly used for data acquisition and control. Data is processed using intelligent algorithms, such as machine learning and deep learning, to make informed irrigation decisions. Fuzzy logic handles uncertainties and inaccuracies by introducing the concept of "degree" of truth, determining irrigation needs based on sensor inputs or predictions. Neuro-fuzzy systems combine neural networks and fuzzy logic, leveraging learning capabilities for prediction and rule-based decision-making for control. The smart irrigation system automatically controls the irrigation process, activating or deactivating solenoid valves or water pumps based on predicted soil moisture values, direct sensor readings, or alerts sent to the user via mobile applications. Some systems also incorporate chatbots for user notification and data collection.
Advances in Precision Agriculture and Crop Monitoring	Machine vision and deep learning algorithms are revolutionizing precision agriculture by detecting and predicting fruit and vegetable maturity. The YOLOv7 architecture and multi-head attention mechanisms are used for drone-based apple detection, while the Segment Anything Model (SAM) enables farmers to identify and analyze crop conditions without extensive data training. The application of machine learning algorithms for mapping cropland extent using Sentinel-2 datasets is crucial for large-scale agricultural monitoring, aiding in resource management and policy-making.
Technological Innovation for Farm Management and Infrastructure	Artificial intelligence is crucial for agricultural sustainability and productivity. A Raspberry-Pi-based IoT system for real-time tractor performance monitoring is a significant advancement in farm machinery management. A ResNet-9 model for detecting blight disease in potato and tomato leaf images reduces chemical use and early detection. Deep learning-based automatic disease detection is poised to revolutionize pest management strategies, reducing reliance on chemicals and lowering production costs.
Digital Outreach and Decision Support Systems (DSS)	In order to close knowledge gaps and spread best practices across farming communities—especially in the area of citizen engagement—e-agriculture projects and digital platforms such as Facebook for agricultural extension services are essential. The potential of social media to provide urban agriculturists with peer support and information access is demonstrated by the example of rooftop gardeners in Bangladesh. Another illustration of how digital technologies are fostering more responsive and interconnected agricultural ecosystems is the AgroAPI platform. Furthermore, by providing tailored advice and services to improve climate adaptation and food sovereignty, decision support algorithms and farming-as-a-service models are opening the door for more resilient and nutrition-secure agricultural systems.

(*Continued*)

TABLE 7.3 (*Continued*)
Application of Smart Farming

Coffee ripeness using multispectral data from UAVs (Nogueira Martins et al., 2023)	This study aimed to predict fruit ripeness using spectral and textural variables and determine the best variables for developing spatiotemporal variability maps of fruit ripeness. An experiment was conducted with six arabica coffee fields during the 2018–2019 and 2020–2021 seasons. Seven flights were conducted using a quadcopter equipped with a five-band multispectral camera, and 12 spectral and 64 textural variables were obtained. The percentage of unripe fruits was determined using an irregular grid on each field. Fruit ripeness was predicted using six machine learning (ML) algorithms, with the random forest algorithm showing higher accuracy. The model using spectral and textural variables performed better than the model based solely on spectral variables. The study demonstrated the feasibility of using spectral and textural variables derived from UAV imagery for mapping and monitoring spatiotemporal changes in fruit ripeness at a fine scale.
Disease diagnosis using mobile capture devices (Johannes et al., 2017)	Early symptom-based disease detection is essential for integrated pest control systems since it reduces production losses and boosts treatment effectiveness. However, the emergence of novel illnesses linked to resistant agricultural varieties makes early detection more difficult and postpones remedial measures. Automated identification systems based on images can be useful, but they don't work well with mobile devices in actual field settings. To address illness diagnosis in natural settings, a novel image processing system that combines statistical inference techniques with candidate hot-spot detection is suggested. Using seven mobile devices and more than 3,500 photos taken at pilot locations in Germany and Spain, this study examines the effectiveness of early detection of three endemic wheat diseases in Europe: rust, tan spot, and septoria.
Variable-Rate Technology (VRT) and Grid Soil Sampling (Batte, n.d.)	Variable-rate technologies (VRTs) are used in farming to predict the delivery rate of inputs based on a predetermined map from GIS. Grid soil sampling is a method used to collect soil from a systematic grid to establish a map for every parameter. These maps are loaded into a variable-rate applicator, which directs and controls changes in the delivery amount or fertilizer product based on map features. New technologies like VRT and grid soil sampling can improve soil fertility management and assess the spatial distribution of nutrients and yields. Samples are collected from sampled sections based on the subdivision of a field into small areas, or cells, and composite samples represent an entire area of each smaller area at the intersections of grid lines. Field-specific variability in phosphorus and potassium requires uniform fertilization to improve precision agriculture.
Geographic Information System (GIS) (Ehlers, 2008)	GIS is a hardware and software system that collects, stores, retrieves, analyzes attributes, and provides location data for maps and statistical analysis. It provides information on field soil types, nutrient status, topography, irrigation, drainage, chemical applications, and crop production. GIS also helps assess current and alternative management by compounding and altering data layers for decision-making. It also establishes relationships between elements affecting crops in specific farming fields.

(*Continued*)

TABLE 7.3 (*Continued*)
Application of Smart Farming

Crop Management (Rembold et al., 2013)	Satellite images offer valuable insights into soil conditions and crop performance, enabling farmers to monitor factors like seeds, fertilizers, and pesticides that contribute to yield increase and efficiency. These images provide near real-time information at a regional scale, with vegetation indices like the normalized difference vegetation index (NDVI) being the most popular indicator for assessing vegetation health and crop production. Crop monitoring methods compare actual crop status to previous or normal seasons, allowing for early estimation of crop yield. Automated field management offers more complex functionalities, including data acquisition, processing, monitoring, decision-making, and management of farm operations, including crop production, profits and losses, farm weather prediction, field mapping, and soil nutrients tracking. This technology allows for more accurate monitoring and decision-making in the agricultural sector.
Hydroponics System Control (Zamora-Izquierdo et al., 2019)	The smart farming IoT platform aims to address the challenges of controlling hydroponics systems, particularly closed recirculation systems using moderately saline water. It consists of a three-tier architecture: local Cyber-Physical Systems (CPS), an edge computing tier, and a cloud platform. At the local level, CPS nodes interact with sensors and actuators, such as solar radiation, humidity, temperature, CO_2, pH, electrical conductivity, and liquid consumption. The edge computing plane monitors and manages main Precision Agriculture tasks near the access network, increasing system reliability against network access failures. The platform comprises virtualized control modules for key hydroponics operations like irrigation and nutrition control, as well as climate and auxiliary tasks. Nutrition Control orchestrates the nutrient solution unit, controlling the discharge of waters and fertilizers, and managing the delivery of the solution to the crops. The edge control modules perform data fusion and aggregation to offload analytics from the cloud, communicating with the CPS nodes using IoT communication protocols like MQTT and CoAP. The cloud platform collects and hosts data analytics modules, serving as the interface between users and the core platform.
Indoor automated system (Chowdhury et al., 2020).	An indoor automated system is a vertical hydroponic farming test-bed designed for indoor cultivation in challenging climates like the Gulf region. This system combines hydroponics and automation technologies to create a self-regulating environment for plant growth, ensuring essential nutrients, minerals, and a stable pH within a certain range. Key features include a vertical Nutrient Film Technique (NFT) setup, a microcontroller acting as the brain, and various sensors monitoring critical parameters. The primary motivation behind developing this system is to enhance food security and reduce dependency on imports, especially in countries like Qatar with limited arable land and challenging climates. The system offers individuals the opportunity to grow their own food in their homes without significantly increasing water and power consumption, with government support for such initiatives. The estimated cost of the system is around $400. Comparing the proposed system with other research projects and commercial solutions, the source notes that cheaper options often lack complete automation, IoT integration, or sufficient food production capacity. More expensive commercial solutions may not be suitable for individual home use. In conclusion, the indoor automated vertical hydroponics system represents a promising approach to sustainable agriculture, particularly in regions with unfavorable outdoor conditions.

Complementing these are low-power leaf sensors using backscatter communication, which detect plant water stress by monitoring the temperature differential between leaf and air. These batteryless sensors operate on microwatts of power using solar energy and communicate through Morse code-based protocols, making them ideal for scalable WSNs in water-constrained regions (Daskalakis et al., 2018). Additionally, general reviews underscore the growing role of WSNs in agriculture, highlighting their contributions to soil monitoring, irrigation control, and sustainable farming by minimizing power consumption while maximizing data precision (Piyar & Baulkani, 2022). Collectively, these technologies exemplify the shift toward intelligent, low-energy agricultural systems that are both environmentally sustainable and economically viable.

7.10 FUTURE TRENDS IN AI AND IOT FOR AGRICULTURE

7.10.1 Future Directions

Limited training data hinders accurate disease identification. Explainable AI (XAI) is needed for crop disease identification to ensure users' confidence in predictions. Visualizing features contributing to the model's decision is a solution. Statistical validation is crucial for analyzing the significance of results when multiple classifiers are used, and cross-validation is essential for estimating model robustness (Coulibaly et al., 2022). The device and platform should be expanded to accommodate a wider range of crops and geographic locations, enabling more farmers to benefit from its monitoring and recommendation functionalities. Future research should also focus on incorporating satellite imagery and weather data into the system to provide more accurate insights into crop health, growth patterns, and environmental conditions, enhancing the device's overall capabilities. The device's scalability and usability should be enhanced to meet the diverse needs of farmers across different regions with varying technological infrastructures. This includes adapting to different communication networks and power sources, and creating user-friendly interfaces accessible to farmers with varying technological literacy levels (Coulibaly et al., 2022).

Improving the understanding and explainability of DL models in agriculture is essential for building trust and facilitating adoption by stakeholders. Future research should incorporate XAI techniques to provide insights into the models' reasoning and predictions. Visualization techniques, such as saliency maps, can help identify influential features in predictions. Enhancing collaboration between agricultural researchers and companies is suggested to further assist farmers in developing advanced systems like vision-based vehicle guidance and autonomous robots. Finally, future work will explore agricultural paradigms related to ecological collapse and climate change, focusing on sustainability and environmental concerns. The special issue on "Smart Agricultural Technology" outlines future directions in data-driven science and smart agricultural technologies. It emphasizes the need to overcome digital literacy and infrastructure deficits for successful implementation and the integration of traditional knowledge with modern methodologies. The editorial calls for investment in technology transfer and collaboration between policymakers and industry stakeholders to support the diffusion of these technologies, especially in developing societies. The overarching direction is to harness the synergy between agriculture and technology to create productive, resilient, and sustainable farming systems. Research areas include advances in PA, technological innovation for farm management and infrastructure, digital outreach and decision support systems, and a focus on sustainability and resilience in agricultural practices. The future directions revolve around broader adoption and accessibility of smart agricultural technologies, enhanced data utilization and governance, and continued innovation in sensing and AI-driven analysis.

Future directions in IoT, edge-cloud computing, and PA include incorporating factors beyond high bandwidth time, such as network capacity and queue pause during data transmission, and focusing on latency and data transmission efficiency. The proposed method, IoT-SCOM, is more practical for microservice distribution problems and should consider delay, data transmission rate, latency requirements, and compatibility for cloud computing. The study also suggests broader adoption and implementation of IoT-based systems for real-time soil health monitoring, assessing reliability in field conditions for long-term use, and integrating advanced data analysis and decision support features. Future research could explore the optimization of communication protocols based on different environmental and infrastructure landscapes, and develop user-friendly web applications for real-time visualization. These directions include enhancing the complexity and practicality of optimization models, focusing on latency and data transmission efficiency, promoting long-term reliability testing, expanding the scope of soil health monitoring systems, optimizing communication strategies, and developing sophisticated data analysis and decision support tools. The future of smart agriculture will involve the integration of IoT, AI, and advanced technologies. Key technologies include interconnecting agriculture with AI and big data services, leading to converged systems managing farm operations. IoT will play a significant role in rural areas due to its improved coverage, low power consumption, and cost-effectiveness. WSNs will provide real-time information for farmers, while remote sensing and UAVs will be integrated for improved weed detection and management. Smart grids will provide opportunities for power sector reform and alleviate rural electricity problems. Fuzzy logic is a significant technology for irrigation control, and future research could focus on overcoming challenges in urban soil cultivation using advanced AI methods.

REFERENCES

Afaqui, M. S., Garcia-Villegas, E., & Lopez-Aguilera, E. (2017). IEEE 802.11ax: Challenges and requirements for future high efficiency WiFi. *IEEE Wireless Communications*, 24(3), 130–137. https://doi.org/10.1109/MWC.2016.1600089WC

Afzaal, H., Farooque, A. A., Abbas, F., Acharya, B., & Esau, T. (2020). Precision irrigation strategies for sustainable water budgeting of potato crop in Prince Edward Island. *Sustainability*, 12(6), 2419. https://doi.org/10.3390/su12062419

Anoop, G. L., Nandini, C., & Naresh, E. (2025). 3TFL-XLnet-CP: A novel transformer-based crop yield prediction framework with weighted loss based 3-tier feature learning model. *SN Computer Science*, *6*(3). https://doi.org/10.1007/s42979-025-03778-9

Arsenovic, M., Karanovic, M., Sladojevic, S., Anderla, A., & Stefanovic, D. (2019). Solving current limitations of deep learning based approaches for plant disease detection. *Symmetry*, 11(7), 939. https://doi.org/10.3390/sym11070939

Asseng, S., Ewert, F., Rosenzweig, C., Jones, J. W., Hatfield, J. L., Ruane, A. C., Boote, K. J., Thorburn, P. J., Rötter, R. P., Cammarano, D., Brisson, N., Basso, B., Martre, P., Aggarwal, P. K., Angulo, C., Bertuzzi, P., Biernath, C., Challinor, A. J., Doltra, J., … Wolf, J. (2013). Uncertainty in simulating wheat yields under climate change. *Nature Climate Change*, 3(9), 827–832. https://doi.org/10.1038/nclimate1916

Awari, A., Bhokare, V., Daundkar, H., Dhas, S., & Take, Prof. A. (2024). Plant disease detection and classification. *International Journal of Scientific Research in Engineering and Management*, 08(10), 1–7. https://doi.org/10.55041/IJSREM38320

Bah, M. D., Hafiane, A., & Canals, R. (2018). Deep learning with unsupervised data labeling for weed detection in line crops in UAV images. *Remote Sensing*, *10*(11), 1690. https://doi.org/10.3390/rs10111690

Batte, M. (n.d.). *Precision Farming-Factors Influencing Profitability*. https://www.researchgate.net/publication/237258730

Benincasa, P., Antognelli, S., Brunetti, L., Fabbri, C. A., Natale, A., Sartoretti, V., Modeo, G., Guiducci, M., Tei, F., & Vizzari, M. (2018). Reliability of NDVI derived by high resolution satellite and UAV compared to in-field methods for the evaluation of early crop N status and grain yield in wheat. *Experimental Agriculture*, 54(4), 604–622. https://doi.org/10.1017/S0014479717000278

Boopathi, S. (2024). *Sustainable Development Using IoT and AI Techniques for Water Utilization in Agriculture* (pp. 204–228). https://doi.org/10.4018/979-8-3693-1722-8.ch012

Cheshkova, A. F. (2022). A review of hyperspectral image analysis techniques for plant disease detection and identification. *Vavilov Journal of Genetics and Breeding*, 26(2), 202–213. https://doi.org/10.18699/VJGB-22-25

Chowdhury, M. E. H., Khandakar, A., Ahmed, S., Al-Khuzaei, F., Hamdalla, J., Haque, F., Reaz, M. B. I., Shafei, A. Al, & Al-Emadi, N. (2020). Design, construction and testing of IoT based automated indoor vertical hydroponics farming test-bed in Qatar. *Sensors (Switzerland)*, 20(19), 1–24. https://doi.org/10.3390/s20195637

Chu, X., Wang, W., Ni, X., Li, C., & Li, Y. (2020). Classifying maize kernels naturally infected by fungi using near-infrared hyperspectral imaging. *Infrared Physics & Technology*, 105, 103242. https://doi.org/10.1016/j.infrared.2020.103242

Coulibaly, S., Kamsu-Foguem, B., Kamissoko, D., & Traore, D. (2019). Deep neural networks with transfer learning in millet crop images. *Computers in Industry*, 108, 115–120. https://doi.org/10.1016/j.compind.2019.02.003

Coulibaly, S., Kamsu-Foguem, B., Kamissoko, D., & Traore, D. (2022). Deep learning for precision agriculture: A bibliometric analysis. *Intelligent Systems with Applications*, 16, 200102. https://doi.org/10.1016/j.iswa.2022.200102

Daskalakis, S. N., Goussetis, G., Assimonis, S. D., Tentzeris, M. M., & Georgiadis, A. (2018). A uW backscatter-morse-leaf sensor for low-power agricultural wireless sensor networks. *IEEE Sensors Journal*, 18(19), 7889–7898. https://doi.org/10.1109/JSEN.2018.2861431

Devalal, S., & Karthikeyan, A. (2018). LoRa technology – An overview. *2018 Second International Conference on Electronics, Communication and Aerospace Technology (ICECA)*, 284–290. https://doi.org/10.1109/ICECA.2018.8474715

Dhanaraju, M., Chenniappan, P., Ramalingam, K., Pazhanivelan, S., & Kaliaperumal, R. (2022). Smart farming: Internet of Things (IoT)-based sustainable agriculture. *Agriculture*, *12*(10), 1745. https://doi.org/10.3390/agriculture12101745

Ehlers, M. (2008). Geoinformatics and digital earth initiatives: A German perspective. *International Journal of Digital Earth*, 1(1), 17–30. https://doi.org/10.1080/17538940701781975

Espejo-Garcia, B., Mylonas, N., Athanasakos, L., Fountas, S., & Vasilakoglou, I. (2020). Towards weeds identification assistance through transfer learning. *Computers and Electronics in Agriculture*, 171, 105306. https://doi.org/10.1016/j.compag.2020.105306

Fezari, M., & Al Dahoud, A. (n.d.). Exploring one-wire temperature sensor "DS18B20" with microcontrollers. https://www.researchgate.net/publication/330854061

Gabrielli, M., Lançon-Verdier, V., Picouet, P., & Maury, C. (2021). Hyperspectral imaging to characterize table grapes. *Chemosensors*, 9(4), 71. https://doi.org/10.3390/chemosensors9040071

Golhani, K., Balasundram, S. K., Vadamalai, G., & Pradhan, B. (2018). A review of neural networks in plant disease detection using hyperspectral data. *Information Processing in Agriculture*, 5(3), 354–371. https://doi.org/10.1016/j.inpa.2018.05.002

Guo, Y., Zhang, J., Yin, C., Hu, X., Zou, Y., Xue, Z., & Wang, W. (2020). Plant disease identification based on deep learning algorithm in smart farming. *Discrete Dynamics in Nature and Society*, *2020*, 1–11. https://doi.org/10.1155/2020/2479172

Kim, H. J., Sudduth, K. A., Hummel, J. W., & Drummond, S. T. (2013). Validation testing of a soil macronutrient sensing system. *Transactions of the ASABE*, 56(1), 23–31. https://doi.org/10.13031/2013.42582

Halonen, T., Romero, J., & Melero, J. (n.d.). *GSM, GPRS and EDGE Performance: Evolution towards 3G/UMTS*, 2nd Edition. ISBN: 978-0-470-86695-5

Huang, H., Fei, X., Hu, X., Tian, J., Ju, J., Luo, H., & Huang, D. (2023). Analysis of the spectral and textural features of hyperspectral images for the nondestructive prediction of amylopectin and amylose contents of sorghum. *Journal of Food Composition and Analysis*, 115, 105018. https://doi.org/10.1016/j.jfca.2022.105018

Husain, S. O., Jagadish, S., Armoogum, V., Siri, D., & Palanivel, R. (2024). Analysis of machine learning and deep learning algorithms for plant disease detection and classification. *2024 Second International Conference on Networks, Multimedia and Information Technology (NMITCON)*, 1–6. https://doi.org/10.1109/NMITCON62075.2024.10698945

Islam, M. R., Oliullah, K., Kabir, M. M., Alom, M., & Mridha, M. F. (2023). Machine learning enabled IoT system for soil nutrients monitoring and crop recommendation. *Journal of Agriculture and Food Research*, 14, 100880. https://doi.org/10.1016/j.jafr.2023.100880

Jarolmasjed, S., Khot, L., & Sankaran, S. (2018). Hyperspectral imaging and spectrometry-derived spectral features for bitter pit detection in storage apples. *Sensors*, *18*(5), 1561. https://doi.org/10.3390/s18051561

Ji, Y., Sun, L., Li, Y., & Ye, D. (2019). Detection of bruised potatoes using hyperspectral imaging technique based on discrete wavelet transform. *Infrared Physics & Technology*, 103, 103054. https://doi.org/10.1016/j.infrared.2019.103054

Johannes, A., Picon, A., Alvarez-Gila, A., Echazarra, J., Rodriguez-Vaamonde, S., Navajas, A. D., & Ortiz-Barredo, A. (2017). Automatic plant disease diagnosis using mobile capture devices, applied on a wheat use case. *Computers and Electronics in Agriculture*, 138, 200–209. https://doi.org/10.1016/j.compag.2017.04.013

Kaushik, P. (2025). *PyCIAT: A Configurable Python Framework for Scalable, Multi-Model Assessment of Climate Change Impacts and Adaptations in Agriculture*. https://doi.org/10.20944/preprints202504.0153.v1

Kerkech, M., Hafiane, A., & Canals, R. (2020). Vine disease detection in UAV multispectral images using optimized image registration and deep learning segmentation approach. *Computers and Electronics in Agriculture*, 174, 105446. https://doi.org/10.1016/j.compag.2020.105446

Khanal, K., Ojha, G., Chataut, S., & Ghimire, U. K. (2024). IoT-based real-time soil health monitoring system for precision agriculture. *International Research Journal of Engineering and Technology*. www.irjet.net

Klaina, H., Guembe, I. P., Lopez-Iturri, P., Campo-Bescós, M. Á., Azpilicueta, L., Aghzout, O., Alejos, A. V., & Falcone, F. (2022). Analysis of low power wide area network wireless technologies in smart agriculture for large-scale farm monitoring and tractor communications. *Measurement*, 187, 110231. https://doi.org/10.1016/j.measurement.2021.110231

Kumar, A., Karn, N., & Sharma, H. (2024). *IoT, AI, and Robotics Applications in the Agriculture Sector* (pp. 243–272). https://doi.org/10.4018/979-8-3693-3583-3.ch014

Leong, Y. M., Lim, E. H., Subri, N. F. B., & Jalil, N. B. A. (2023). Transforming agriculture: Navigating the challenges and embracing the opportunities of artificial intelligence of things. *2023 IEEE International Conference on Agrosystem Engineering, Technology & Applications (AGRETA)*, 142–147. https://doi.org/10.1109/AGRETA57740.2023.10262747

Li, R., Jia, X., Hu, M., Zhou, M., Li, D., Liu, W., Wang, R., Zhang, J., Xie, C., Liu, L., Wang, F., Chen, H., Chen, T., & Hu, H. (2019). An effective data augmentation strategy for CNN-based pest localization and recognition in the field. *IEEE Access*, 7, 160274–160283. https://doi.org/10.1109/ACCESS.2019.2949852

Martre, P., Wallach, D., Asseng, S., Ewert, F., Jones, J. W., Rötter, R. P., Boote, K. J., Ruane, A. C., Thorburn, P. J., Cammarano, D., Hatfield, J. L., Rosenzweig, C., Aggarwal, P. K., Angulo, C., Basso, B., Bertuzzi, P., Biernath, C., Brisson, N., Challinor, A. J., … Wolf, J. (2015). Multimodel ensembles of wheat growth: Many models are better than one. *Global Change Biology*, 21(2), 911–925. https://doi.org/10.1111/gcb.12768

Mim, M. I., Sultana, F., & Hasan, M. R. (2025). AI-powered autonomous farming: The future of sustainable agriculture. *European Journal of Theoretical and Applied Sciences*, 3(1), 11–31. https://doi.org/10.59324/ejtas.2025.3(1).02

Mostafa, A. M., Kumar, S. A., Meraj, T., Rauf, H. T., Alnuaim, A. A., & Alkhayyal, M. A. (2021). Guava disease detection using deep convolutional neural networks: A case study of guava plants. *Applied Sciences*, 12(1), 239. https://doi.org/10.3390/app12010239

Nogueira Martins, R., de Assis de Carvalho Pinto, F., Marçal de Queiroz, D., Sárvio Magalhães Valente, D., Tadeu Fim Rosas, J., Fagundes Portes, M., & Sânzio Aguiar Cerqueira, E. (2023). Digital mapping of coffee ripeness using UAV-based multispectral imagery. *Computers and Electronics in Agriculture*, 204, 107499. https://doi.org/10.1016/j.compag.2022.107499

O'Grady, M. J., & O'Hare, G. M. P. (2017). Modelling the smart farm. *Information Processing in Agriculture*, 4(3), 179–187.

Okasha, A. M., Ibrahim, H. G., Elmetwalli, A. H., Khedher, K. M., Yaseen, Z. M., & Elsayed, S. (2021). Designing low-cost capacitive-based soil moisture sensor and smart monitoring unit operated by solar cells for greenhouse irrigation management. *Sensors*, 21(16), 5387. https://doi.org/10.3390/s21165387

Partel, V., Charan Kakarla, S., & Ampatzidis, Y. (2019). Development and evaluation of a low-cost and smart technology for precision weed management utilizing artificial intelligence. *Computers and Electronics in Agriculture*, 157, 339–350. https://doi.org/10.1016/j.compag.2018.12.048

Pinheiro Lisboa, I., Melo Damian, J., Roberto Cherubin, M., Silva Barros, P. P., Ricardo Fiorio, P., Cerri, C. C., & Eduardo Pellegrino Cerri, C. (2018). Prediction of sugarcane yield based on NDVI and concentration of leaf-tissue nutrients in fields managed with straw removal. Agronomy, 8(9), 196. https://doi.org/10.3390/agronomy8090196

Purushotham, P., Rakhonde, G., Reddy, N. K., Nayak, A. M., Pravalika, K. M., & Sahana, G. (2024). *Harvesting the Future: Navigating the Challenges and Limitations of AI in Agriculture* (pp. 547–564). https://doi.org/10.1007/978-981-97-0341-8_25

Qiu, Z., Ma, F., Li, Z., Xu, X., Ge, H., & Du, C. (2021). Estimation of nitrogen nutrition index in rice from UAV RGB images coupled with machine learning algorithms. *Computers and Electronics in Agriculture*, 189, 106421. https://doi.org/10.1016/j.compag.2021.106421

Razfar, N., True, J., Bassiouny, R., Venkatesh, V., & Kashef, R. (2022). Weed detection in soybean crops using custom lightweight deep learning models. *Journal of Agriculture and Food Research*, 8, 100308. https://doi.org/10.1016/j.jafr.2022.100308

Rembold, F., Atzberger, C., Savin, I., & Rojas, O. (2013). Using low resolution satellite imagery for yield prediction and yield anomaly detection. *Remote Sensing*, 5(4), 1704–1733. https://doi.org/10.3390/rs5041704

Ruigrok, T., van Henten, E., Booij, J., van Boheemen, K., & Kootstra, G. (2020). Application-specific evaluation of a weed-detection algorithm for plant-specific spraying. *Sensors*, 20(24), 7262. https://doi.org/10.3390/s20247262

Piyar, S. N., & Baulkani, Dr. S. (2022). A Review of Wireless Sensor Networks in Agriculture. *International Journal for Research in Applied Science and Engineering Technology*, 10(12), 577–580. https://doi.org/10.22214/ijraset.2022.47936

Shuaibu, M., Lee, W. S., Schueller, J., Gader, P., Hong, Y. K., & Kim, S. (2018). Unsupervised hyperspectral band selection for apple Marssonina blotch detection. *Computers and Electronics in Agriculture*, 148, 45–53. https://doi.org/10.1016/j.compag.2017.09.038

Singh, P. P., Kanth, D. R., Madhuri, G. S., Yadav, A., Chauhan, S. S., Kundu, P., Bhattacharyya, U. K., Banoo, S., . R., & Rajput, U. S. (2025). Deep learning techniques for plant disease detection and classification: A comprehensive review. *International Journal of Advanced Biochemistry Research*, 9(1S), 187–200. https://doi.org/10.33545/26174693.2025.v9.i1Sc.3457

Wang, L., Chen, S., Li, D., Wang, C., Jiang, H., Zheng, Q., & Peng, Z. (2021). Estimation of paddy rice nitrogen content and accumulation both at leaf and plant levels from UAV hyperspectral imagery. *Remote Sensing*, 13(15), 2956. https://doi.org/10.3390/rs13152956

Wang, Z., Fan, S., Wu, J., Zhang, C., Xu, F., Yang, X., & Li, J. (2021). Application of long-wave near infrared hyperspectral imaging for determination of moisture content of single maize seed. *Spectrochimica Acta Part A: Molecular and Biomolecular Spectroscopy*, 254, 119666. https://doi.org/10.1016/j.saa.2021.119666

Weraikat, D., Šorič, K., Žagar, M., & Sokač, M. (2024). Data analytics in agriculture: Enhancing decision-making for crop yield optimization and sustainable practices. *Sustainability*, 16(17), 7331. https://doi.org/10.3390/su16177331

Yan, S., & Li, S. (2024). Smart agricultural management scheme based on low-power wireless sensor network. *Journal of Multimedia Information System*, 11(1), 101–108. https://doi.org/10.33851/JMIS.2024.11.1.101

Zamora-Izquierdo, M. A., Santa, J., Martínez, J. A., Martínez, V., & Skarmeta, A. F. (2019). Smart farming IoT platform based on edge and cloud computing. *Biosystems Engineering*, 177, 4–17. https://doi.org/10.1016/j.biosystemseng.2018.10.014

Zhang, Y., Xia, C., Zhang, X., Cheng, X., Feng, G., Wang, Y., & Gao, Q. (2021). Estimating the maize biomass by crop height and narrowband vegetation indices derived from UAV-based hyperspectral images. *Ecological Indicators*, 129, 107985. https://doi.org/10.1016/j.ecolind.2021.107985

Zhou, S., Sun, L., Xing, W., Feng, G., Ji, Y., Yang, J., & Liu, S. (2020). Hyperspectral imaging of beet seed germination prediction. *Infrared Physics & Technology*, 108, 103363. https://doi.org/10.1016/j.infrared.2020.103363

Section 3

Leveraging Real-Time Environmental Data Analysis for Precision Agriculture

8 Employing Integrated Data to Study the Impact of Climate Change on Agriculture

Prateek Gupta, Priyanka Gupta, Gitika Sharma, Bhavna Thakur, and Vijay Prakash

8.1 INTRODUCTION

The ongoing progression of climate change is exerting profound influences on agricultural systems worldwide by altering rainfall distribution, raising ambient temperatures, and increasing both the frequency and intensity of extreme climatic events. These climatic transformations introduce considerable challenges to crop productivity and contribute to growing concerns over global food security. Conventional methodologies for examining these impacts often fail to encapsulate the full scope of these complex and interrelated processes, primarily due to their limited spatial granularity and the inadequate integration of dynamic variables.

In contrast, the development of integrated data frameworks presents a more holistic approach for exploring the intricate relationships between agricultural performance and climate variability. Recent scholarly efforts emphasize the importance of incorporating diverse, multidimensional datasets that can effectively capture both the immediate outcomes—such as yield fluctuations—and broader indirect factors, including ecological service disruption and socio-economic transitions. Scholars have highlighted the benefits of linking agricultural simulation models with climate forecasting systems to enable more accurate predictions and to support the formulation of resilient adaptation strategies [1]. Figure 8.1 depicts the conceptual framework used in this study to demonstrate the integration of climate variables with agricultural system modelling.

The relationship between agriculture and climate is inherently interdependent. Increasing global temperatures, shifting rainfall regimes, improved occurrence of droughts as well as floods, and changes in agroecological zones are disrupting cropping patterns and reducing yields in many areas. While climate modelling has seen substantial development, accurately predicting localized agricultural impacts remains tough due to the complex interplay between environmental, biophysical, and socio-economic variables. To effectively report this challenge, it is vital to integrate diverse and heterogeneous data sources. Traditional agricultural assessments often rely on localized field observations or standalone simulations, which may not reflect broader spatial or temporal trends. In contrast, integrated datasets allow for multi-scale monitoring, identifying

DOI: 10.1201/9781003637264-11

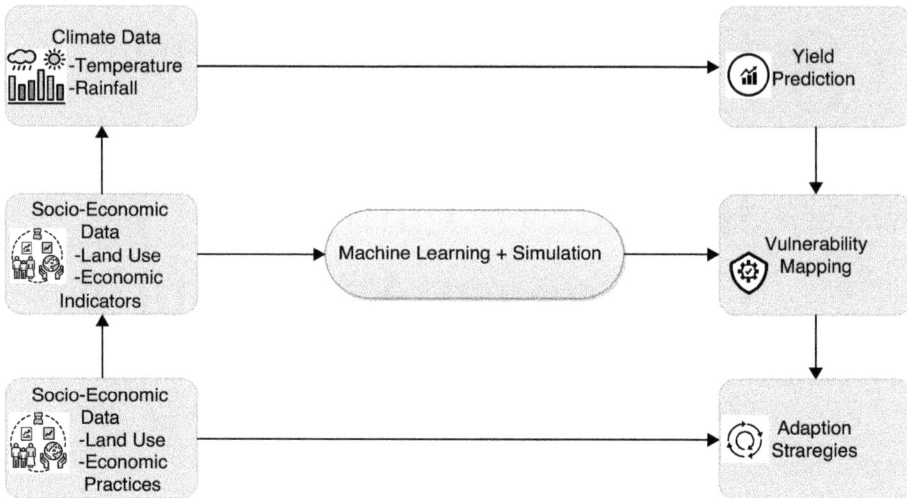

FIGURE 8.1 Conceptual framework of climate-agriculture integration.

spatial patterns and scenario-based modelling under many climate projections, thus offering a more robust context for understanding future risks. Agricultural efficiency is particularly sensitive to climate variability. Shifts in weather patterns and environmental stressors directly influence crop responses, which in turn affect yield stability. As greenhouse gas (GHG) concentrations continue to rise at an unprecedented rate, the urgency to understand and forecast their effects on crop systems has become a central concern within climate-agriculture research. While climate models can simulate future scenarios at regional scales, typically using grids ranging from tens to hundreds of kilometres, these coarse resolutions often oversimplify the local heterogeneity current in farming environments.

This spatial generalization constrains the precision of impact assessments, especially where localized factors such as soil diversity, irrigation practices, and farming techniques significantly shape agricultural outcomes. In contrast, field-scale crop models are designed to incorporate such site-specific details and are particularly useful for evaluating the effects of environmental stressors at a granular level. These models are increasingly being employed in scientific inquiry and decision-making processes to assess vulnerability and guide adaptation. This study focuses on integrating crop and climate models as a strategy for evaluating the potential impacts of climate change on agricultural productivity, which remains a cornerstone of global food security. While regional-scale modelling offers broader insights, a growing need exists to bridge the spatial gap between coarse climate projections and fine-scale agricultural realities. The review examines the techniques used to couple climate and crop models, estimates the challenges of working across different spatial and temporal scales, and highlights recent innovations in modelling approaches that progress the representation of climate-crop interactions.

Additionally, this study explores how these integrated models contribute to the development of targeted adaptation strategies and the enhancement of agricultural resilience.

It also discusses existing limitations—such as the lack of standardized data frameworks, the challenges of uncertainty quantification, and emerging trends that promise to advance the field. Eventually, this study presents recommendations for refining simulation tools as well as data systems, aiming to support more sustainable and climate-resilient agricultural planning in the face of an indeterminate future.

8.2 LITERATURE REVIEW

Several studies have attempted to model agricultural efficiency in the context of climate change. For example, Lobell et al. [2] used remote sensing to distinguish yield anomalies and related them with climatic variables across sub-Saharan Africa. Similarly, Ray et al. [3] suggested temperature and precipitation effects on crop yields globally. Previous studies examining the impacts of climate change on agriculture frequently lacked the incorporation of vital datasets, such as socio-economic indicators and soil health parameters, which are essential for holistic and meaningful analysis. Contemporary research has shifted towards integrating remote sensing data with climate forecasts and machine learning techniques to better evaluate agricultural risks. Organizations like the Food and Agriculture Organization (FAO) and the World Bank have actively supported the development of advanced tools such as the Agricultural Stress Index System (ASIS) and remote crop monitoring platforms based on Earth Observation technologies. Building on this foundation, the present study introduces a comprehensive framework designed to analyse the interplay between climate dynamics and agriculture through the use of integrated and diverse data sources.

Agriculture remains one of the most vulnerable sectors to the adverse effects of climate change, which increasingly threatens both productivity and long-term sustainability. Rising sea levels are reducing the extent of arable land by inundating coastal farming zones, while excessive rainfall is expanding flood-prone areas and constraining the availability of cultivable terrain [4]. Simultaneously, rising temperatures—both locally and globally—have triggered widespread crop failures and are jeopardizing future agricultural stability [5]. If proactive adaptation mechanisms are not implemented, ongoing climatic warming is expected to further hinder agricultural output. This is particularly concerning given the mounting challenges such as erratic rainfall patterns, extended periods of extreme heat, and recurrent flooding events. These factors interfere with optimal planting cycles, weaken resilience in fertile regions, and diminish both crop quality and harvest efficiency [5,6].

Climate model projections indicate that key crops like maize and wheat may suffer substantial yield reductions, particularly in tropical zones [7]. Meanwhile, the escalating global demand for food places immense pressure on agricultural systems to expand into previously underutilized environments, including coastal and low-quality lands. Climate change also intensifies biotic pressures by modifying the relationships between crops and harmful agents like pests, diseases, and invasive weeds. This disruption is further aggravated by declining pollinator populations, increased atmospheric ozone concentrations, and limited freshwater resources, all of which collectively threaten agricultural productivity [7]. In densely populated urban regions, the overlapping impacts of climate variability and demographic pressures significantly increase the risk to food systems and nutritional security [5,8].

In response to these urgent issues, the CGIAR's Climate Change, Agriculture and Food Security (CCAFS) programme has laid out four strategic areas for action: reshaping agricultural pathways, enhancing adaptive capabilities, curbing emissions from food production systems, and transforming institutional support structures [9]. These strategies form the backbone of the Climate-Smart Agriculture (CSA) approach—a multifaceted framework that aims to strengthen agricultural resilience, boost productivity, lower GHG emissions, and uphold food security amidst evolving climatic challenges [10].

Given the highly dynamic nature of agricultural ecosystems, achieving CSA objectives demands a nuanced understanding of the interconnections between ecological, biological, and socio-economic components [11]. Modern digital tools—including Internet of Things (IoT) devices, machine learning models, and big data platforms—are becoming essential in providing such insights. These technologies enable continuous environmental monitoring and real-time data generation, offering the analytical power needed to support data-driven decision-making in agriculture [12].

However, leveraging such large, heterogeneous datasets poses challenges in terms of data acquisition, storage, processing, modelling, and interpretation [11]. Although climate science is now among the most data-intensive fields, its integration of big data tools has lagged behind other domains like genomics and e-commerce [11,13]. While earlier climate observations required manual verification, modern techniques now employ sensors and satellites to deliver accurate, real-time data [13]. This change has significantly advanced the study of climate impacts on agriculture.

Big data applications previously addressed key issues in agriculture, such as weather prediction, pest and disease surveillance, and yield forecasting [14]. Despite its relative infancy, agricultural big data has the potential to drive precision farming, support predictive models, and promote climate-resilient agricultural practices [15–17]. Nevertheless, the design and implementation of big data infrastructure remain complex tasks—mainly in the context of climate change—due to the essential need for interdisciplinary collaboration across fields such as agronomy, physiology, molecular biology, social sciences, and economics [18–21]. Addressing the multifaceted nature of climate-related agricultural challenges requires a System of Systems method, termed "climate computation," that facilitates integration across diverse domains [22]. This method aims to generate holistic solutions reflecting the complex interdependencies between natural systems, human behaviour, and technological interferences.

Agricultural big data emerges as a vital solution for adapting to and mitigating the effects of climate change. The present study explores how big data technologies are applied in managing crops, water resources, climate conditions, and soil. The study also emphasizes the technological tools employed for data processing, analytical modelling, and visualization. It explores how these tools facilitate the development of adaptive and resilient agricultural practices. To support ongoing research, this work introduces a structured schema that consolidates critical data dimensions while simultaneously identifying current knowledge gaps and potential avenues for future investigation.

To establish this foundation, a Systematic Literature Review (SLR) was undertaken following the PRISMA framework [15], with a specific focus on literature situated at the convergence of agricultural big data and climate change. A synthesized summary of prior research contributing to understanding climate change impacts and adaptive strategies in agriculture is presented in Table 8.1.

TABLE 8.1
Key Studies on Climate Change Impacts and Adaptation in Agriculture

Author & Ref.	Region	Dataset Used	Modelling/Analytical Framework	Climate Challenge Addressed	Adaptation Contribution
Lobell et al. [2]	Global	Weather + crop yield	Statistical	Temperature variability	Yield sensitivity to temperature
Ray et al. [3]	Global	Yield and climate data	Regression	Temperature and rainfall shifts	Climate-driven yield attribution
Zhang and Cai [4]	Global	Climate projections + land use	Spatial IAM model	Land availability	Climate impact on land suitability
Nguyen et al. [5]	Asia	Climate, policy, regional data	Integrated climate policy modelling	Flood and drought risk	Assessment of policy barriers to adaptation
Gaál et al. [6]	Hungary	Regional crop + land datasets	Statistical panel models	Drought and flood risk	Irrigation-based yield strategy
Zougmoré et al. [7]	Sub-Saharan Africa	Institutional and socio-economic data	Climate-smart agriculture (CSA) framework	Climate adaptation policy gaps	CSA integration into policy
Malhi et al. [8]	India	Secondary datasets (Literature review)	Narrative climate-agriculture review	Temperature and rainfall stress	Adaptive practices for India
Byerlee and Lynam [9]	Global	Synthetic crop-climate dataset	Naïve Bayes classifier	Disease risk from climatic variability	Forecasting crop disease using ML
Steiner et al. [10]	Global	CCAFS reports + scenarios	Narrative review	Food system resilience	Food system resilience
Cravero et al. [12]	Africa	Peer-reviewed studies (SLR)	Systematic review	Climate change impact on food security in Africa	Drought-tolerant crops, irrigation, insurance, agroforestry

(Continued)

TABLE 8.1 (*Continued*)
Key Studies on Climate Change Impacts and Adaptation in Agriculture

Author & Ref.	Region	Dataset Used	Modelling/Analytical Framework	Climate Challenge Addressed	Adaptation Contribution
Mangal et al. [13]	Global	Climate + satellite + sensor data	Big data analytics framework	Heterogeneous climate data	Data-driven climate insights
Hassani et al. [14]	Global	Climate sensor data + big data tools	Big data analytics	Climate variability	Data-driven adaptation
Fenu and Malloci [15]	Italy	Weather station data	ANN, SVM, SimCast	Crop disease due to climate variability	Data-driven disease forecasting
Sadiku et al. [16]	Global	Sensor, satellite, climate, socio-economic data	Big data analytics	Environmental shifts and resource stress	Real-time agri-response
Talari et al. [17]	Europe	Climate and food safety databases	Big data analytics, DSS, ML, GIS	Climate-driven food safety risks	Real-time risk-informed decisions
Tiwari and Rana [20]	Global	Feature engineering + big data	Narrative review	Climate data heterogeneity	Big data scalability
Del Pozo et al. [21]	Mediterranean	Secondary data	Narrative review	Drought and warming	Policy-oriented adaptation synthesis

8.3 CONCEPTUAL FRAMEWORK

The following subsections provide a detailed examination of the conceptual framework adopted in this study.

8.3.1 Integrated Agricultural Systems (IAS)

Gil et al. [23] examined the benefits of diversified agricultural practices that integrate crop, livestock, and forestry components. Such integrated systems demonstrate enhanced resilience to climatic fluctuations by minimizing reliance on external inputs, improving nutrient recycling, and distributing risk across a range of production activities.

8.3.2 Ecosystem Services Perspective

Runting et al. [24] emphasized integrating climate variables with ecosystem service assessments. Their findings show that ignoring drivers like land-use change, socio-economic pressures, or invasive species can undermine agricultural adaptation and sustainability strategies.

8.3.3 Simulation-Based Integrated Assessment

Rivington et al. [25] proposed an Integrated Modelling Framework (IMF) combining simulation tools with stakeholder participation. This methodology assesses climate impacts at the whole-farm level, accounting for socio-economic trade-offs and adaptive capacity.

8.4 CONTEXT

8.4.1 Big Data in Agriculture

Big data is commonly characterized by four key dimensions, often called the "Four Vs": **volume, velocity, variety**, and **veracity**.

- **Volume** pertains to the massive amount of data being generated, stored, and processed.
- **Velocity** reflects the rapid rate at which data is produced, collected, and exchanged and the speed of data transmission.
- **Variety** addresses the heterogeneous nature of data sources, encompassing structured, semi-structured, and unstructured formats.
- **Veracity** focuses on the trustworthiness and quality assurance of the data used in analyses [26].

Big data analytics allows researchers and practitioners to uncover patterns and insights by analysing vast, diverse datasets. This has led to the emergence of big data science—a multidisciplinary field integrating database systems, machine learning, distributed systems, statistics, mathematics, AI, and information technology [27,28].

Table 8.2 summarizes the use of selected big data platforms and analytical techniques for agricultural applications such as crop monitoring, water resource management, and pest/disease identification.

The Big Data Ecosystem consists of various components that work together to manage data life cycles, including storage, processing, analysis, and visualization. The typical process begins with identifying data sources, which are then stored using appropriate models based on the data's structure. The next stage involves filtering and categorizing the data according to the analysis needs. Processing can occur in batch or stream mode, and analysis may utilize ad hoc queries, deep learning, or broader data science approaches. Visualization tools present the analysed data, ultimately supporting informed decision-making. In agriculture, big data is applied to enhance several domains [29–35]:

- Weather and climate forecasting
- Crop and soil monitoring
- Pest and disease detection
- Land-use and biodiversity assessment
- Food availability and farmer decision-making
- Finance and insurance services
- Supply chain optimization
- Remote sensing and precision farming [11]

It also supports creating platforms that provide actionable insights to stakeholders, helping improve yield, forecast demand, optimize fertilizer use, and enable data-driven farm management.

Figure 8.2 shows the 4Vs—Volume, Velocity, Variety, and Veracity—that are important for big data in agriculture. It also shows some of the most important uses of big data in agriculture, such as predicting the weather, keeping an eye on pests, and digital supply chains.

TABLE 8.2
Big Data Platforms and Analytical Techniques in Agriculture

Application Area	Big Data Tool/ Platform	Analytical Technique	Data Source	Outcome	Reference
Crop monitoring	Google Earth Engine (GEE)	NDVI-based analysis	MODIS, Landsat	Early detection of crop stress	[11,43]
Water management	GEE	Time-series satellite analysis	Sentinel, Landsat	Improved irrigation and basin planning	[44,45]
Pest and disease detection	Sensor integration platforms	Classification models (ML)	Sensor + weather data	Early outbreak alerts	[15,16]

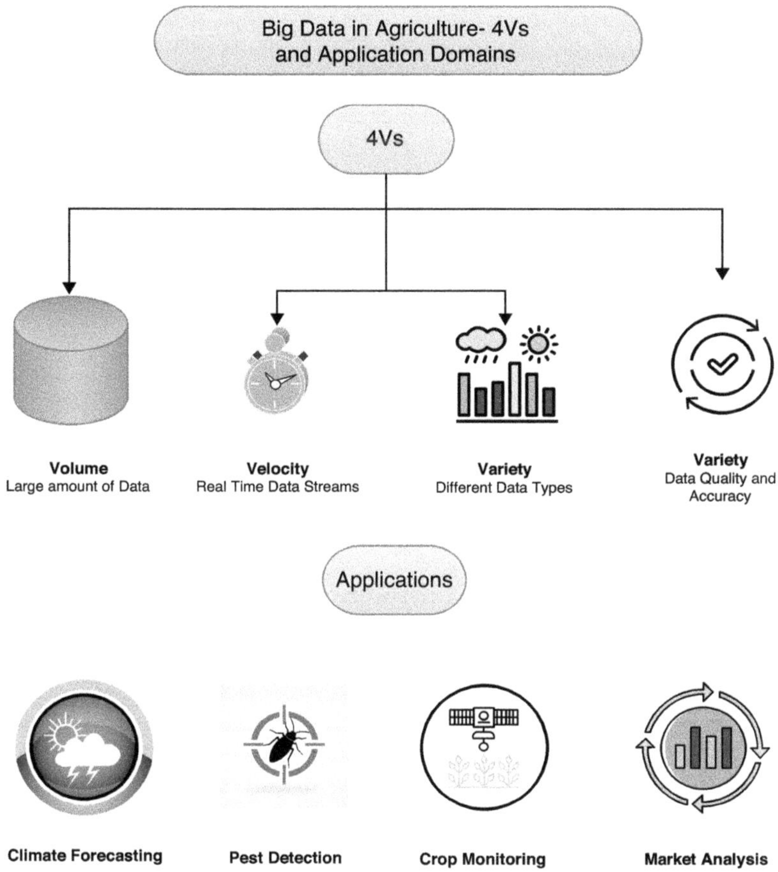

FIGURE 8.2 Big data application in agriculture.

8.4.2 Climate Change

Climate change is one of the most pressing global challenges today. It refers to significant alterations in long-term meteorological variables such as temperature and precipitation [28]. Human activity has substantially altered the atmospheric composition, triggering sharp climatic shifts in recent decades [29]. Since 1750, concentrations of key GHGs have risen dramatically—CO_2 by 150%, N_2O by 40%, and CH_4 by 20% [36]. CO_2 emissions alone increased from 22,150 million metric tonnes in 1990 to 36,140 million metric tonnes by 2014. This escalation in GHG emissions has led to global warming, with temperatures rising by 0.15–0.20°C per decade since 1975. By 2021, temperature rises ranged from 1.4°C to 5.8°C. The primary drivers contain fossil fuel combustion (CO_2), as well as non-CO_2 gases like methane (CH_4), nitrous oxide (N_2O), and chlorofluorocarbons [8].

Agriculture is mainly vulnerable to climate change due to its dependence on climatic conditions [8]. Climate variability affects crop yield as well as sustainability through changes in temperature, rainfall, and CO_2 levels. The impression of these changes varies by crop type, location, and the magnitude of environmental fluctuations [37].

As well-defined by the Intergovernmental Panel on Climate Change (IPCC), adaptation includes adjusting systems to minimize harm or exploit opportunities from climate change impacts [38]. Resilience means a system's capacity to absorb shocks as well as maintain functionality [39]. Adaptation helps farmers cope with evolving conditions, but mitigation efforts—such as reducing GHG emissions—are significant in preventing unmanageable changes. Big data plays a pivotal role in both adaptation and mitigation. It can offer real-time insights into ongoing climate impacts, guide resource allocation during disasters, enable early warning systems, and support policy decisions [40]. However, realizing its full potential needs overcoming challenges related to data quality, privacy, coordination, and infrastructure readiness. Investments in tools, technologies, and expertise are vital for the effective application of big data in climate-related agricultural strategies [40].

8.5 RESEARCH APPROACH

This study accepts a **Systematic Literature Review (SLR)** procedure, guided by the **PRISMA** framework (Preferred Reporting Items for Systematic Reviews and Meta-Analyses) [15]. The PRISMA process comprises four key stages: identification, screening, eligibility, and inclusion. Additionally, the review follows the **Kitchenham and Charters** protocol [37] to define its goals and formulate research questions (RQs). The primary RQs include:

i. What types of agricultural challenges are addressed through big data in the context of climate change?
ii. What are the core components of agricultural big data architectures?
iii. What technological tools are used to implement these architectures?
iv. What are the trends and challenges in deploying big data solutions in agriculture?

To structure the search, the **PICOC** framework was applied:

- **Population**: Agricultural big data
- **Intervention**: Climate change impact on agriculture
- **Outcomes**: Big data technologies, solutions, and implementation issues
- **Context**: Agricultural systems under climate pressure

A comprehensive search was conducted across six major databases: **Web of Science, IEEE, ACM, MDPI, Springer**, and **Scopus**. Search terms included combinations of:

- "Big Data"
- "Climate change" OR "climate-change"
- "Agriculture" OR "farm"

- Supplementary terms: "global-change," "land use change," "drought"

Searches were limited to the **title, abstract, and keywords** of publications written in English and published between **2015 and 2021**. The initial query yielded the following:

- 357 papers from Web of Science
- 38 from IEEE
- 2 from MDPI
- 5,108 from ACM
- 1,000 from Springer

After removing duplicates, non-peer-reviewed sources, and irrelevant content (e.g., books, news, grey literature), a total of **33 publications** were selected as relevant for final analysis [41].

8.6 CONNECTING CLIMATE AND CROP MODELS

8.6.1 Integrating Crop and Climate Models for Regional-Scale Assessment

Simulation models are vital tools for exploring the complex interactions between climate and agricultural systems. Given that only one physical realization of climate exists, the scope for observation-based evaluation is inherently limited—making simulation models particularly valuable for investigating climate change impacts. Unlike weather forecasting, which benefits from frequent validation against real-time observations, climate change research must rely heavily on modelled projections. In this context, coupling crop and climate models provides a makeshift laboratory for assessing potential future scenarios and adaptation strategies. Table 8.3 compares commonly used climate-crop model integration strategies, highlighting their benefits, limitations, and regions of applicability.

Various techniques are employed to integrate crop models with climate model outputs. These methods span from using raw climate simulation data directly to post-processing techniques that adjust or downscale data for compatibility with crop models. The climate models range in scope from short-term, high-resolution weather predictions to long-term, global climate change simulations. Despite their differences, these models are based on the same fundamental physical principles, and recent developments aim to present climate and weather modelling as a seamless continuum. This harmonized approach enhances user understanding and supports consistent methodologies across time scales. Crop models can either utilize unprocessed climate outputs or rely on post-processed data. Post-processing techniques—such as the application of weather generators or statistical transformations—are often employed to refine climate datasets, enhancing their suitability for agricultural modelling applications. Although raw climate data maintains internal consistency, it frequently exhibits systematic biases. These may include the overestimation of the number of wet days or inaccuracies in the representation of rainfall intensity, both of which can introduce significant errors into crop growth simulations. The complexity of bias correction varies, with some discrepancies being more straightforward to adjust than others. Moreover, the influence of these biases is highly dependent on the specific crop type and regional context under examination. The workflow for climate-crop modelling is illustrated in Figure 8.3.

TABLE 8.3
Summary of Climate-Crop Model Integration Techniques

Technique Used	Description	Pros	Limitations	Application Examples
Direct use of raw GCM output	Crop models fed with raw climate data	High consistency	Biases may distort crop simulations	Early regional assessments
Statistical downscaling	Adjusts climate data statistically	Tailored local relevance	Requires historical data	India, sub-Saharan Africa
Weather generators	Synthesizes realistic weather data	Captures variability	Parameter sensitive	Used in ensemble runs
Bias correction methods	Correct systematic errors	Improves realism	May overfit	Widespread in ensemble studies

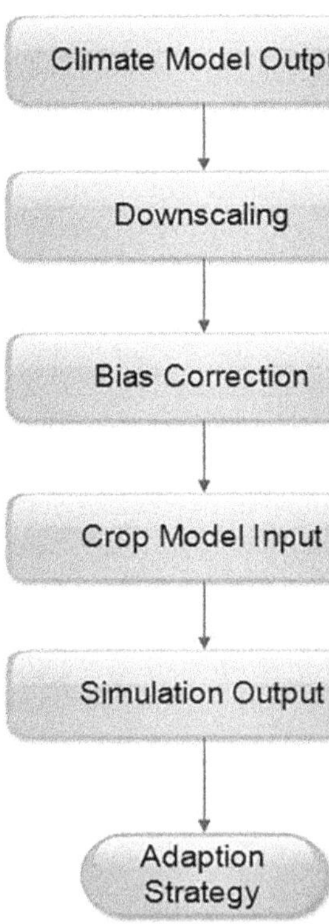

FIGURE 8.3 Crop-climate integration flow.

To overcome the spatial resolution gap between coarse-scale climate models and fine-grained crop models, researchers often utilize downscaling strategies or develop region-specific modelling frameworks. Downscaling techniques enable the translation of broad-scale climate outputs into localized projections, thereby enhancing their applicability for field-level agricultural planning. In some cases, alignment between climate and crop models is achieved through methods such as aggregation of inputs or outputs, calibration adjustments, or subgrid sampling approaches.

Additionally, empirical methods have been adopted for estimating crop yields at regional scales. These rely on statistical regression models that correlate observed weather parameters with historical crop performance. However, despite their computational efficiency and data-driven nature, empirical models often fall short in dynamic environments. Their dependency on historical data ranges limits their predictive capacity under future or unprecedented climate scenarios. Moreover, their reliance on linear relationships and failure to incorporate intra-seasonal climate variability can result in substantial forecasting errors. As a result, there is a growing preference for process-based modelling techniques, especially for applications at regional and larger spatial scales, due to their ability to simulate underlying physiological and environmental mechanisms more effectively. These models integrate biophysical processes, capture temporal variability, and offer greater explanatory power, making them more suitable for studying long-term climate change impacts and informing adaptation strategies. In sum, linking crop and climate models requires careful consideration of scale, data quality, and modelling objectives. Whether through downscaling, bias correction, aggregation, or regional model development, these integrative approaches enable more accurate and context-sensitive assessments of agricultural vulnerability and resilience in a changing climate [1].

8.6.2 Crop Modelling for Large Areas

The development of large-area crop modelling arose from the need to use climate model outputs directly—without applying downscaling techniques—to evaluate the impacts of climate variability and change on agricultural systems through a process-based lens. These approaches combine the strengths of detailed field-scale models, which offer biophysical accuracy under diverse environments, with the broader applicability of empirical methods that require minimal input data and can be scaled over large regions. At the core of this modelling strategy is recognizing that responses to climate variability, as observed in empirical studies, can be simulated at a broader scale using process-oriented models. For instance, research in India has shown that crop yield responses to climate variability can be predicted at the same spatial resolution used in climate models. However, future modelling must be grounded in physiological principles and maintain appropriate complexity to ensure reliability and avoid misinterpretation of causality.

One of the key considerations in large-area modelling is choosing the right level of model complexity. Increasing the number of processes and parameters raises the risk of error due to the greater potential for inaccurate parameterization and unintended interactions. Hence, models should focus on simulating only those processes that are immediately relevant to the primary variable of interest—typically yield.

This targeted approach reduces the likelihood of overfitting and enhances the model's transferability to new environments. Another fundamental principle in designing large-scale crop models is the preference for observable parameters wherever possible. Using empirically validated or directly measurable parameters enhances model transparency and reliability. Semi-empirical approaches, such as linking a physiologically based plant water stress index to yield, help balance empirical observations with mechanistic understanding. However, care must be taken when applying parameters derived from current climatic conditions to future climate scenarios, as these values may change under different environmental conditions.

Large-area models are generally simpler than field-scale models and contain fewer non-observable parameters. For instance, rather than simulating individual leaf development, some models estimate leaf area growth using parameters that define the maximum rate of change in the leaf area index (LAI). Similarly, instead of relying on detailed photosynthesis equations at the leaf level, biomass accumulation may be modelled using aggregate physiological indicators such as transpiration efficiency or radiation-use efficiency (RUE)—metrics that are more suitable for regional applications, especially in water-limited environments.

Despite their advantages, large-area crop models should be applied with caution. They are particularly well-suited for evaluating the influence of weather and climate on crop yields. Still, they often exclude non-climatic yield determinants such as pests, diseases, weeds, nutrient limitations, and air pollutants like tropospheric ozone. These factors contribute to the "yield gap"—the difference between actual farm yields and potential yields achievable under optimal conditions. In regions where such non-climatic influences dominate, or where there is high spatial variability within climate model grid cells, the utility of large-area models may be reduced. Nonetheless, climate remains a dominant driver of yield variability at the regional scale, especially when analysing trends over larger geographic areas. This reinforces the value of large-area crop models in climate impact assessments. Their demonstrated capacity to replicate observed yields under current conditions and to maintain predictive skill under projected climate scenarios lends credibility to their use in

TABLE 8.4
Comparison of Crop Modelling Approaches

Approach	Scale/Resolution	Strengths	Limitations	Example Usage
Empirical models	Local to regional	Easy to implement, data-driven	Cannot simulate unseen conditions, lacks causality	Regression-based yield forecasting
Process-based models	Field-scale or regional	Captures biophysical mechanisms, scalable	Complex, requires calibration and detailed inputs	DSSAT, APSIM
Large-area models	National/global	Regional yield estimation with fewer inputs	Simplified, excludes non-climatic stressors	Predicting wheat yield under RCP8.5

guiding adaptation strategies. Table 8.4 presents a comparative analysis of various crop modelling methodologies.

In summary, large-scale crop modelling presents a robust and scalable approach grounded in process-based principles to assess the impacts of climate change on agricultural productivity. When appropriately calibrated and applied with an awareness of their inherent constraints, these models serve as valuable instruments for guiding regional climate adaptation strategies and evaluating food security risks [11].

8.7 ADVANCES IN UNDERSTANDING AND MODELLING THE EFFECTS OF CLIMATE CHANGE ON AGRICULTURE

8.7.1 Ensemble Modelling and Quantifying Uncertainty

Assessing the effects of climate change on agricultural systems remains a complex task, largely due to uncertainties inherent in both climate forecasting and crop response mechanisms—particularly those associated with elevated atmospheric CO_2 levels. The wide disparity in yield projections, such as the broad range of simulated maize outcomes across African regions, underscores how methodological differences in model construction can significantly influence results. To address these uncertainties, ensemble modelling has emerged as a widely adopted approach. By utilizing multiple climate models or varying model parameters, researchers are able to explore structural uncertainties and produce a spectrum of plausible scenarios.

These climate model ensembles are frequently integrated with crop simulation tools and weather generators, thereby enhancing the robustness of yield projections by capturing variability across different climate trajectories. In addition, altering key agricultural inputs—such as selecting different crop genotypes or adjusting planting dates—expands the ensemble's scope, offering a more comprehensive representation of potential agricultural responses under diverse future climate conditions. This approach has proven effective in identifying key variables influencing agricultural productivity and assessing adaptation strategies' robustness.

8.7.2 Biophysical Processes and Elevated CO_2

Among the direct effects of atmospheric changes on crops, increased carbon dioxide concentrations have received considerable attention. Elevated CO_2 generally enhances biomass production in C3 plants and improves water use efficiency in C3 and C4 crops. While experimental setups—including Free Air CO_2 Enrichment (FACE) trials—have demonstrated yield improvements, there is concern that models may overestimate these benefits, particularly when parameter values are drawn from controlled environments. Recent studies, however, suggest that many crop models can reproduce crop responses observed in FACE trials, lending credibility to their predictive ability. Continued improvements in simulating canopy development and RUE are essential for increasing the reliability of field-scale CO_2 response models. At larger spatial scales, better integrating CO_2

effects with variability in soil, climate, and management practices is critical to improving model accuracy.

8.7.3 The Role of Ozone in Crop Performance

In addition to CO_2, surface-level ozone is another atmospheric factor that substantially affects crop yield. Ozone is a reactive gas formed through photochemical reactions involving nitrogen oxides and volatile organic compounds. Exposure to high ozone levels can impair photosynthesis, damage leaf tissue, and accelerate senescence, resulting in significant yield losses. Experimental evidence also indicates that elevated ozone levels can reduce the nutritional quality of crops, such as lowering protein content in wheat. Global-scale simulations have attempted to project future ozone impacts on agriculture, but these estimates are highly uncertain. This uncertainty stems from inconsistencies in ozone emission data, variability in exposure-response relationships, and limited understanding of how ozone interacts with other environmental stressors. More accurate projections will require process-based models that compute ozone uptake through stomata and incorporate environmental factors such as radiation, temperature, and soil moisture.

8.7.4 Indirect Effects of Climate Change on Agriculture

Warming associated with GHG emissions is expected to indirectly impact crop growth by altering development rates, growing seasons, and heat stress exposure. Although longer growing seasons have been observed, there is limited evidence suggesting these changes translate into higher yields. Thermal time response functions—used to model development rates—suggest that modest warming could reduce yield in tropical regions and provide limited benefits at higher latitudes. Sub-seasonal weather variations, such as the timing of rainfall, can also significantly affect yields. For example, even if total seasonal rainfall remains similar across years, differences in distribution can influence critical growth stages such as pod filling. This highlights the importance of including intra-seasonal dynamics in crop models, especially for large-area simulations.

8.7.5 Complex Interactions between Environmental Stressors

Crop yields result from many interacting biophysical processes, many of which are non-linear and highly sensitive to environmental context. While elevated CO_2 is expected to enhance crop productivity under water stress, the relationship is inconsistent across all models and observations. The effectiveness of crop models in capturing these interactions varies by spatial scale and model complexity. Interactions between elevated CO_2, ozone exposure, and other stressors—such as nutrient availability and temperature extremes—add further complexity. For example, ozone-induced stress can compromise plant morphology and physiological function, thereby increasing crop susceptibility to pests, pathogens, and extreme climatic events. Such interactive effects must be thoroughly accounted for when evaluating prospective risks to agricultural systems.

8.7.6 Modelling Approaches and Future Research Directions

Current modelling frameworks are beginning to incorporate some of the complex physiological interactions between crops and atmospheric variables, particularly with respect to ozone and carbon dioxide. Models that simulate stomatal ozone uptake—rather than relying solely on ambient ozone concentrations—offer more accurate predictions of crop responses. Integrating these physiological mechanisms into coupled process-based models is vital for advancing the scientific understanding of climate-agriculture interactions. However, progress in this area is hindered by the limited availability of high-quality experimental data, especially across diverse crop types and climatic regions. There is an urgent need for expanded field experiments conducted under elevated ozone and CO_2 conditions, as these would help refine existing model parameters and significantly enhance the credibility and reliability of predictive outcomes.

8.7.7 Producing Reliable and Valuable Information

To generate actionable insights for climate change adaptation in agriculture, the credibility and contextual relevance of the information produced must be ensured. Process-based crop models, while valuable, often struggle to accurately represent the complex interactions found in real-world agricultural systems. Oversimplified models risk omitting critical dynamics, whereas overly intricate models may introduce an excessive number of parameters—many of which are difficult to validate with available observational data. This can result in *overtuning*, where models accurately replicate past observations but do so for incorrect reasons, thereby limiting their predictive validity under future climate conditions.

Enhancing the reliability of such models requires rigorous calibration against a broad range of observational datasets captured under diverse environmental and management conditions. Calibration strategies should utilize directly measurable variables such as LAI, RUE, or indicators of soil fertility to constrain model behaviour. To avoid overfitting and improve generalizability, internal validation mechanisms and independent variables should be incorporated into the calibration process.

Moreover, the adoption of ensemble modelling approaches—anchored by empirical observations—has proven effective in capturing a range of possible outcomes. These ensembles help quantify uncertainty in simulation results and enhance confidence in model projections, particularly when applied to assess agricultural responses under changing climatic scenarios.

8.7.8 Informing Adaptation Strategies

Once the outputs of simulation models are validated for reliability, they can effectively inform adaptive agricultural strategies. For instance, simulating crop genotype responses under projected climate scenarios enables the identification of traits that are favourable for breeding climate-resilient cultivars. This modelling approach has been applied in the Indian context to assess whether existing germplasm possesses characteristics compatible with adaptation to rising atmospheric CO_2 levels.

Interestingly, findings from these simulations revealed increased uncertainty in model projections for genotypes adapted to elevated CO_2 conditions, underscoring the complexity of such interactions.

Successful adaptation planning also hinges on aligning model outputs with practical considerations, including forecast lead times and their relevance to end-users. Without ensuring temporal alignment and contextual applicability, even robust simulations may fall short in supporting timely and actionable decision-making at the farm or policy level. Short-term forecasts support operational decisions, while long-term projections inform infrastructure and policy planning. Presenting forecasts in accessible formats and aligning them with the capacities of stakeholders—ranging from high-tech systems to community-level observational networks—enhances their utility.

8.7.9 Future Directions: Complex Effects and Integrated Modelling

Improving model resolution, complexity, and ensemble size is a key opportunity made possible by advances in computational capacity. Higher spatial resolution improves precision, while larger ensembles enhance robustness by better capturing uncertainty. However, increasing model complexity must be balanced against data availability and parameterization limitations. Inter-model comparisons have revealed considerable differences in how crop models simulate development and yield responses, depending on whether they emphasize sink strength, source strength, or resource-use efficiency. Consequently, models are often region-specific, shaped by local crop-environment interactions and data availability. No single model is universally superior; instead, multiple approaches should be combined to improve predictions across different contexts.

8.8 INTEGRATING SOCIO-ECONOMIC AND BIOPHYSICAL DIMENSIONS

Agricultural productivity is influenced by more than just physical conditions; socio-economic factors such as labour, capital, infrastructure, and institutional support also play pivotal roles. Drawing from development studies, it is evident that households often diversify income sources to reduce vulnerability to climate shocks. However, upscaling such qualitative insights into regional or global models remains a significant challenge. Studies in Europe have linked variability in crop yields to farm characteristics, including size, intensity, and land-use practices. These factors can obscure the climate signal, making it necessary to account for farm-level diversity when simulating crop responses. Some attempts have been made to develop socio-climatic indices that identify regions particularly vulnerable due to combined climatic and socio-economic stressors. Integrating socio-economic constraints into crop models at the same spatial scale as biophysical models could significantly improve the representation of climate impacts on food security. Doing so enables a more comprehensive assessment of the capacity to produce food and adapt to changing conditions.

8.9 BRIDGING SIMULATION WITH ADAPTATION PLANNING

For simulations to meaningfully support climate adaptation, closer collaboration with practical adaptation efforts is essential. One example is using simulations to guide plant breeding programmes. Since breeding for climate-resilient traits can take years, models can help identify the geographic regions where new cultivars are likely to perform well, improving targeting and deployment strategies. Additionally, models can simulate various adaptation interventions, such as changing sowing dates, adopting new crop varieties, or altering irrigation regimes. These simulations inform local decision-making while maintaining a broader perspective on climate resilience. True success in these efforts depends on interdisciplinary and transdisciplinary collaboration—integrating climatology, plant science, agronomy, and social sciences to reflect the multifaceted nature of agricultural systems under climate change.

8.10 METHODOLOGICAL APPROACHES

The subsections below discuss the detailed methodology used in this study.

8.10.1 Linking Crop and Climate Models

Challinor et al. [1] described regional-scale climate simulations with large-area crop models to assess yield impacts.

8.10.2 Multi-source Data Integration

Incorporating long-term meteorological data, satellite observations, soil health indices, and socio-economic indicators into farm-level vulnerability assessments.

8.10.3 Handling Uncertainty

Climate change assessments must incorporate multiple uncertainties—from model assumptions to external socio-economic pressures. Only then can adaptation policies be both robust and realistic.

8.10.4 Data Sources

The study employs multiple data layers:

- **Remote Sensing Data**: MODIS NDVI, Landsat-8 imagery for vegetation monitoring.

TABLE 8.5
Data Sources Used in the Study

Data Type	Source	Resolution	Format	Application
Climate data	NASA POWER, IMD	Daily, Grid	CSV/NetCDF	Crop-climate simulation input
Remote sensing	MODIS NDVI, Landsat-8	250 m–30 m	Raster	Vegetation index, drought monitoring
Soil data	ISRIC SoilGrids, FAO HWSD	250 m–1 km	Raster	Soil mapping, yield potential
Socio-economic	World Bank, NSSO, Agri Surveys	-	CSV	Income, insurance, and cropping analysis

- **Climate Data**: Temperature, rainfall, and humidity from NASA POWER and IMD.
- **Soil Data**: FAO's Harmonized World Soil Database and ISRIC SoilGrids.
- **Socio-economic Data**: Crop yield statistics, irrigation patterns, and farmer income from national agricultural surveys and World Bank databases.

Table 8.5 presents the key data sources used in this study, including type, source, resolution, format, and application.

8.10.5 Analytical Framework

The analysis involves the following steps:

a. **Data Preprocessing**: Standardizing spatial and temporal resolutions across datasets using Python and QGIS.
b. **Correlation and Regression Analysis**: Examining relationships between climatic variables and yield trends.
c. **Geospatial Analysis**: Using GIS tools to map vulnerable zones.
d. **Machine Learning Modelling**: We Employ the Random Forest and XGBoost algorithms to predict yield under climate scenarios (RCP4.5 and RCP8.5).
e. **Validation**: Comparing model outputs with actual yield data from ground-truth sources.

8.11 KEY FINDINGS

- **Climate-Smart Integration**: Crop modelling integrated with real climate projections enables early identification of yield variability under different scenarios.
- **System Resilience**: IAS buffer climate shocks better than monoculture systems, although evidence is limited and context-dependent.

- **Socio-Economic Linkages**: Effective adaptation strategies depend on understanding ecological and socio-economic contexts, highlighting the need for cross-disciplinary integration.
- **Stakeholder Involvement**: Engaging farmers and policymakers in simulation processes leads to more relevant and actionable outputs.

8.12 RESULTS AND DISCUSSION

Preliminary findings indicate a strong negative correlation between temperature anomalies and wheat yields in northern India and parts of sub-Saharan Africa. Regions with poor irrigation infrastructure and low organic carbon content in soil show greater vulnerability to climate stress. Geospatial risk maps highlight "hotspots" of vulnerability, especially in rainfed agricultural zones. Integrating socio-economic data revealed that districts with diversified income sources and access to crop insurance showed better resilience.

8.12.1 Implications

The use of integrated data enables:

- **Early Warning Systems**: Real-time monitoring of crop stress using remote sensing.
- **Policy Insights**: Identifying regions in need of targeted subsidies or infrastructure investment.
- **Climate Adaptation Planning**: Tailoring strategies such as crop diversification, modified sowing dates, and resilient varieties.

This approach also supports global Sustainable Development Goals (SDGs), particularly Goal 2 (Zero Hunger) and Goal 13 [11].

Table 8.6 outlines the key findings of study that influence crop resilience and their implications and suggests corresponding policy recommendations.

TABLE 8.6
Key Findings and Policy Implications

Key Finding	Implication	Suggested Policy Action
High temperature anomalies reduce wheat yield.	Increased climate vulnerability in wheat-growing zones	Invest in heat-tolerant cultivars
Poor irrigation is linked with yield instability.	Infrastructure gaps increase exposure to climate shocks	Expand micro-irrigation and canal modernization
Multi-income districts show better adaptation.	Socio-economic diversification improves resilience	Promote off-farm employment and crop insurance schemes
NDVI-based hotspots align with drought-prone zones.	Satellite data is effective for early warning systems	Develop remote sensing-based advisory platforms for farmers

8.12.2 Using Big Data to Manage Water

Adaptation: Five key studies were identified that explore the role of big data in supporting climate change adaptation strategies within water management systems. These studies illustrate how data-driven insights enhance decision-making across various hydrological and agricultural applications.

Coro et al. employed big data frameworks integrating NASA and AquaMaps data sources to examine the greenhouse effect and climate variability across 10 marine regions. These environmental parameters were processed using a cloud-based computing platform to develop intermediate climate scenarios and generate derived variables. The study found that the Mediterranean Sea may respond differently to climate stressors compared to other regions, potentially affecting local fisheries and related economic inputs [42]. Senay et al. used remote-sensing data from Landsat satellites and meteorological big data to evaluate the relative impact of climate variability and water management practices on crop productivity. Their findings help inform agricultural decisions, hydrological modelling, and commercial water use by clearly understanding the interactions between climate and irrigation systems [43]. Alencar et al. described how collaboration between the private sector and governmental institutions, including hydrographic river basin agencies, has led to the development of decision-support tools. These tools integrate big data systems to track the combined effects of climate change, irrigation, and municipal/industrial water extraction. For example, such tools have shown how these factors influence river flow rates. The authors emphasized that big data facilitates a paradigm shift from reactive to proactive strategies in river basin modelling and management [44]. Kandekar et al. investigated how the availability of surface water resources significantly influences climate change and agricultural development in semi-arid regions. Using a cloud-based geospatial analysis platform, they applied advanced algorithms—particularly those supported by Google Earth Engine (GEE)—to process remote-sensing imagery. Vector data on surface water bodies, derived from big data systems, were used as a basis for extracting hydrological parameters, enabling enhanced water resource planning [45]. Reddy and Sureshbabu developed and tested a big data-driven early warning system for precipitation, aimed at supporting farmers in flood-prone areas. Their approach significantly enhances the capacity for timely and informed water and crop management decisions. By leveraging predictive analytics, the system contributes not only to improved agricultural productivity but also to increased resilience against extreme weather events [46].

8.13 CONCLUSION

Effectively responding to the multifaceted impacts of climate change on agricultural productivity necessitates an integrated, multidimensional approach that accommodates both scientific complexity and real-world applicability. The interactions among climate variables—such as rising temperatures, shifting precipitation regimes, increased atmospheric CO_2 concentrations, and the frequency of extreme weather events—are intricately linked with socio-economic, geographic, and land management factors. As a result, the development of meaningful crop-climate modelling frameworks must go beyond conventional biophysical simulations and incorporate interdisciplinary perspectives that encompass ecological, economic, and social dimensions across appropriate spatial and temporal scales.

This study underscores the importance of leveraging integrated big data systems—merging satellite remote sensing, climate projections, soil profiles, hydrological datasets, and socio-economic indicators—to evaluate crop vulnerabilities and guide the design of adaptive strategies. To ensure the robustness and practical utility of such modelling systems, three foundational pillars are essential:

a. **Transparent Quantification of Uncertainty**
   Uncertainty must be explicitly addressed through ensemble modelling and sensitivity analysis. These techniques enable the exploration of a range of plausible futures without presenting misleading precision, thereby facilitating more robust and informed decision-making.
b. **Integration of Observational and Simulation Tools**
   Combining empirical observations with simulation models helps capture non-linear dynamics and context-specific interactions that are crucial for understanding agricultural responses. This hybrid approach enhances the reliability of simulations by grounding them in real-world conditions.
c. **Contextual Calibration and Model Customization**
   Models must be locally adapted and appropriately parameterized, taking into account region-specific environmental conditions, management practices, and socio-economic realities. Calibration using diverse, high-quality observational datasets improves model relevance and accuracy.

Looking ahead, crop-climate modelling must evolve beyond theoretical exercises to become integral components of adaptive management systems. These models should inform a broad spectrum of climate-smart strategies, including cultivar development, water resource planning, risk mitigation, and evidence-based policy formulation. Importantly, the outputs from simulation models must be translated into actionable insights that are accessible and useful to farmers, extension agents, and policymakers.

This work advocates for a paradigm shift in agricultural climate research—one that embraces interdisciplinary collaboration, user-centred design, and real-time responsiveness. Future modelling efforts should prioritize the integration of farmer feedback, improvement of data interoperability, and development of open-access, scalable platforms that connect scientific insights with ground-level impact. Through such coordinated efforts, crop-climate modelling can become a cornerstone of climate-resilient, sustainable, and equitable food systems worldwide.

REFERENCES

1. Challinor, A.J.; Ewert, F.; Arnold, S.; Simelton, E.; Fraser, E. Crops and climate change: Progress, trends, and challenges in simulating impacts and informing adaptation. *J. Exp. Bot.* 2009, 60(10), 2775–2789. doi:10.1093/jxb/erp062
2. Lobell, D.B.; Schlenker, W.; Costa-Roberts, J. Climate trends and global crop production since 1980. *Science* 2011, 333(6042), 616–620.

3. Ray, D.K.; West, P.C.; Clark, M.; Gerber, J.S.; Prishchepov, A.V.; Chatterjee, S. Climate change has likely already affected global food production. *Nat. Food*. 2019, 1(1), 1–8.
4. Zhang, X.; Cai, X. Climate change impacts on global agricultural land availability. *Environ. Res. Lett.* 2011, 6, 014014.
5. Nguyen, T.H.; Sahin, O.; Howes, M. Climate change adaptation influences and barriers impacting the Asian agricultural industry. *Sustainability* 2021, 13, 7346.
6. Gaál, M.; Quiroga, S.; Fernandez-Haddad, Z. Potential impacts of climate change on agricultural land use suitability of the Hungarian counties. *Reg. Environ. Chang.* 2013, 14, 597–610.
7. Zougmoré, R.B.; Läderach, P.; Campbell, B.M. Transforming food systems in Africa under climate change pressure: Role of climate-smart agriculture. *Sustainability* 2021, 13, 4305.
8. Malhi, G.; Kaur, M.; Kaushik, P. Impact of climate change on agriculture and its mitigation strategies: A review. *Sustainability* 2021, 13, 1318.
9. Byerlee, D.; Lynam, K.L. The invention of the international agricultural research center model and the CGIAR System. *World Dev.* 2020, 135, 105080.
10. Steiner, A.; Aguilar, G.; Bomba, K.; Bonilla, J.P.; Campbell, A.; Echeverria, R. Actions to Transform Food Systems under Climate Change. 2020. Available online: https://ageconsearch.umn.edu/record/310920/files/CCAFS%20Actions%20to%20Transform%20Food%20Systems%20Under%20Climate%20Change.pdf (accessed on 22 May 2022).
11. Kamilaris, A.; Kartakoullis, A.; Prenafeta-Boldú, F.X. A review on the practice of big data analysis in agriculture. *Comput. Electron. Agric.* 2017, 143, 23–37.
12. Cravero, A.; Pardo, S.; Sepúlveda, S.; Muñoz, L. Challenges to use machine learning in agricultural big data: A systematic literature review. *Agronomy* 2022, 12, 748.
13. Mangal, P.; Rajesh, A.; Misra, R. Big data in climate change research: Opportunities and challenges. In *Proceedings of the 2020 International Conference on Intelligent Engineering and Management (ICIEM)*, London, UK, 17–19 June 2020; pp. 321–326.
14. Hassani, E.; Huang, H.; Silva, X. Big data and climate change. *Big Data Cogn. Comput.* 2019, 3, 12.
15. Fenu, G.; Malloci, F.M. An application of machine learning technique in forecasting crop disease. In *Proceedings of the 2019 3rd International Conference on Big Data Research*, New Delhi, India, 20–23 December 2014; Volume 143, pp. 23–37.
16. Sadiku, M.N.O.; Ashaolu, T.J.; Musa, S.M. Big data in agriculture. *Int. J. Adv. Res. Comput. Sci. Softw. Eng.* 2020, 9, 70–72.
17. Talari, G.; Cummins, E.; McNamara, C.; O'Brien, J. State of the art review of Big Data and web-based Decision Support Systems (DSS) for food safety risk assessment with respect to climate change. *Trends Food Sci. Technol.* 2022, 126, 192–204.
18. BenSaid, F.; Alimi, A.M. Online feature selection system for big data classification based on multi-objective automated negotiation. *Pattern Recognit.* 2020, 110, 107629.
19. Guo, Y.; Zhang, B.; Sun, Y.; Jiang, K.; Wu, K. Machine learning based feature selection and knowledge reasoning for CBR system under big data. *Pattern Recognit.* 2020, 112, 107805.
20. Tiwari, S.R.; Rana, K.K. Feature selection in big data: Trends and challenges. In *Data Science and Intelligent Applications*; Editors by K. Kotecha, V. Piuri, H. N. Shah, R. Patel; Springer: Berlin/Heidelberg, Germany, 2021; pp. 83–98.
21. Del Pozo, A.; Brunel-Saldias, N.; Engler, A.; Ortega-Farias, S.; Acevedo-Opazo, C.; Lobos, G.A.; Molina-Montenegro, M.A. Climate change impacts and adaptation strategies of agriculture in Mediterranean-climate regions (MCRs). *Sustainability* 2019, 11, 2769.
22. Sebestyén, V.; Czvetkó, T.; Abonyi, J. The applicability of big data in climate change research: The importance of system of systems thinking. *Front. Environ. Sci.* 2021, 9, 70.

23. Gil, J.D.; Cohn, A.S.; Duncan, J.; Newton, P.; Vermeulen, S. The resilience of integrated agricultural systems to climate change. *WIREs Climate Change* 2017. doi:10.1002/wcc.461
24. Runting, R.K.; Bryan, B.A.; Dee, L.E.; Maseyk, F.J.; Mandle, L.; Hamel, P.; Wilson, K.A.; Yetka, K.; Possingham, H.P.; Rhodes, J.R. Incorporating climate change into ecosystem service assessments and decisions: A review. *Global Change Biol.* 2016. doi:10.1111/gcb.13457
25. Rivington, M.; Matthews, K.B.; Bellocchi, G.; Buchan, K.; Stöckle, C.O.; Donatelli, M. An integrated assessment approach to conduct analyses of climate change impacts on whole-farm systems. *Environ. Model. Softw.* 2007. doi:10.1016/j.envsoft.2005.07.018
26. Sassi, I.; Ouaftouh, S.; Anter, S. Adaptation of classical machine learning algorithms to big data context: Problems and challenges. In *Proceedings of the 2019 1st International Conference on Smart Systems and Data Science (ICSSD)*, Rabat, Morocco, 3–4 October 2019; pp. 1–7.
27. Elshawi, R.; Sakr, S.; Talia, D.; Trunfio, P. Big data systems meet machine learning challenges: Towards big data science as a service. *Big Data Res.* 2018, 14, 1–11.
28. Haig, B.D. *Big Data Science: A Philosophy of Science Perspective*; American Psychological Association: Washington, DC, USA, 2020; pp. 15–33.
29. De Mauro, A.; Greco, M.; Grimaldi, M. A formal definition of Big Data based on its essential features. *Libr. Rev.* 2016, 65, 122–135.
30. Demchenko, Y.; de Laat, C.; Membrey, P. Defining architecture components of the Big Data Ecosystem. In *Proceedings of the 2014 International Conference on Collaboration Technologies and Systems (CTS)*, Minneapolis, MN, USA, 19–23 May 2014; pp. 104–112.
31. Santos, M.Y.; e Sá, J.O.; Costa, C.; Galvão, J.; Andrade, C.; Martinho, B.; Lima, F.V.; Costa, E. A big data analytics architecture for Industry 4.0. In *A World Conference Information Systems and Technologies*; Editors: Á. Rocha, A.M. Correia, H. Adeli, L.P. Reis, and S. Costanzo; Springer: Berlin/Heidelberg, Germany, 2017; pp. 175–184.
32. Salma, C.A.; Tekinerdogan, B.; Athanasiadis, I.N. Domain-driven design of big data systems based on a reference architecture. In *Software Architecture for Big Data and the Cloud*; Editors: I. Mistrik, R. Bahsoon, N. Ali, M. Heisel, and B. Maxim; Morgan Kaufmann: Burlington, MA, USA, 2017; pp. 49–68.
33. Sowmya, R.; Suneetha, K. Data mining with big data. In *Proceedings of the International Conference on Intelligent Systems and Control (ISCO)*, Coimbatore, India, 5–6 January 2017; IEEE: Piscataway, NJ, USA; pp. 246–250.
34. Torky, M.; Hassanein, A.E. Integrating blockchain and the internet of things in precision agriculture: Analysis, opportunities, and challenges. *Comput. Electron. Agric.* 2020, 178, 105476.
35. WMO, G.; OMM, G. *International Meteorological Vocabulary*; World Meteorological Organization: Geneva, Switzerland, 1966.
36. Team, X.W. Contribution ofWorking Groups I, II and III to the Fourth Assessment Report of the Intergovernmental Panel on Climate Change: Working group II contribution. *Russian Meteorology and Hydrology*; Springer: Berlin/Heidelberg, Germany, 2007; Volume 104.
37. Karimi, V.; Karami, E.; Keshavarz, M. Climate change and agriculture: Impacts and adaptive responses in Iran. *J. Integr. Agric.* 2018, 17, 1–15.
38. de Cambio Climático, I.P.I. Cambio Climático. Base Física; Suiza IPCC: Geneva, Switzerland, 2013.
39. Sarker, M.N.; Yang, B.; Yang, L.; Huq, M.E.; Kamruzzaman, M.M. Climate change adaptation and resilience through big data. *Int. J. Adv. Comput. Sci. Appl.* 2020, 11, 533–539.

40. Keele, S. Guidelines for Performi Systematic Literature Reviews in Software Engineering. Version 2.3, Technical Rep. EBSE' 2007. Available online: https://www.researchgate.net/profile/Barbara Kitchenham/publication/302924724_Guidelines_for_performing_Systematic_Literature_Reviews_in_Software_Engineering/links/61712932766c4a211c03a6f7/Guidelines-forperforming-Systematic-Literature-Reviews-in-Software-Engineering.pdf (accessed on 22 May 2022).
41. Cravero, A.; Bustamante, A.; Negrier, M.; Galeas, P. Agricultural big data architectures in the context of climate change: A systematic literature review. *Sustainability* 2022, 14(13), 7855.
42. Coro, G.; Pagano, P.; Ellenbroek, A. Detecting patterns of climate change in long-term forecasts of marine environmental parameters. *Int. J. Digit. Earth* 2020, 13, 567–585.
43. Senay, G.B.; Schauer, M.; Friedrichs, M.; Velpuri, N.M.; Singh, R.K. Satellite-based water use dynamics using historical landsat data (1984–2014) in the Southwestern United States. *Remote Sens. Environ.* 2017, 202, 98–112.
44. Alencar, P.; Cowan, D.; McGarry, F.; Palmer, R.M. An open and big data platform for cumulative environmental analysis and management. In *Proceedings of the 2015 3rd International Conference on Future Internet of Things and Cloud*, Rome, Italy, 24–26 August 2015; IEEE: Piscataway, NJ, USA, 2015; pp. 412–417.
45. Kandekar, V.U.; Pande, C.B.; Rajesh, J.; Atre, A.A.; Gorantiwar, S.D.; Kadam, S.A.; Gavit, B. Surface water dynamics analysis based on sentinel imagery and Google Earth Engine Platform: A case study of Jayakwadi dam. *Sustain. Water Resour. Manag.* 2021, 7, 44.
46. Reddy, P.C.; Sureshbabu, A. An adaptive model for forecasting seasonal rainfall using predictive analytics. *Int. J. Intell. Eng. Syst.* 2019, 12, 22–32.

9 Hybrid Gradient Boosting and Gaussian Process Regression for Crop Yield Prediction Using Feature Engineering

Mohanapriya C. and C. Radhiya Devi

9.1 INTRODUCTION

The increasing pressure on global food systems—driven by climate variability, rising population, and shrinking arable land—has made it imperative to adopt data-driven strategies for maximizing agricultural productivity. Accurate crop yield prediction is one such strategy, empowering stakeholders such as farmers, agronomists, and policymakers to make informed decisions regarding crop selection, irrigation planning, fertilizer usage, and supply chain logistics.

However, conventional statistical models used in yield forecasting often fail to cope with the complexities of real-world agriculture. These models typically rely on linear assumptions and require high-quality, structured data. For example, consider a scenario where a farmer grows two crops—rice and chickpea—under fluctuating temperature and rainfall conditions. A traditional linear regression model might predict similar yield responses for both crops when temperature increases by 2°C. In reality, rice might tolerate the shift due to high water availability, while chickpea yield could decline sharply due to sensitivity to heat stress. Such nuances in crop behaviour are difficult for conventional models to capture, especially in the presence of noisy or sparse data.

This gap motivates the need for advanced, flexible, and scalable machine learning (ML) approaches that can handle

- Nonlinear relationships between climatic and soil variables
- Noisy or synthetic datasets
- Crop-specific environmental responses
- Uncertainty in prediction outcomes

DOI: 10.1201/9781003637264-12

In this study, we present a hybrid ML framework that combines Gradient Boosting (GB) and Gaussian Process Regression (GPR) for crop yield prediction. This dual-model approach addresses the limitations of single-model systems by combining the strength of GB in capturing complex patterns in large datasets with GPR's ability to model uncertainty and fine-tune predictions through probabilistic reasoning.

The methodology involves:

Using a well-known crop recommendation dataset augmented with a synthetic yield simulation function to mimic real-world variability. Applying advanced feature engineering techniques such as interaction terms, polynomial transformations, and logarithmic adjustments.

Training models separately for each crop type to capture crop-specific yield dynamics.

Evaluating performance using both regression metrics (R^2, root mean square error (RMSE)) and classification metrics (confusion matrix based on yield bins).

9.1.1 Key Contributions

This work makes the following major contributions:

Hybrid Prediction Architecture: Proposes a novel hybrid model (GB-GPR) that improves crop yield prediction accuracy through a weighted ensemble strategy, leveraging both deterministic and probabilistic strengths.

Advanced Feature Engineering: Introduces a comprehensive feature transformation pipeline tailored to agricultural data, enhancing the model's ability to capture nonlinear interactions.

Synthetic Yield Simulation: Develops a noise-augmented yield simulation technique that enables robust training even when actual yield data is unavailable.

Crop-wise Model Training: Implements individualized models for each crop, improving specificity and generalizability over traditional generalized approaches.

Comprehensive Evaluation: Achieves state-of-the-art performance (e.g., R^2 up to 0.9942 for grapes and RMSE as low as 0.2564 for apple) and demonstrates practical reliability via 100% accurate yield category classification.

Overall, this work presents a scalable, interpretable, and high-accuracy yield prediction framework with strong applicability in real-world precision agriculture scenarios, particularly where environmental variability and limited yield data are significant challenges.

By combining the strengths of both ensemble learning and probabilistic modelling, this framework offers a scalable, interpretable, and highly effective solution for yield forecasting. It holds great potential for applications in precision agriculture, climate-resilient farming, and resource optimization. ML methods such as decision trees, Support Vector Machines (SVMs), and ensemble models are reviewed with a focus on feature selection and model performance across crop datasets (Halder et al. 2023). Key challenges identified include inconsistent data availability and lack of standardization. Bassine et al. (2023) emphasize the role of remote sensing, Internet of Things (IoT), and ML in precision agriculture for yield prediction. Integrating satellite data and ground-based observations offer significant improvements in accuracy, especially in variable climates.

Random Forest (RF) and SVMs are highlighted for their superior handling of nonlinear data patterns in yield prediction (Sujatha et al. 2022). The comparative analysis shows improved performance over traditional statistical techniques (Yuan et al. 2024).

Combining UAV imagery with ML models enhances yield estimation by capturing real-time field variability. The review outlines preprocessing techniques and optimal feature extraction methods from aerial datasets. Coupling crop simulation models with ML algorithms reduces prediction error and increases robustness (Shahhosseini et al. 2021). Using model-generated variables as ML inputs leads to better generalization, particularly for corn yield forecasting. A hybrid approach unifying physics-based and data-driven models provides more accurate yield predictions in data-scarce conditions (Kallenberg et al. 2023). The meta-modelling strategy effectively balances interpretability and predictive power. A smart farming framework uses ML models trained on historical weather and soil parameters (Islam et al. 2024). Emphasis is placed on climate-resilient strategies and scalable architectures for large-scale deployment. Early fusion of multi-modal data such as Normalized Difference Vegetation Index (NDVI), temperature, and rainfall enhances the accuracy of yield prediction models (Pathak et al. 2023). The method supports different temporal resolutions, enabling broader field-level application. Deep learning architectures, especially LSTM networks, outperform traditional ML methods by modelling temporal dependencies in weather and genotype data (Shook et al. 2020). Results show high accuracy in predicting maize yield under fluctuating conditions. ML optimizes input usage by predicting yield responses to water and fertilizer levels (Ersin et al. 2023). The model contributes to reduced resource wastage while maintaining or increasing crop output. Application of UAVs and ML facilitates detailed spatial monitoring of crop development stages (Yuan et al. 2024). Results indicate substantial gains in prediction precision compared to ground-only datasets. Simulation-enhanced ML models demonstrate improved adaptability across regions and planting seasons (Shahhosseini et al. 2021). Feature importance analysis supports integration into real-time advisory systems. Meta-modelling techniques show promise in combining theoretical crop growth principles with real-world data trends (Kallenberg et al. 2023). The approach is particularly useful when empirical training data is limited or inconsistent. Emphasis is placed on cost-effective ML integration in smart agriculture, balancing prediction quality with deployment feasibility (Islam et al. 2024). The system architecture allows real-time adaptation to environmental changes. A robust ML pipeline is built for early-stage yield forecasting using combined environmental and spectral data (Pathak et al. 2023). The approach supports real-time predictions in dynamic weather conditions.

Conventional models often rely on linear assumptions and lack the capacity to model the intricate interactions between climatic factors, soil conditions, and crop genetics. In contrast, ML techniques can identify nonlinear relationships and complex patterns in large datasets. Among these, ensemble methods and probabilistic models stand out for their predictive power and flexibility. This chapter presents a hybrid approach that combines two powerful ML techniques GB and GPR to predict crop yield using environmental and soil attributes.

The study uses a crop recommendation dataset containing key features such as temperature, humidity, rainfall, soil pH, and macronutrient levels (nitrogen, phosphorus, and potassium). Since real-world yield values are not part of the dataset, yields are synthetically generated using a weighted formula reflecting the contribution of each feature, combined with random noise to mimic natural variability.

Feature engineering plays a central role in enhancing model **GB** performance. Interaction terms (e.g., temperature × humidity), polynomial transformations (e.g., $temperature^2$), and logarithmic adjustments are introduced to capture hidden dependencies among variables. These derived features help the model better represent the underlying agronomic conditions affecting crop productivity.

To address variability across crops, the dataset is grouped by crop type and models are trained separately for each group. GB, known for its high accuracy and ability to reduce bias through sequential learning, serves as the primary predictive model. It is complemented by GPR, which offers probabilistic predictions and handles uncertainty well, particularly with smaller datasets or noisy inputs. Final yield predictions are obtained by combining both models using a weighted average, giving more influence to GB while refining outputs with GPR.

Model evaluation is performed using two common regression metrics: the coefficient of determination (R^2) and RMSE. These metrics provide insights into how well the model generalizes to unseen data. Additionally, visual tools such as bar charts, scatter plots, and heat maps are used to compare actual versus predicted yields across different crops and to assess prediction accuracy.

A unique aspect of this study is the crop-specific modelling approach. Instead of building a one-size-fits-all model, we isolate each crop's data, allowing the hybrid model to tailor its predictions to the specific environmental requirements and responses of each crop. This strategy enhances accuracy and reflects real-world agricultural practices more closely.

The proposed framework demonstrates that integrating diverse ML models can provide a more balanced and accurate yield estimation system. It also highlights the value of domain-informed synthetic labelling and thoughtful feature engineering, even when actual yield data is unavailable.

The rest of this chapter elaborates on the methodology, including data preprocessing, feature construction, model design, and performance evaluation. The results section presents findings on crop-wise model performance, followed by a discussion on potential applications in agricultural forecasting and precision farming.

9.1.2 Related Works

Crop yield prediction has received significant attention in recent years with advancements in ML and remote sensing technologies. Most prior studies have employed standalone ML models or combinations with simulation-based techniques to forecast yield outcomes based on soil, weather, or spectral data. However, limitations remain in terms of generalizability, feature adaptability, and uncertainty modelling.

Sujatha et al. (2022) compared several classic ML algorithms including RF, SVM, and K-Nearest Neighbours (KNN). While RF achieved strong performance with R^2 above 0.90, the study relied only on basic preprocessing and lacked hybrid or probabilistic modelling strategies.

TABLE 9.1
Comparative Summary of Crop Yield Prediction Approaches

Study	Approach	Feature Engineering	Dataset Used	Performance (R^2/RMSE)	Remarks
Sujatha et al. (2022)	RF, SVM, KNN	Basic preprocessing	Soil and weather data (India)	>0.90 (RF)/moderate	No hybrid modelling, no uncertainty quantification
Shahhosseini et al. (2021)	Crop simulation + ML	Simulation variables	U.S. Corn Belt (simulation outputs)	0.85–0.90/low–moderate	High accuracy but complex integration of domain-specific models
Yuan et al. (2024)	UAV + ML (NDVI features)	Spectral feature extraction	UAV field images (various crops)	~0.93/moderate	Spectrally rich, but high computational cost
Pathak et al. (2023)	Early fusion ML	Multi-modal data (NDVI, temperature, rainfall)	Remote sensing + weather datasets	~0.92/not reported	Requires multi-source synchronization
Kallenberg et al. (2023)	Meta-model (physics + ML)	Simulation + sensor fusion	Simulated + empirical mixed datasets	High accuracy/not reported	Balanced interpretability and performance, high setup complexity
Proposed GB-GPR (This work)	Hybrid ML (GB + GPR)	Product, polynomial, log transformations	Crop recommendation dataset (synthetic yield)	**Up to 0.9942/as low as 0.2564**	High performance, lightweight, interpretable, uncertainty-aware, crop-wise

Shahhosseini et al. (2021) proposed coupling crop simulation models with ML to improve generalization and robustness. Although this method improved accuracy, it required domain-specific simulations, which limit its scalability across diverse crop types and environments.

Several other studies have explored deep learning (e.g., LSTM), spectral imaging (e.g., UAV-based NDVI), and process-based meta-models. While powerful, these often demand high computation, extensive field-level calibration, or multi-source data integration, making them less suitable for flexible, low-resource applications.

Table 9.1 presents a comparative overview of various crop yield prediction approaches from existing literature, highlighting their core methodologies, feature engineering strategies, datasets used, and performance metrics. Sujatha et al. (2022) employed traditional ML algorithms such as RF, SVM, and KNN, achieving reasonably high R^2 scores (above 0.90 for RF) but relying only on basic preprocessing, which limited their capacity to model complex interactions.

Shahhosseini et al. (2021) introduced a hybrid approach by coupling crop simulation models with ML. This integration improved robustness and accuracy (R^2: 0.85–0.90) but came with increased system complexity and dependence on domain-specific models, making it less flexible across varied crops or regions.

Yuan et al. (2024) explored the combination of UAV-based remote sensing data with ML, focusing on spectral indices like NDVI. Their approach achieved high accuracy (R^2 ~0.93), but required sophisticated UAV systems and heavy computational resources, restricting its accessibility for general or low-resource applications. Similarly, Pathak et al. (2023) proposed a multi-modal ML pipeline that fuses diverse inputs—such as temperature, rainfall, and NDVI—early in the process. While this early fusion strategy enhanced predictive accuracy (R^2 ~0.92), it required synchronized data streams from different sensors, adding logistical complexity.

In contrast, the proposed hybrid GB-GPR framework distinguishes itself by combining the strengths of GB and GPR within a lightweight, fully data-driven pipeline. It incorporates advanced feature engineering techniques (interaction, polynomial, and logarithmic transformations) and uses synthetically generated yield values to simulate real-world agricultural variability. Unlike previous approaches, it avoids dependency on simulation models or expensive sensors. This results in state-of-the-art performance—achieving R^2 values up to 0.9942 and RMSE as low as 0.2564—while remaining scalable, interpretable, and applicable across a wide range of crop types and environmental conditions

9.2 WORKING METHODOLOGY

9.2.1 Dataset

The Crop Recommendation Dataset from Kaggle, compiled by Atharva Ingle, is a structured dataset designed to support the advancement of precision agriculture through ML. It enables users to build predictive models that recommend the most suitable crops for cultivation based on key environmental and soil-related parameters. Precision agriculture focuses on maximizing yield and resource efficiency through data-driven decisions and this dataset provides the necessary foundation for such innovations.

The dataset Crop_recommendation.csv is loaded, containing environmental and soil-related features including:

- **Climatic Factors**: Temperature, Humidity and Rainfall
- **Soil Chemistry**: pH, Nitrogen (N), Phosphorus (P) and Potassium (K)
- **Label**: Crop name (indicating the most suitable crop for given conditions)

These attributes influence plant physiology, nutrient absorption, and overall growth, making them essential for accurate crop prediction. The dataset has been curated by augmenting multiple Indian datasets related to climate, rainfall, and fertilizer usage, reflecting real-world agricultural scenarios. Each entry in the dataset represents a sample environment with the corresponding ideal crop label, enabling supervised learning approaches.

This dataset is widely used in agricultural analytics, aiding researchers, agronomists, and data scientists in developing tools that optimize land use, reduce input costs, and promote sustainable farming practices. The dataset can be accessed via Kaggle's Crop Recommendation Dataset.

9.2.2 Yield Simulation

In the absence of real crop yield data, a pseudo-yield function is employed to simulate crop yields. This simulated yield is crucial for training ML models that predict actual crop yields based on environmental and soil-related features. The purpose of using a simulated yield function is to mimic the real-world impact of various factors on crop productivity while allowing for model training and testing.

The yield is simulated by using a linear combination of key environmental and soil features such as temperature, humidity, rainfall, soil pH, and the levels of essential nutrients like nitrogen, phosphorus, and potassium. These factors are known to influence crop growth and productivity. For example, an increase in temperature or humidity can impact crop health and nutrient levels directly affecting growth potential. The relationship between these factors and crop yield is assumed to be linear in this case, meaning the overall yield is the weighted sum of these features.

However, real-world crop yields are subject to various unpredictable elements, such as fluctuations in weather conditions or variations in soil quality. To account for this uncertainty and make the simulated yield more realistic, Gaussian noise is added to the model. This noise introduces randomness into the predictions, reflecting the inherent variability found in actual agricultural systems. The noise term ensures that the simulated yield is not perfectly deterministic, providing a more accurate representation of real-world unpredictability.

Thus, the pseudo-yield function provides a synthetic yet realistic approximation of crop yields, which can be used for model development in the absence of actual yield data. It also helps assess how environmental and soil conditions can impact agricultural productivity under various scenarios, aiding in crop recommendation and yield prediction tasks.

9.2.3 Feature Engineering

Feature engineering plays a crucial role in improving the performance of ML models, especially when dealing with agricultural data that often involve complex relationships between variables. To enhance model learning and capture intricate dependencies between environmental factors and crop yield, several transformations are applied to the dataset. These transformations provide the model with a richer set of features, allowing it to learn more effectively.

9.2.3.1 Product Features

To capture the interaction between different environmental factors, product features are introduced. These features represent the combined effects of two variables on crop yield, which can reveal deeper insights into how environmental conditions jointly influence agricultural outcomes. For example:

- **Temperature × Humidity**
- **Humidity × Rainfall**

These interactions can represent phenomena such as the amplification of temperature effects under high humidity conditions, or how rainfall might interact with humidity levels to influence crop growth. Including product features allows the model to better understand these complex interactions, which might not be captured by individual features alone.

9.2.3.2 Polynomial Features

Polynomial features are introduced to model nonlinear relationships between environmental factors and crop yield. Some relationships in agriculture are inherently nonlinear, meaning that a simple linear approach might fail to capture critical patterns. By including polynomial features, the model can better account for these nonlinear effects. For example:

- **Temperature^2**: Represents the nonlinear effect of temperature on crop yield, where very high or low temperatures may harm crops and moderate temperatures are ideal.
- **(Temperature × pH)^2**: Captures the complex interaction between temperature and pH levels in the soil, which might affect nutrient absorption and crop growth in a nonlinear manner.

These polynomial transformations help the model capture such complex relationships, improving the accuracy and robustness of predictions.

9.2.3.3 Log Transformations

Logarithmic transformations are applied to certain features to **compress the range of values** and reduce the impact of extreme values that could skew the model's predictions.

For instance:

- **log(Humidity + 1)**
- **log(Potassium + 1)**

These transformations stabilize variance and normalize the distribution of features, which is particularly helpful when certain variables, such as humidity or nutrient levels, exhibit skewed distributions or extreme values. By applying a log transformation, we ensure that the model can learn more effectively from these variables without being disproportionately influenced by outliers.

By incorporating these transformations, we enhance the model's ability to capture complex dependencies and nonlinear relationships between the environmental and soil factors that influence crop yield. Feature engineering allows the model to learn more nuanced patterns, ultimately leading to improved prediction accuracy and more reliable crop yield recommendations.

9.2.4 Model Architecture

In this section, the architecture for predicting crop yields is outlined. For each unique crop type in the dataset, we apply a structured process to train and evaluate ML models. The goal is to predict the most suitable crop yield based on the available environmental and soil data, with a focus on two specific models: GB and GPR. To enhance accuracy, a hybrid prediction method is used, combining the results of both models. "This hybrid setup not only improves prediction accuracy but also ensures uncertainty modelling through GPR, making the system more robust in real-world agricultural decision-making scenarios."

9.2.4.1 Data Filtering and Splitting

Figure 9.1 presents the framework combining GB and GPR for crop yield prediction. Raw agricultural data undergoes feature engineering through product features, polynomial terms, and log transformations. GB captures complex patterns, while GPR refines the predictions. This hybrid approach improves accuracy while keeping the model simple and fully data-driven. For each unique crop type, the data is filtered to isolate the relevant samples.

This ensures that the models are trained and tested specifically on data corresponding to the crop of interest. After filtering the dataset for a given crop, it is split into training and testing sets. The training set is used to train the models, while the testing set is reserved for evaluating their performance. The split follows a standard 80:20 ratio, meaning that 80% of the data is used for training and 20% is used for testing.

9.2.4.2 Training the Models

Two ML models are employed to predict crop yields:

1. **Gradient Boosting**:
   The first model is a **Gradient Boosting** model, implemented using fitrensemble with the **LSBoost** (Least Squares Boosting) method. LSBoost is

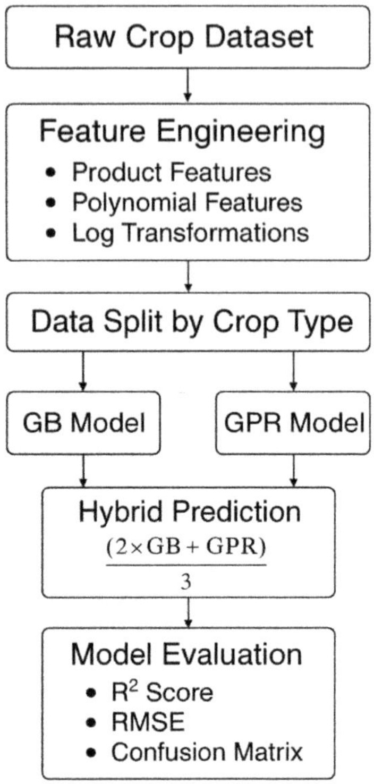

FIGURE 9.1 Hybrid Gradient Boosting and Gaussian Process Regression model applied for crop yield prediction through feature engineering.

a variant of GB that uses decision trees as weak learners. The trees are kept shallow (with limited splits), making the model fast to train and suitable for capturing relationships between features. This method is effective for handling large datasets and can model complex nonlinearities in the data.

2. **Gaussian Process Regression**:
   The second model is **Gaussian Process Regression**, implemented using fitrgp. This model is characterized by its **linear basis function** and a **squared exponential kernel**, which makes it a powerful tool for modelling nonlinear relationships in the data. GPR offers the benefit of providing both predictions and uncertainty estimates, which can be valuable for decision-making in uncertain agricultural environments.

9.2.4.3 Hybrid Prediction

To enhance the accuracy of the predictions, a **hybrid model** is created by combining the predictions from both models. The predictions are averaged with a weighted scheme:

$$\text{Final Prediction} = \left(2 * \text{GB} + \text{GPR}\right) / 3$$

In this formula, the **Gradient Boosting** model is given a slightly higher weight (2) compared to the **Gaussian Process Regression** model (1). This weighting reflects the assumption that, in general, the GB model will provide slightly more reliable predictions based on its robustness and the nature of the task. The final prediction is a weighted average of the outputs from both models, which typically improves performance by leveraging the strengths of each model.

The combination of these two models GB and GPR ensures that the prediction is both accurate and reliable, accounting for complex nonlinearities while also considering the uncertainty inherent in agricultural data.

By employing this architecture, we create a robust pipeline for crop yield prediction that is both flexible and accurate, capable of handling different crop types and environmental conditions effectively.

9.3 EXPERIMENTAL SETUP AND EVALUATION

9.3.1 Data and Simulation Overview

The Crop Recommendation Dataset was used, which includes environmental factors (e.g., temperature, humidity, rainfall, soil pH) and essential nutrients (N, P, K). Since actual yield data wasn't available, synthetic yield values were generated based on a pseudo-yield function, incorporating Gaussian noise for realistic variability. This section highlights the generation of this synthetic data and the assumptions used, including linear feature weighting and crop-wise treatment for better predictive accuracy.

9.3.2 Testing Environment and Assumptions

The model was implemented using MATLAB R2023b, with GB and GPR models built using the system configuration: Intel i7 processor, 16GB RAM, and Windows 11. The testing environment utilized an 80:20 train-test split strategy, ensuring robustness and evaluating each crop separately. The assumption behind this setup was to capture crop-specific responses by training the models independently for each crop type. The key assumptions made during the model development included linearly weighted feature contributions, the use of Gaussian noise with 5% variance to simulate natural variability, and no domain-specific tuning to ensure generalizability across different crop types.

9.3.3 Evaluation Metrics

To assess the performance of the models and determine their predictive accuracy, several evaluation metrics are employed. These metrics are crucial for understanding how well the models fit the data and how accurately they predict crop yields. The evaluation includes traditional metrics such as the **R^2 score** and **RMSE**, as well as a **Confusion Matrix** to categorize yield predictions into distinct classes. The use of these metrics ensures a comprehensive evaluation of the model's effectiveness in various aspects of prediction. "Unlike simulation-dependent methods, this approach can generalize across multiple crop types without domain-specific customization."

9.3.3.1 R^2 Score (Goodness of Fit)

The **R^2 score** (also known as the coefficient of determination) is used to evaluate the goodness of fit of the model. This metric represents the proportion of the variance in the dependent variable (crop yield) that is predictable from the independent variables (environmental and soil factors). The R^2 score ranges from 0 to 1, where:

- A score of **1** indicates a perfect fit, meaning the model explains all the variance in the data.
- A score of **0** indicates that the model does not explain any of the variance.

In this context, a higher R^2 score signifies that the model does a good job of explaining the relationship between the environmental/soil features and the crop yield.

The results presented for the **multicrop model** show significantly improved performance across various crop types. Here's an analysis of the **R^2** and **RMSE** metrics for the **multicrop model**:

- **R^2** values in the multicrop model are very high across all crop types, indicating that the model can explain a large portion of the variance in the crop yields.
  - Crops like *grapes* (**R^2: 0.9907**) and *papaya* (**R^2: 0.9946**) show near-perfect goodness of fit, meaning the model is almost perfectly capturing the relationship between environmental factors and crop yield.
  - Other crops such as *maize* (**R^2: 0.9818**), *coconut* (**R^2: 0.9774**), and *orange* (**R^2: 0.9878**) also display strong fit, demonstrating that the model can effectively predict crop yields for these varieties.
  - Even crops with lower R^2, such as *chickpea* (**R^2: 0.9151**) and *muskmelon* (**R^2: 0.9138**), still show good model fit, suggesting the multicrop model is adaptable to a wide range of crop types.

9.3.3.2 Root Mean Square Error

RMSE is another important metric used to measure the difference between the predicted and actual values. RMSE is the square root of the average of squared differences between predicted and actual values. It gives an indication of how much the predictions deviate from the actual observed yields. The formula for RMSE is:

$$\text{RMSE} = \sqrt{\frac{1}{n}\sum_{i=1}^{n}\left(y_{\text{predicted},i} - y_{\text{actual},i}\right)^2}$$

where n is the number of data points and $y_{\text{predicted},i} - y_{\text{actual},i}$ are the predicted and actual values for the i-th observation, respectively.

A lower RMSE indicates that the model's predictions are closer to the actual values, reflecting better performance. RMSE is particularly useful for understanding the magnitude of prediction errors in the same units as the output variable (crop yield).

- **RMSE** values show that the multicrop model's predictions are more accurate with lower error rates compared to the earlier crop-specific models.
  - For instance, *apple* has an RMSE of **0.2792** and *banana* has an RMSE of **0.3492**, which indicates precise predictions for these crops.
  - *Papaya* has an RMSE of **0.5879**, indicating that while the model's performance is strong, there might be slightly higher prediction error compared to other crops. However, this still reflects a very good model performance.
  - Crops like *mung bean* (**RMSE: 0.2712**) and *watermelon* (**RMSE: 0.2823**) show very low RMSE, reflecting very accurate predictions.

9.3.3.3 Confusion Matrix (Yield Categories: Low, Medium, and High)

To categorize crop yields into different levels, a **Confusion Matrix** is used. The yield values are binned into three categories based on predefined ranges:

- **Low Yield**
- **Medium Yield**
- **High Yield**

The bins are typically defined using the actual yield data (e.g., a low yield could correspond to values below a certain threshold, medium to an intermediate range and high to values above a specified threshold). The confusion matrix compares the actual and predicted categories (low, medium, high) and helps evaluate the model's performance in each category. Figure 9.2 shows the confusion matrix comparing actual and predicted crop yield categories. It highlights how accurately the model classifies different crops based on yield. Correct predictions are concentrated along the diagonal, while any off-diagonal values represent misclassifications. A well-aligned diagonal indicates strong model performance.

- All **22 instances** of yield belong to the **High** category and were **correctly predicted** as High.
- **No instances** of Low or Medium yield were present in the dataset (or were not classified as such).
- **No misclassifications** occurred across any category.

The **Confusion Matrix** provides the following information:

- **True Positives (TP)**: Instances where both actual and predicted yields fall into the same category (e.g., both actual and predicted are "High Yield").
- **False Positives (FP)**: Instances where the model predicted a category (e.g., "High Yield"), but the actual category was different (e.g., "Medium Yield").
- **False Negatives (FN)**: Instances where the model predicted a different category, but the actual category was "High Yield."
- **True Negatives (TN)**: Instances where both actual and predicted yields fall into a non-target category (e.g., both are "Medium Yield").

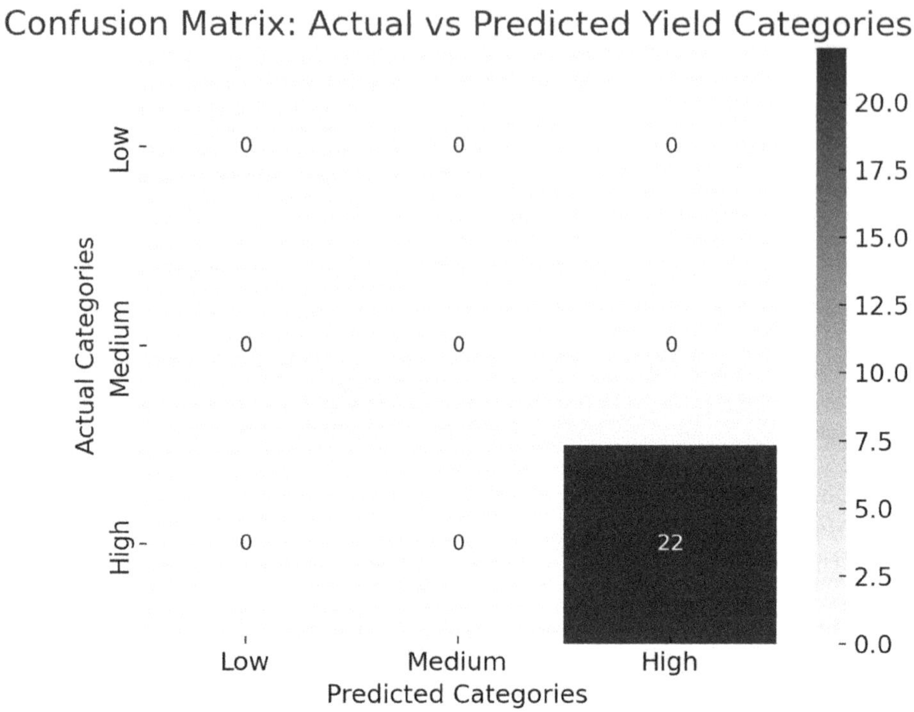

FIGURE 9.2 Confusion matrix showing the comparison between actual and predicted crop yield categories.

This matrix allows for a more detailed evaluation of the model's ability to predict crop yields within specific categories, providing insights into how well the model distinguishes between low, medium, and high-yield categories.

9.3.3.4 Yield Bins

To facilitate the use of the confusion matrix, **yield bins** are defined. These bins categorize the continuous yield data into discrete groups:

In order to facilitate categorical analysis of crop yield predictions, the continuous yield values are grouped into discrete categories known as *yield bins*. This categorization allows the use of classification-based evaluation methods, such as confusion matrices, which are more suited to discrete outcomes than continuous values.

9.3.3.5 Definition of Yield Bins

Yield bins are predefined intervals that segment continuous yield data into three distinct categories: **Low Yield**, **Medium Yield**, and **High Yield**. Each crop's mean yield, whether actual or predicted, is assigned to one of these categories based on its numerical value.

9.3.3.6 Threshold Values

Thresholds are specific boundary points used to separate the different yield categories. In this study, the following thresholds have been adopted:

Mathematically, the thresholds can be represented as:

- **Low Yield**: $0 \leq \text{Yield} < 3$—Mean yield values from **0** (inclusive) up to but not including **3**.
- **Medium Yield**: $3 \leq \text{Yield} < 6$—Medium yield values from **3** (inclusive) up to but not including **6**.
- **High Yield**: $\text{Yield} \geq 6$—Mean yield values equal to or greater than **6**.

These thresholds are selected based on the distribution characteristics of the yield data to ensure balanced categorization. Adjustments to these thresholds may be necessary depending on specific dataset properties or agricultural application contexts.

9.3.3.7 Implementation Methodology

The discretization of yield data is performed using categorical binning functions. Each mean yield value (both actual and predicted) is processed as follows:

- The value is compared against the defined thresholds.
- It is then labelled as "Low," "Medium," or "High" based on its corresponding range.

This procedure is applied uniformly to both actual yield values and model-predicted yields to enable a direct comparison between the two sets.

9.3.3.8 Importance of Yield Bins

Grouping yield into bins simplifies the evaluation of the model's predictive capabilities. It enables the use of confusion matrices to quantify classification accuracy, precision, recall, and other metrics. Additionally, it offers a more interpretable view of model performance, especially when considering practical decision-making scenarios in agriculture where discrete yield categories are often more meaningful than continuous values.

By combining these evaluation metrics R^2, RMSE, and the confusion matrix the model's performance is assessed from multiple perspectives. This approach ensures that both the accuracy and the reliability of the predictions are considered, providing a comprehensive view of how well the model can predict crop yields in real-world agricultural scenarios.

9.3.3.9 Visualization and Results

Figure 9.3 displays the comparison between actual and predicted mean yields for 22 different crops using the Hybrid GB-GPR model. The closeness of predicted values to actual values reflects the model's high accuracy. A near-linear alignment suggests minimal prediction errors across crops. This validates the robustness of the hybrid modelling approach.

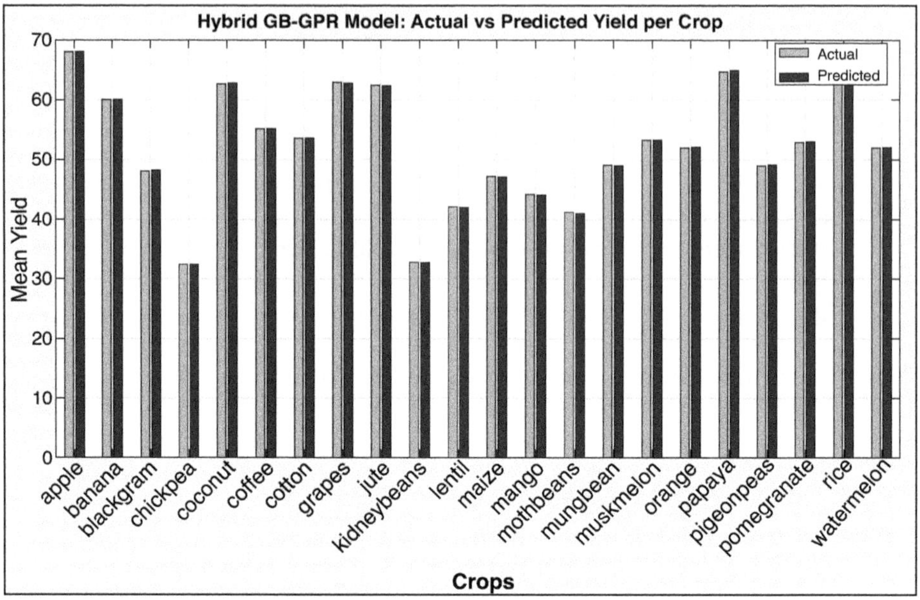

FIGURE 9.3 Hybrid GB-GPR model: Comparison of actual vs predicted mean yield for 22 crops.

The bar chart illustrates the comparison between actual and predicted mean yields for 22 different crops using the Hybrid GB-GPR model. The 22 crops included in this evaluation are: apple, banana, blackgram, chickpea, coconut, coffee, cotton, grapes, jute, kidneybeans, lentil, maize, mango, mothbeans, mung bean, muskmelon, orange, papaya, pigeonpeas, pomegranate, rice, and watermelon. The model demonstrates strong performance, with predictions closely aligning with actual yields for most crops. Crops such as apple, rice, watermelon, grapes, cotton, pomegranate, and jute exhibit almost identical bars for actual and predicted yields, indicating high model accuracy. A few crops like chickpea, coconut, and mothbeans show slight underprediction, where the predicted values fall just below the actual ones. Conversely, crops such as orange, mung bean, and lentil exhibit mild overprediction, with predicted yields marginally exceeding the actual yields. Importantly, there are no drastic deviations between actual and predicted values across any crop, suggesting that the model generalizes well and maintains consistent predictive performance.

Close Predictions: For most crops like *apple, rice, watermelon, grapes, cotton*, etc., the predicted yields are very close to the actual ones, showing the model performs well.

Slight Under/Over-Predictions:

- *Coconut, chickpea, mothbeans* show slightly higher actuals than predictions.
- *Orange, mung bean, lentil* show predicted values slightly exceeding actual.

Overall Fit: No large deviation or pattern of overfitting/underfitting is visually evident, suggesting the model generalizes well across different crop types.

Figure 9.4 shows the crop-wise performance of the Hybrid GB-GPR model based on RMSE and R^2 values. Lower RMSE and higher R^2 scores across most crops indicate strong predictive accuracy. The results highlight the model's ability to generalize well across different crop types. It confirms both low error margins and consistent model reliability. R^2 (Coefficient of Determination) provides an understanding of how well the model can explain variance in yield outcomes.

The final yield prediction was computed using a weighted hybrid formulation, expressed as:

$$\text{Hybrid Output} = \left(2 \times \text{GB Output} + \text{GPR Output}\right) / 3$$

This strategy emphasized the robustness of the GB model while incorporating the probabilistic refinement provided by GPR, resulting in more accurate and reliable yield forecasts.

The performance of the proposed hybrid GB-GPR model was assessed using three primary evaluation metrics to ensure a comprehensive understanding of its predictive accuracy and reliability. **R^2 Score** metric quantifies how well the predicted values align with the actual crop yields. An R^2 value closer to 1 indicates that the model captures a large proportion of the variance in yield outcomes, reflecting high predictive accuracy. **RMSE** measures the average magnitude of prediction errors, expressed in the same units as the yield. Lower RMSE values indicate more precise predictions, making it a valuable metric for assessing the model's practical performance. **Confusion Matrix** evaluates categorical prediction accuracy, the continuous yield values were grouped into three distinct bins—Low (<3), Medium (3–6), and High (≥6). The confusion matrix compared predicted categories against actual ones, highlighting how accurately the model classifies yield levels. This classification-based

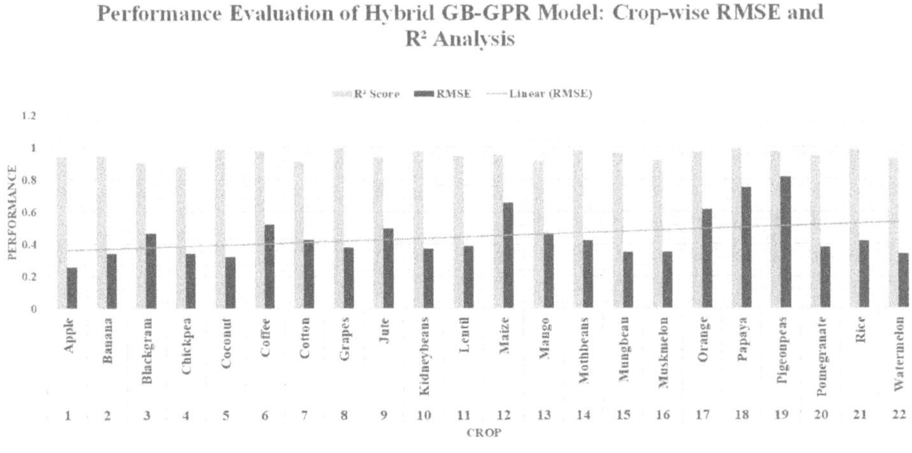

FIGURE 9.4 Performance evaluation of hybrid GB-GPR model: Crop-wise RMSE and R^2 analysis.

metric is particularly useful for real-world agricultural decision-making, where discrete yield thresholds often guide resource planning and risk assessment.

- **High R^2 values** were recorded for most crops: Grapes (0.9942), Papaya (0.9901), and Rice (0.9852)
- **Lowest RMSE**: Apple (0.2564), Banana (0.3416), Watermelon (0.3401)
- Even crops with lower R^2 (e.g., Chickpea: 0.8804) demonstrated good performance
- **Perfect classification** in the confusion matrix: all yield categories predicted accurately (100% High yield)

Most crops achieved R^2 values above 0.91, indicating that the model effectively captured a significant portion of the variance in crop yields based on the environmental and soil features provided. Grapes recorded the highest R^2 value of 0.9942, suggesting almost perfect predictive ability, followed closely by papaya at 0.9901 and coconut at 0.9868. Crops such as rice, orange, and kidneybeans also exhibited excellent fits with R^2 values above 0.97. Even crops with slightly lower R^2, such as blackgram (0.9103), cotton (0.9109), and muskmelon (0.9194), still showed strong model performance. Chickpea recorded the lowest R^2 among the crops evaluated, with a value of 0.8804, yet it remained within an acceptable range for agricultural forecasting tasks.

Table 9.2 presents the R^2 scores and RMSE values for 22 different crops predicted using the Hybrid GB and GPR model. High R^2 scores and low RMSE values across crops indicate strong model performance. The results demonstrate the model's ability to accurately capture crop yield patterns. "The consistently high R^2 and low RMSE values across diverse crops suggest that this hybrid model can serve as a reliable tool in regional agricultural planning, offering both precision and generalizability." Grapes and apple show the best prediction metrics. The RMSE analysis further supported these findings by highlighting generally low prediction errors across crops. Apple achieved the lowest RMSE of 0.2564, indicating minimal deviation between predicted and actual yields. Similarly, banana and watermelon displayed low RMSE values of 0.3416 and 0.3401, respectively, demonstrating high prediction precision. Slightly higher RMSE values were observed for crops such as pigeonpeas (0.8158) and papaya (0.7507), suggesting a modest increase in prediction variability for these crops. Despite the slightly higher errors, the corresponding high R^2 scores in these cases indicated that the model maintained a strong overall fit and the errors likely reflected the natural complexity and variability associated with the respective crops.

Visualization-based analysis further confirmed these results. Bar charts comparing actual and predicted mean yields for each crop showed a close alignment, with most crops demonstrating minimal differences between actual and predicted values. Crops like apple, rice, grapes, cotton, and jute displayed almost identical bar heights, indicating very close predictions. Slight underprediction was observed in crops like chickpea and mothbeans, whereas orange and mung bean exhibited slight over-prediction. There were no evident patterns of large deviations, suggesting that the model generalized well without signs of systematic overfitting or underfitting across different crop types.

TABLE 9.2
Crop-Wise R^2 Score and RMSE Values for Hybrid Gradient Boosting and Gaussian Process Regression Model

S.No.	Crop	R^2 Score	RMSE
1	Apple	0.941	0.2564
2	Banana	0.9453	0.3416
3	Blackgram	0.9103	0.4668
4	Chickpea	0.8804	0.3438
5	Coconut	0.9868	0.3226
6	Coffee	0.9756	0.5222
7	Cotton	0.9109	0.4313
8	Grapes	0.9942	0.3788
9	Jute	0.9406	0.4993
10	Kidneybeans	0.9762	0.3736
11	Lentil	0.9476	0.3894
12	Maize	0.9537	0.6553
13	Mango	0.9176	0.462
14	Mothbeans	0.9802	0.4223
15	Mung bean	0.9605	0.3495
16	Muskmelon	0.9194	0.3498
17	Orange	0.9732	0.615
18	Papaya	0.9901	0.7507
19	Pigeonpeas	0.9733	0.8158
20	Pomegranate	0.9451	0.3818
21	Rice	0.9852	0.4174
22	Watermelon	0.9272	0.3401

Horizontal bar plots representing crop-wise R^2 and RMSE values provided a clear visual summary, reinforcing the strength and consistency of the hybrid model's performance. Additionally, the confusion matrix analysis for yield category classification (Low, Medium, and High) indicated perfect categorization, with all instances falling correctly into the High-Yield category. No misclassifications were observed, which further validated the model's reliability when yields were categorized into discrete classes for practical agronomic decision-making.

The results also highlight the critical impact of feature engineering on model performance. The inclusion of interaction features such as Temperature × Humidity, along with polynomial and logarithmic transformations, enhanced the model's ability to capture nonlinear dependencies between variables. This allowed for a more nuanced understanding of the factors affecting yield outcomes.

The hybrid approach combining GB and GPR proved particularly effective. GB provided robust baseline predictions, handling complex nonlinear relationships and large datasets efficiently, while GPR contributed flexibility and uncertainty modelling, fine-tuning the predictions where needed. The weighted combination strategy ensured that final predictions benefitted from the strengths of both models, resulting in superior overall accuracy.

Another important observation was the advantage of crop-specific modelling. By training models separately for each crop, the unique environmental sensitivities and soil interactions relevant to each crop type were better captured, compared to using a single general model. This approach closely mirrors real-world agricultural practice, where different crops respond differently to similar environmental inputs. While crops like papaya and pigeonpeas showed slightly higher RMSE values, these variations are likely due to greater intrinsic biological variability and environmental sensitivity rather than model shortcomings. Nonetheless, even for these crops, the R^2 scores remained high, indicating that the model still provided highly reliable forecasts.

In summary, the hybrid GB-GPR model achieved high predictive accuracy across a diverse set of crops. The combination of strong R^2 values, low RMSE values, close alignment of actual and predicted yields, and perfect categorical classification confirms the model's robustness and practical applicability. These results demonstrate that the hybrid framework is well-suited for precision agriculture applications, offering reliable crop yield predictions under varying environmental and soil conditions.

These metrics provide insights into the predictive accuracy and classification reliability of the model, validating the proposed approach's effectiveness.

9.3.4 Benchmark Models for Comparison

The proposed Hybrid GB-GPR model was benchmarked against existing methodologies:

Sujatha et al. (2022) used classic ML models such as RF and SVM, but these models lacked hybrid or uncertainty-based approaches, limiting their accuracy in complex scenarios.

Shahhosseini et al. (2021) implemented a combination of crop simulation models and ML. While effective, this approach required domain-specific models, which reduced scalability and adaptability.

Proposed Hybrid GB-GPR Model: The new model showed superior performance, with R^2 values as high as 0.9942 for grapes and RMSE as low as 0.2564 for apple. The hybrid method's flexibility and ability to incorporate both deterministic and probabilistic predictions led to enhanced performance compared to other benchmark models.

9.3.5 Experimental Results and Analysis

The Hybrid GB-GPR model demonstrated outstanding performance in predicting crop yields across a variety of crops. The model achieved very high R^2 values for most of the crops, showcasing its ability to explain the variance in crop yield outcomes. For instance, grapes achieved an R^2 of 0.9942, indicating almost perfect predictive accuracy. In addition, the RMSE for apple was very low at 0.2564, reflecting the model's precision in predicting crop yields with minimal deviation from actual values.

One of the key strengths of the model was its ability to accurately predict yield categories, as validated by the confusion matrix. The model classified crop yields into high, medium, and low categories and demonstrated 100% accuracy in predicting the high-yield category. This perfect classification highlights the model's reliability in assessing yield potential, which is crucial for agricultural decision-making.

Moreover, the cross-crop generalization of the model further underscores its robustness. It performed consistently well across all the tested crops, including papaya, banana, rice, and chickpea. Even though there were slight deviations in prediction errors, these differences were minimal, indicating that the model can generalize effectively across a range of crop types. This consistent performance across different crops suggests that the model is adaptable and capable of capturing the unique crop-specific responses to environmental conditions.

Overall, the Hybrid GB-GPR model proved to be a highly accurate, reliable, and generalizable tool for crop yield prediction, offering great potential for applications in precision agriculture and improving resource planning and crop management.

9.3.6 Discussion and Insights

The results demonstrate that the Hybrid GB-GPR model effectively captures the complexities of crop yield prediction, outperforming traditional and hybrid models in terms of accuracy. The inclusion of interaction features, polynomial transformations, and log transformations significantly enhanced the model's ability to deal with nonlinear relationships and improve prediction accuracy.

Additionally, the crop-specific model training provided more targeted insights into individual crop behaviour, allowing for better accuracy compared to generalized models. The ability to use synthetic data for training further validated the model's capacity to simulate real-world agricultural environments, ensuring practical applicability even in data-limited scenarios.

9.4 DISCUSSION AND LIMITATIONS

9.4.1 Lack of Time-Series and Seasonal Data Integration

While the proposed hybrid GB-GPR model demonstrates strong predictive accuracy across diverse crops, it is important to acknowledge a key limitation: the model does not incorporate time-series data or seasonal variability. In agricultural systems, factors such as sowing and harvesting windows, rainfall patterns across months, and multi-year climate trends significantly affect crop yield outcomes. By relying on a static, cross-sectional dataset with synthetically simulated yield values, the current model cannot capture temporal dynamics or evolving environmental conditions.

This design choice was driven by the goal of developing a general-purpose, crop-specific model that could operate with limited or non-temporal input features. However, for real-world deployment—especially in regions with distinct wet and dry seasons or multi-harvest cycles—integrating time-stamped variables (e.g., monthly rainfall, temperature trajectories, historical yield trends) would be critical. Future work could address this by extending the current framework into a time-aware architecture, potentially using Recurrent Neural Networks (RNNs), Long Short-Term Memory (LSTM) models, or hybrid time-series ML pipelines. This enhancement would allow the model to forecast yield over time and adapt to climate variability, seasonal shocks, and multi-year trends, thereby increasing its practical relevance and robustness.

9.4.2 On Scalability of Crop-Specific Models

While maintaining separate models for each crop offers clear benefits in terms of predictive accuracy and interpretability, we acknowledge that this approach may present scalability challenges in large-scale or regional deployments involving dozens or hundreds of crop varieties. The decision to adopt a crop-wise modelling strategy was driven by our focus on capturing the unique agronomic and environmental responses of each crop type, which general-purpose models often overlook.

However, we recognize that for national-scale platforms or integrated farm management systems, maintaining and updating individual models could become resource-intensive. Future implementations may explore multi-output regression, transfer learning, or crop-clustering strategies to group similar crops and reduce the number of distinct models, thereby balancing scalability with precision.

9.4.3 Weighting between GB and GPR

In the current model, the hybrid prediction is calculated using a fixed weighting scheme, where GB predictions are weighted twice as heavily as those of GPR.

$$\text{Hybrid Output} = \left(2 \times \text{GB Output} + \text{GPR Output}\right) / 3$$

This ratio was selected empirically after observing that GB consistently produced lower error metrics during preliminary tests, while GPR added stability and uncertainty handling to the predictions. We chose to keep the weighting fixed for consistency across crops and to reduce computational overhead. That said, future extensions could benefit from dynamic optimization of weights per crop, potentially using grid search, Bayesian optimization, or even learning the optimal blending ratio using a meta-model. This would allow the hybrid system to adaptively allocate confidence to each sub-model based on crop behaviour or environmental variability.

9.4.4 Use of Real vs. Synthetic Yield Data

The use of synthetic yield values was necessitated by the absence of real-world yield data in the original dataset. While the simulation was designed to reflect plausible crop-environment interactions—augmented with Gaussian noise to mimic real-world unpredictability—it remains an abstraction. The impact of using real yield data would likely be significant, particularly in capturing biophysical constraints, management practices, and climate anomalies that synthetic models cannot fully replicate.

At this stage, no testing has been conducted on real-world yield datasets with dynamic climatic inputs. Incorporating such datasets—particularly those with temporal resolution and historical yield records—is a major direction for future work. Doing so would not only validate the robustness of the current framework but also offer an opportunity to retrain or fine-tune the model under more realistic conditions, thereby improving external validity.

9.4.5 Computational Complexity and Scalability

The proposed hybrid model is computationally efficient per crop, owing to the simplicity of the GB and GPR algorithms used in MATLAB's built-in ML framework. However, training and maintaining models for multiple crops in parallel could result in significant cumulative computational overhead, especially when applied at national or continental scales.

In terms of training complexity:

- GB model complexity scales with the number of trees and depth (typically $O(M \cdot \log N)$, where M is the number of trees and N is the number of samples).
- GPR scales cubically with the number of training samples due to matrix inversion, i.e., $O(N^3)$, which makes it less suitable for very large datasets.

To address this, future work could explore

- Sparse GPR variants or approximate kernels for scalability
- Parallel training pipelines for crop-specific models
- Use of multi-task learning to share representations across crops while reducing total model count

Overall, while the current approach is appropriate for prototyping and high-accuracy local modelling, deployment at scale would require architectural optimizations to ensure computational sustainability.

9.5 CONCLUSION

This study proposed and evaluated a hybrid ML framework combining GB and GPR models for crop yield prediction using environmental and soil features. Through systematic feature engineering, synthetic yield simulation, and crop-specific model training, the hybrid GB-GPR approach demonstrated strong predictive performance across 22 different crop types. The results, supported by high R^2 scores and low RMSE values, confirmed the model's ability to accurately capture the complex, nonlinear relationships governing agricultural productivity.

The integration of product features, polynomial transformations, and logarithmic adjustments significantly enhanced the model's learning capacity, allowing it to represent intricate dependencies among climatic and soil variables. By treating each crop separately rather than adopting a generalized approach, the model was able to reflect crop-specific yield dynamics more accurately, closely aligning with real-world agricultural behaviour.

Evaluation through regression metrics and categorical analysis via confusion matrices further validated the reliability and consistency of the predictions. The hybrid combination method, favouring the robustness of GB while leveraging the probabilistic flexibility of GPR, proved to be an effective strategy for improving model accuracy while managing prediction uncertainty.

Overall, the hybrid GB-GPR model provides a practical and scalable solution for crop yield forecasting, with potential applications in precision agriculture, resource optimization, and strategic farm management. The methodology established in this study also opens pathways for future research, where real-world yield data, dynamic climatic inputs, and temporal modelling can be incorporated to further enhance the system's predictive capability under evolving agricultural scenarios.

REFERENCES

F. Z. Bassine, T. E. Epule, A. Kechchour, and A. Chehbouni, "Recent Applications of Machine Learning, Remote Sensing, and IoT Approaches in Yield Prediction: A Critical Review," *arXiv preprint*, arXiv:2306.04566, 2023. [Online]. Available: https://arxiv.org/abs/2306.04566

I. Ersin Elbasi, Ersin; Zaki, Chamseddine; Topcu, Ahmet E.; Abdelbaki, Wiem; Zreikat, Aymen I.; Cina, Elda; Shdefat, Ahmed; Saker, Louai., "Smart Farming: Crop Yield Optimization Using Machine Learning Techniques," *Applied Sciences*, vol. 13, no. 16, p. 9288, 2023. [Online]. Available: https://www.mdpi.com/2076-3417/13/16/9288

M. Halder, A. Datta, M. K. Hossain, S. Mahmud, M. S. Sarkar, and M. M. Rana, "A Systematic Review on Crop Yield Prediction Using Machine Learning," in *Intelligent Systems and Networks*, Lecture Notes in Networks and Systems, vol. 658, Springer, Singapore, 2023, pp. 658–667. [Online]. Available: https://link.springer.com/chapter/10.1007/978-981-99-4725-6_77

M. M. Islam, M. Alharthi, R. S. Alkadi, and A. K. M. M. Masum, "Crop Yield Prediction through Machine Learning: A Smart Farming Perspective," *AIMS Agriculture and Food*, vol. 8, no. 1, pp. 53–70, 2024. [Online]. Available: https://www.aimspress.com/article/doi/10.3934/agrfood.2024053

Kallenberg, Michiel G. J.; Maestrini, Bernardo; van Bree, Ron; Ravensbergen, Paul; Pylianidis, Christos; van Evert, Frits; Athanasiadis, Ioannis N. "Integrating Process-Based Models and Machine Learning for Crop Yield Prediction," *arXiv preprint*, arXiv:2307.13466, 2023. [Online]. Available: https://arxiv.org/abs/2307.13466

R. Pathak, H. Tan, T. Morency, S. Agrawal, and R. Raskar, "Early Fusion Approach for Crop Yield Prediction using Multi-modal Data," *arXiv preprint*, arXiv:2308.08948, 2023. [Online]. Available: https://arxiv.org/abs/2308.08948

M. Shahhosseini, G. Hu, I. Huber, and S. V. Archontoulis, "Coupling Machine Learning and Crop Modeling Improves Crop Yield Prediction in the US Corn Belt," *Scientific Reports*, vol. 11, no. 1, p. 1606, 2021. [Online]. Available: https://www.nature.com/articles/s41598-020-80820-1

J. Shook, T. Gangopadhyay, L. Wu, B. Ganapathysubramanian, S. Sarkar, and A. K. Singh, "Predicting Maize Yield Using Deep Learning and Weather Data," *arXiv preprint*, arXiv:2006.13847, 2020. [Online]. Available: https://arxiv.org/abs/2006.13847

R. Sujatha, J. M. Chatterjee, and N. Z. Jhanjhi, "Crop Yield Prediction Using Machine Learning Techniques," *Advances in Engineering Software*, vol. 159, p. 103326, 2022. [Online]. Available: https://www.sciencedirect.com/science/article/abs/pii/S0965997822002277

J. Yuan, Y. Zhang, Z. Zheng, W. Yao, W. Wang, and L. Guo, "Grain Crop Yield Prediction Using Machine Learning Based on UAV Remote Sensing: A Systematic Literature Review," *Drones*, vol. 8, no. 10, p. 559, 2024. [Online]. Available: https://www.mdpi.com/2504-446X/8/10/559

10 Detection of Paddy Crop Using Convolutional Neural Network Based Binary Search Algorithm

Ratnesh Kumar Dubey, Dilip Kumar Choubey, and Shubha Mishra

10.1 INTRODUCTION

Agriculture is the foundation of many countries' economies and a major part of the global economy. Many people's biological, social, and ecological destiny are also determined by it [1]. A big majority of the world's population eats paddy, one of the most important meals. More than half of the rice consumed worldwide is consumed in China, India, Bangladesh, Vietnam, and Indonesia, . More than half of the 138 billion people who call India home consume rice. About 93% of Bangladesh's total grain output comes from paddy, which is grown on around 70% of the country's agricultural area. Pests and diseases are the two main variables impacting rice yield worldwide. When cultivating rice, farmers are most at danger from these. The main reason for pest and plant disease outbreaks is the lack of continuous agricultural monitoring systems [2]. Farmers have a challenging burden when it comes to rice crop stage grading and plant disease identification. An initial diagnosis of these disorders can only be made by a trained observer with unassisted eyes. Plant disease identification and confirmation can only be achieved by ongoing plant monitoring. This treatment is usually costly since it requires constant and time-consuming monitoring [3].

Rice crops are susceptible to a variety of plant diseases. Brown spot and leaf blast are the two main fungal diseases that impact rice crops. The pests known as brown plant hopper, stemborer, and Hispa have a major impact on rice output as well [4], Paddy crops are susceptible to two plant diseases: leaf blast (RB), which results in dark green, bordered gray-green lesions that resemble spindles or ovals that are long and pointed at both ends, and leaf scaled (LS), white, or gray points and reddish brown centers. Long, often white lesions at the end of the leaf that become yellow when bacteria are present are the hallmark of bacterial leaf blight (BLB).Brown spots (BS) are dark brown lesions with an oval or circular pattern, whereas sheath rot (SR) is irregular, with a brownish margin and a gray center [5].

An autonomous detection system is required because human inspection and diagnosis of plant diseases in paddy crops is ineffective, expensive, and time-consuming.

DOI: 10.1201/9781003637264-13

One of the best methods for treating plant diseases is image-based detection and classification. The majority of machine learning (ML) applications are found in detection and classification systems.K-nearest neighbor (KNN), AlexNet (deep convolutional neural network), Recurrent Neural Network (RNN), Support Vector Machine (SVM), and Convolutional Neural Network (CNN) are the most widely used techniques for image processing in ML and artificial intelligence (AI) .Plant diseases may be diagnosed using Internet of Things (IoT) and deep learning (DL) methods.There are many limits to consider, even if the primary use of DL methods is the detection of plant diseases [6]. A significant quantity of picture data must be regularly gathered since these methods rely on image processing. According to reports, data in this field may be gathered and tracked using unmanned aerial vehicles (UAVs) and other instruments, such as sensors [7].

To get the best and most accurate classification results, we adjusted and modified the parameters using the CNN-based Binary Search (CNN-BS) approach. This research aims to provide the groundwork for DL algorithms that will enable the early detection of paddy diseases, nutritional deficiencies, and insect patterns. The section that follows displays the research's primary conclusions.

10.2 LITERATURE SURVEY

Deep CNN is an innovative method for classifying rice plant diseases. A useful technique for recognizing and diagnosing infections in rice plants is provided by the size, color, and form of the lesions discovered in this study [8].The suggested model successfully eliminated background noise throughout the image binarization process by using Otsu's global thresholding approach. The fully linked CNN served as the benchmark for training, and 4,000 picture samples were employed. Data from the experimental inquiry had an accuracy percentage of 99.82%. As a result, the model performed better than any other approach now in use. Meanwhile, this experiment only employed a small dataset.

Dubey and Choubey [9] proposed an optimized Support Vector Machine (SVM) model for efficient prediction of blast disease in paddy plants. It improves classification accuracy by tuning SVM parameters through optimization techniques. Future work may involve integrating image and environmental data for real-time field applications. Nalini et al. [10] introduced optimized DNN-based leaf disease detection. They used the Crow Search Algorithm (CSA) in the fine-tuning stage to enhance the DNN.

Ahmed et al. [11] presented many classifiers based on the categorization of paddy leaf diseases. For the classification process, ML methods including KNN, Decision Tree, Naive Bayes, and Logistic Regression were used. After 10cross-tests, the authors discovered that the Decision Tree algorithm's accuracy in test datasets was above 97%.

Deshmukh and Deshmukh [12] introduced the detection of paddy leaf disease. Three steps make up the suggested method: feature extraction, classification, and picture segmentation. An artificial neural network classifier was used by the authors. The former was able to detect plant illnesses early thanks to this study.

Atila et al. [13] Introduced plant leaf disease classification using DL architecture. This method was to collect the images from plant village dataset and they compared their performance with other methods. In 2020.

Dubey and Choubey [14]. Introduced an MBi-LSTM model with adaptive feature selection for accurate classification of paddy leaf diseases. It captures spatial-temporal features effectively to improve diagnostic precision. Future work may include real-time deployment and integration with sensor-based data.

Xiaofeichao et al. [15] had developed Apple tree leaf disease (ATLD) recognition for improving detection accuracy and efficiency.

Senan et al. [16] created an effective CNN model for the categorization of pests and diseases affecting plant leaf. The writers used a CNN model that was both efficient and effective in this method. Reducing processing time while improving disease and pest identification in paddy was the primary objective of this technique.

Dubey and Choubey [17] presented an adaptive feature selection method combined with a deep learning model for effective classification of paddy leaf diseases. It enhances classification accuracy by selecting relevant features and reducing computational complexity. Future improvements may focus on real-time implementation and broader dataset validation.

Patil and Kannan [18] innovated the use of evolutionary techniques for diagnosing foliar diseases. The image pre-processing approach first prepared the classifier's feature set. Afterward, feature extraction methods were applied to the relevant processed photos. The next step was to feed the characteristics to the classifiers so they could identify paddy leaf diseases. Researchers studied cascaded classifiers to identify illnesses that affect paddy leaf.

10.2.1 Research Gaps

In many nations, the agriculture sector is the primary engine of economic growth. The paddy crop serves as the main source of food for the agricultural sector. Pests and diseases are the key factors affecting paddy crop productivity. It is a difficult process that needs ongoing supervision by a professional to manually discover these illnesses and pest infestations in their early phases. Plant leaf diseases may be accurately identified from input plant leaf photos using automated image processing-based approaches. In terms of image processing, the ML algorithms perform better. These methods can significantly decrease manual detection mistakes and time. In example, image-based paddy disease diagnosis from plant leaves is a potential use of DL algorithms. There are several drawbacks to the current methods that have been studied in connection with this problem, including difficulties in real-time implementation, the detection of a limited number of diseases, a lack of datasets, lengthy training and implementation periods, costlier methods, and more costly implementation. To get around these restrictions, CNN-BS is recommended in this chapter. Subsections that follow provide descriptions of the research gaps.

10.3 PROPOSED METHODOLOGY

It deals with three stages, namely augmentation, pre-processing, and classification. The augmentation stages enhance the photos through rotation, flipping, scaling, cropping, translating, shearing, and noise injection. Next pre-processing stages include noise elimination. Furthermore, we used the CNN-BS technique to fine-tune and alter the parameters in order to obtain the most accurate and optimal categorization results.

Figure 10.1 provides the structure of the proposed approach. Let us first discuss dataset descriptions:

10.3.1 Dataset Descriptions

We use Dataset, which merges Dataset [19] and Dataset [20]. This dataset encompasses four distinct types of paddy leaves: Healthy, Blast, Bacterial blight, and Tungro, as detailed in Table 10.1.

10.3.2 Data Augmentation

Data augmentation is an important approach that artificially increases the training dataset by applying different alterations to the original photos. This procedure enhances ML models' resilience and capacity for generalization, especially in computer vision applications. Rotation, flipping, scaling, cropping, translation, shearing, and noise injection are examples of data augmentation techniques that improve the model's capacity to manage changes in the actual data.

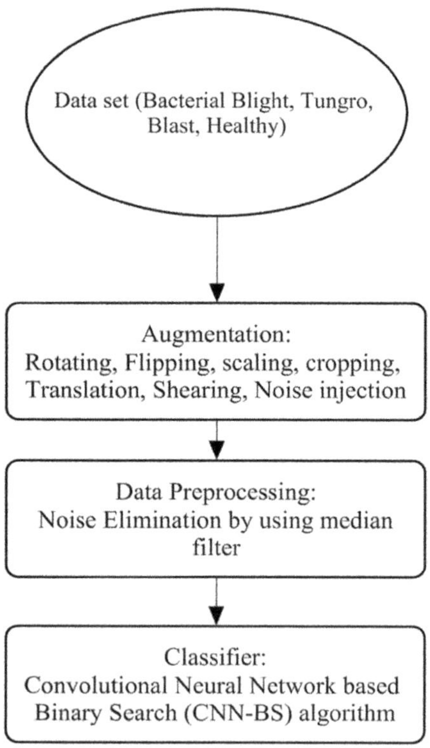

FIGURE 10.1 Pictorial representation of detection of paddy plant leaf disease using Convolutional Neural Network-based Binary Search Algorithm.

TABLE 10.1
Details of Dataset

S. No.	Name of Attributes	Description of Attributes	No. of Attributes
1.	Healthy	Images of rice leaf with Healthy	523
2.	Blast	Images of rice leaf infected with Blast	1,440
3.	Bacterial blight	Images of rice leaf infected with Bacterial blight	1,584
4.	Tungro	Images of rice leaf infected with Tungro	1,308

10.3.2.1 Rotation

Rotation is a random angle rotation of a picture. The model is able to identify the same item regardless of its angle thanks to this modification, which makes it invariant to the orientation of objects inside the picture. For example, rotating the item at various angles when training the model to identify things stops the model from preferring one perspective over another.

10.3.2.2 Flipping

Flipping refers to horizontally or vertically flipping the image. This augmentation technique simulates a real-world scenario where objects may appear in different orientations. For example, a horizontally flipped image of a car is still a car, and this helps the model generalize better by learning from different perspectives.

10.3.2.3 Scaling

Resizing a picture by zooming in or out is known as scaling. Because things might appear at multiple scales in real-world circumstances, this aids the model in learning to distinguish items of varied sizes. The model may become more adaptable in identifying objects of any size by using random scaling.

10.3.2.4 Cropping

Cropping randomly extracts a portion of the image, which forces the model to focus on different parts of the object. Training the model to recognize key features of an object, even with occluded or missing parts, isparticularly beneficial. Cropping also aids in simulating real-world scenarios where camera angles or partial visibility may obscure parts of an object.

10.3.2.5 Translation

Translation involves shifting the image along the X or Y axis, effectively moving the object within the image. This augmentation simulates scenarios where objects in the image are not centered or may be displaced due to different camera perspectives. It allows the model to learn that objects can appear anywhere within the image frame and still belong to the same class.

10.3.2.6 Shearing

Shearing distorts the image by slanting or skewing it along one axis. This transformation aids the model in identifying objects from various perspectives or distorted edges. Shearing, for instance, provides a stretched effect when observing objects from oblique angles.

10.3.2.7 Noise Injection

Adding random noise, such as Gaussian noise, to a picture is known as noise injection. This method makes the model more resilient to changes and flaws in the data, such as sensor noise or hazy photos. The model learns to focus on the important aspects of the item while minimizing the effect of extraneous information caused by noise by training on noisy photos.

10.3.4 Leaf Disease Classification by Using CNN-BS Algorithm

The CNN model categorizes leaf images into various disease categories. For each hyperparameter, perform a binary search to identify the optimal value. The process involves training the CNN with different values of the hyperparameter and evaluating its performance on a validation dataset.

10.3.4.1 Convolutional Neural Network

CNNs may be classified as AI neural networks. It uses the convolution process to extract the best features and has strong generalization capabilities. CNN performs better in tasks like voice recognition and picture recognition. The CNN training method is divided into two phases: forward propagation and reverse propagation for parameter updates.

The procedure of forward propagation: There are three layers in this process, such as the convolution layer, pooling layer, and fully connected layer.

10.3.4.1.1 Convolution Layer

A convolution layer is an essential component of CNNs, which are meant to extract spatial and hierarchical characteristics from the data that is input. Through the use of a sliding window mechanism, it performs its functions by applying learnable filters, which are also referred to as kernels, to the acquired data. Each filter is responsible for carrying out a convolution operation, which involves calculating dot products between the kernel values and portions of the input that overlap. This procedure results in the creation of feature maps that draw attention to certain patterns or qualities that are present within the data. The convolution layer is responsible for gathering localized information, which helps the model to recognize characteristics like edges, textures, and forms. These features are critical for tasks such as picture identification and classification.

10.3.4.1.2 Pooling Layer

A pooling layer is an important part of CNNs. The responsibility lies in diminishing the spatial dimensions of feature maps generated by convolutional layers. The operation is carried out by summarizing information in local areas of the feature map, which assists in the retention of critical features while rejecting data that is redundant or less relevant.

During the pooling procedure, the input feature map is first divided into areas that do not overlap or regions that overlap (depending on the stride and kernel size), and then a pooling operation is applied to each of these regions. These exemplify common pooling operations:

Max pooling is a technique that identifies the maximum value from each region, highlighting the most significant feature. This technique calculates the average value of each area, resulting in a more consistent representation of the data. This technique is known as average pooling. Layers that pool resources make a contribution to the architecture as a whole by: Pooling decreases the dimensions of the feature maps, hence enhancing the computational efficiency of the network. Reducing the spatial dimensions accomplishes this. Pooling helps to preserve important features by concentrating on dominant or average values. This allows for the retention of essential information while simultaneously removing noise or details that are less important. Improving Translation Invariance: Pooling makes sure that even small changes or translations in the input data don't have a big effect on the features that are derived, which makes the model more resilient.

10.3.4.1.3 Fully Connected Layer

In a neural network, a fully connected layer connects each neuron to every neuron in the previous layer. It is an important part of feed forward neural networks and is often used at the end of a CNN or other architectures to do things like regression or classification. The input data in this layer is usually a flattened feature vector from a previous layer. It is changed by a linear combination of weights and biases, and then an activation function is applied. The fully connected layer enables the network to learn complex relationships between features and map them to output classes or values. The fully connected layer plays the following roles:

Feature Combination: It combines the extracted features from previous layers to make predictions or classifications.

Non-linearity: By applying activation functions, it introduces non-linearity, allowing the model to learn more complex patterns. Mapping to Output because they have so many connections and parameters, fully connected layers use a lot of computing power. That's why they're usually put after convolutional and pooling layers, which reduce the size of the input and pull out high-level features. Fully connected layers are computationally intensive due to the large number of connections and parameters, which is why they are often placed after convolutional and pooling layers that reduce the input size and extract high-level features.

10.3.4.2 Binary Search Optimization Algorithm

It splits the search area in half multiple times to find an item more quickly. It is handy when finding a solution requires us to narrow the search as we go along. Let's see how binary search is used for hyperparameter optimization.

1. Initialization of Hyperparameter Search Range: The initial thing to do is choose a range for each hyper parameter. As mentioned earlier, this range could be something like ⟦10⟧^(-6) to ⟦10⟧^(-2) for the learning rate or 16 to 128 for the batch size. These values are the endpoints of the search interval.

2. Evaluating the Midpoint: The algorithm determines the midpoint of the section it has set as the first step after initialization. Using the learning rate search space as 〚10〛^(-6), 〚10〛^(-2), the midpoint is 〚10〛^(-4).

   Training for the CNN is performed using the hyperparameter value selected as our midpoint. To do this, we set the learning rate or batch size that we picked and review the model's results on a separate validation dataset. Common performance indicators are usually precision, loss or metrics that apply to each project.

   After measuring the model's performance in the middle, the range of possible values is reduced to half. If the observations from the model become better with the value at the midpoint, the space being investigated is made smaller by focusing on values close to the original midpoint. In that case, if performance is not as good, the search is directed to look at higher numbers. Specifically:

   - If a midpoint value works the best, you may choose to test values in between that midpoint and the small end of the search interval.
   - When the performance lowers, you'll change the range to cover the upper half of the interval to see if bigger values help.
3. Repeat until Convergence: Items in the list are always split around the middle, until the range becomes so narrow that the value must be present. As a result, the algorithm can slowly approach the perfect hyperparameter value.

   The process goes on until the separation between the two bounds is nearly nothing or the solution falls within a certain range we set.

10.3.5 Results and Discussion of Proposed Methodology

This section discusses the efficacy of the proposed paddy leaf disease classification using the CNN-BS algorithm. The proposed approach is implemented on the MATLAB version 20a tool. The system configuration is CPU Intel® Pentium 1.9 GHz, 64-bit operating system, Microsoft® Windows 10, and 4 GB of RAM. This section discusses the efficiency of the proposed paddy leaf disease classification using the CNN-BS algorithm.

Table 10.2 presents the performance of the proposed techniques in the context of Precision, Specificity, Sensitivity, F1-score, and Accuracy.

Each classifier's performance is analyzed based on different metrics. In Table 10.2, it may be observed that the CNN-BS technique achieved good performance.

Table 10.3 compares the proposed technique with existing ones in the context of accuracy. The same Table 10.3is presented in graphical form in Figure10.2.

TABLE 10.2
Performance of Proposed Techniques

Techniques	Precision	Specificity	Sensitivity	F1-score	Accuracy
Proposed	97.48%	96.50%	97.50%	97.70%	97.40%

TABLE 10.3
Comparison with Existing Work in Terms of Accuracy for Dataset [19,20]

Source	Plant name	Techniques	Accuracy (%)
Rajendran et al. [21]	Paddy plant	ANN	89.50
Nalini et al. [10]	Paddy plant	DNN	92.30
Pant et al. [22]it is vulnerable to many diseases. One of the trickiest issues in agriculture is the early detection of plant diseases. The diseases must be detected early otherwise, they may negatively impact the overall production, lowering the farmers' profits. Many researchers have presented various cutting-edge frameworks relying on Neural Networks to address this issue. However, the majorities of these processes either have limited prediction performance or use massive amounts of input variables. In this system, a hybrid approach of Long Short Term memory (LSTM	Paddy plant	GRU	91.20
Venkatakrishnan and Natarajan [23]	Paddy plant	LSTM	94.50
Chinna et al. [24]	Paddy plant	Bi-LSTM	93.00
Our Study	Paddy plant	**CNN-BS**	97.40%

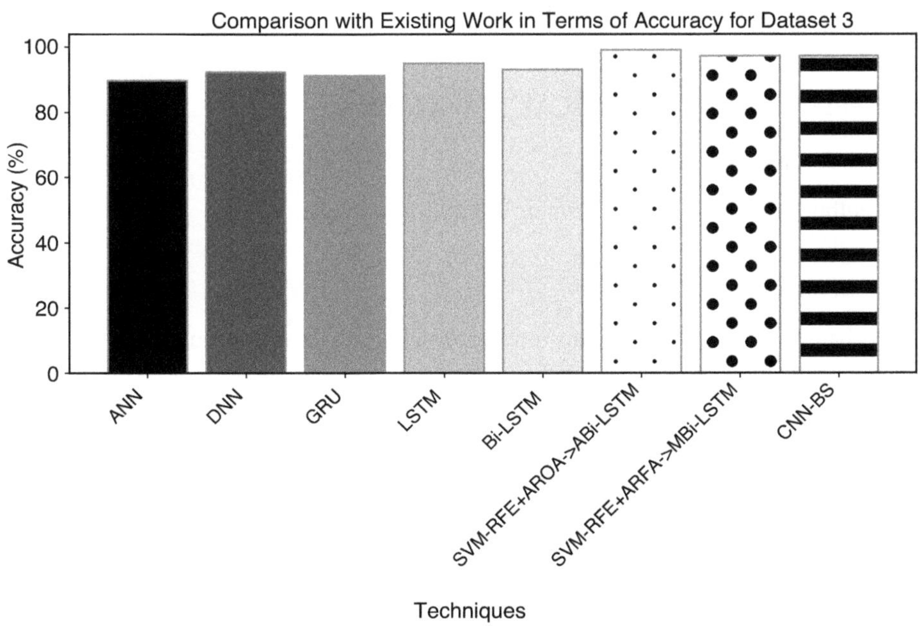

FIGURE 10.2 Comparison with existing work in terms of accuracy.

Each classifier's performance is analyzed based on different metrics. In Table 10.3, it may be observed that the SVM-RFE+AROA→ABi-LSTM technique achieved the highest performance, whereas the ANN technique achieved the worst performance among all. The performance techniques from increasing to decreasing are SVM-RFE+AROA → ABi-LSTM, CNN-BS, SVM-RFE+ARFA→MBi-LSTM, LSTM, Bi-LSTM, DNN,GRU,ANN.

10.4 CONCLUSION

The symptoms of plant diseases may be identified by their appearance in different plant regions. Researchers are motivated to design a variety of methods for accurately identifying plant diseases by the advancements in computer vision technology. In order to efficiently classify the illnesses affecting paddy crops, we propose the CNN-BS algorithm, an ideal practical method for paddy leaf disease identification. To get the best and most accurate classification results, we proposed the CNN-BS approach. The highest accuracy of 97.54% was attained using the proposed CNN-BS algorithm.

FUNDING

Not applicable.

ETHICS APPROVAL

This chapter does not contain any studies with human participants.

COMPLIANCE WITH ETHICAL STANDARDS

Conflict of Interest

The authors declare that they have no conflict of interest.

Human and Animal Rights

This chapter does not contain any studies with human or animal subjects performed by any of the authors.

REFERENCES

1. P. V. Richard and D. S. Radha, "A study on determinants of yield of paddy cultivation in Tamil Nadu with special reference to Madurai district," *Int. J. Curr. Adv. Res.*, vol. 6, no. 8, pp. 5246–5249, 2018.
2. M. A. Islam, N. Rahman Shuvo, M. Shamsojjaman, S. Hasan, S. Hossain, and T. Khatun, "An automated convolutional neural network based approach for paddy leaf disease detection," *Int. J. Adv. Comput. Sci. Appl.*, vol. 12, no. 1, pp. 280–288, 2021, doi: 10.14569/IJACSA.2021.0120134.

3. M. Sibiya and M. Sumbwanyambe, "Automatic fuzzy logic-based maize common rust disease severity predictions with thresholding and deep learning," *Pathogens*, vol. 10, no. 2, pp. 1–17, 2021, doi: 10.3390/pathogens10020131.
4. S. Chakraborty and A. C. Newton, "Climate change, plant diseases and food security: an overview," *Plant Pathol.*, vol. 60, no. 1, pp. 2–14, 2011.
5. D. I. Wisaputra, "Rice Leafs Disease Dataset,"2021. [Online]. Available: https://www.kaggle.com/datasets/dedeikhsandwisaputra/rice-leafs-disease-dataset
6. P. K. Sethy, N. K. Barpanda, A. K. Rath, and S. K. Behera, "Image processing techniques for diagnosing rice plant disease: a survey," *Procedia Comput. Sci.*, vol. 167, pp.516–530, 2020.
7. H. B. Prajapati, J. P. Shah, and V. K. Dabhi, "Detection and classification of rice plant diseases," *Intell. Decis. Technol.*, vol. 11, no. 3, pp. 357–373, 2017.
8. X. Wang and others, "Early detection of rice diseases using deep learning-based multiscale fusion of visible and near-infrared images," *Remote Sens.*, vol. 12, no. 16, p. 2572, 2020.
9. R.K. Dubey and D.K. Choubey, "Efficient prediction of blast disease in paddy plant using optimized support vector machine," *IETE J. Res.*, vol. 70, pp. 3679–3689, 2024.
10. N. Venkatakrishnan and M. Natarajan, "Automated plant disease recognition using tasmanian devil optimization with deep learning model," *Int. J. Recent Innov. Trends Comput. Commun.*, vol. 11, no. 11, pp. 359–366, 2023, doi: 10.17762/ijritcc.v11i11s.8163.
11. K. Ahmed, T. R. Shahidi, S. M. I. Alam, and S. Momen, "Rice leaf disease detection using machine learning techniques," in *2019 International Conference on Sustainable Technologies for Industry 4.0 (STI)*, 2019, pp. 1–5.
12. R. Deshmukh and M. Deshmukh, "Detection of paddy leaf diseases," *Int. J. Comput. Appl.,* vol. 975, no. 8887, 2015.
13. Ü. Atila, M. Uçar, K. Akyol, and E. Uçar, "Plant leaf disease classification using Efficient Net deep learning model," *Ecol. Inform.*, vol. 61, pp. 101182, 2021.
14. R. K. Dubey and D. K. Choubey, "Adaptive feature selection with deep learning MBi-LSTM model-based paddy plant leaf Disease classification,"*Multimed. Tools Appl.*, vol. 83, no. 9, pp. 25543–25571, 2024. Springer (ISSN NO: 1380-7501/1573-7721).
15. X. Chao, G. Sun, H. Zhao, M. Li, and D. He, "Identification of apple tree leaf diseases based on deep learning models," *Symmetry (Basel)*, vol. 12, no. 7, pp. 1–17, 2020, doi: 10.3390/sym12071065.
16. N. Senan, M. Aamir, R. Ibrahim, N. S. A. M. Taujuddin, and W. H. N. W. Muda, "An efficient convolutional neural network for paddy leaf disease and pest classification," *Int. J. Adv. Comput. Sci. Appl.*, vol. 11, no. 7, pp. 116–122, 2020, doi: 10.14569/IJACSA.2020.0110716.
17. R. K. Dubey and D. K. Choubey, "An efficient adaptive feature selection with deep learning model-based paddy plant leaf disease classification," *Multimed. Tools Appl.*, vol. 83, no. 8, pp. 22639–22661, 2024. Springer (ISSN NO: 1380-7501/1573-7721).
18. N. S. Patil, "Identification of paddy leaf diseases using evolutionary and machine learning methods," *Turkish J. Comput. Math. Educ.*, vol. 12, no. 2, pp. 1672–1686, 2021.
19. J. R. J. Pereira, "Rice Disease Dataset,"2021. [Online]. Available: https://www.kaggle.com/datasets/jonathanrjpereira/rice-disease
20. N. Sankalana, "Rice Leaf Disease Image Dataset." [Online]. Available: https://www.kaggle.com/datasets/nirmalsankalana/rice-leaf-disease-image
21. C. G. Simhadri, H. K. Kondaveeti, V. K. Vatsavayi, A. Mitra, and P. Ananthachari, "Deep learning for rice leaf disease detection: A systematic literature review on emerging trends, methodologies and techniques," *Inf. Process. Agric.*, vol. 12, no. 2, pp. 151–168, 2024, doi: 10.1016/j.inpa.2024.04.006

22. T. Rajendran, T. V. Ramana, M. Nalini, R. S. Subramanian, D. C. Devi, and M. Anuradha, "Paddy plant leaf disease classification based on artificial intelligence," in *2023 4th International Conference on Electronics and Sustainable Communication Systems (ICESC)*, Coimbatore, India, from July 6th to 8th, 2023, pp. 1002–1009.
23. S. Nalini, N. Krishnaraj, T. Jayasankar, K. Vinothkumar, A.S. Francis Britto, K. Subramaniam, and C. Bharatiraja, "Paddy leaf disease detection using an optimized deep neural network," Comput. Mater. Contin., vol. 68, no. 1, pp. 1117–1128, 2021, doi: 10.32604/cmc.2021.012431
24. K. Pant, B. Pant, and D. Bordoloi, "Automatic classification of plant leaf disease using a novel hybrid approach of LSTM-CNN" vol. 20, no. 3, pp. 4220–4228, 2021, doi: https://ilkogretim-online.org/index.php/pub/article/view/2851/2783

11 A Hybrid Machine Learning Approach for Accurate Crop Yield Forecasting

Avi Chaurasia and Amandeep Kaur

11.1 INTRODUCTION

Agriculture sustains the livelihoods of over two billion people worldwide and plays a critical role in ensuring global food security [1,2]. However, the sector faces escalating challenges due to climate variability, pest outbreaks, and soil degradation [3,4]. Traditional crop yield prediction techniques—often based on historical averages or limited environmental variables such as rainfall—fail to capture the multifaceted interactions between soil health, microclimatic conditions, and biotic stressors. For instance, linear regression models can incur prediction errors as high as 30%–40% in drought-prone regions due to their inability to model nonlinear and dynamic agricultural systems [5].

Recent advancements in machine learning (ML), Internet of Things (IoT) technologies, and remote sensing present new opportunities for precision agriculture [6,7]. Despite their promise, many existing studies rely on siloed datasets—such as satellite-derived vegetation indices or localized soil readings—limiting their effectiveness in real-world scenarios that demand a more holistic view of farm ecosystems [1,8].

To bridge this gap, this study proposes a hybrid ML framework that integrates historical agronomic records (e.g., past crop performance, soil health, and regional climate trends) with real-time data collected from high-resolution satellite imagery, drone surveillance, and IoT-enabled soil sensors [9]. The aim is to build a comprehensive, multimodal system capable of modeling complex patterns and interactions in agricultural environments.

The research evaluates six supervised ML algorithms—Linear Regression, Support Vector Machines (SVMs), K-Nearest Neighbors (KNNs), Decision Tree, Random Forest (RF), and Ensemble Learning—using publicly available crop yield prediction datasets from Kaggle [10]. These models are assessed not only for accuracy but also for their ability to detect early warning signals and optimize decision-making.

DOI: 10.1201/9781003637264-14

Validated using publicly available Kaggle datasets covering major crops like wheat, maize, and soybean [10], our approach delivers three major innovations:

1. **Early Stress Detection**: By examining spectral band data, the model can spot early signs of crop stress—such as *Fusarium* wilt—up to two weeks before any visible symptoms appear [6].
2. **Pest Outbreak Prediction**: The integration of drone-captured thermal images with soil moisture patterns enables the prediction of pest infestations, such as *Helicoverpa armigera*, in cotton fields [4].
3. **Enhanced Prediction Accuracy**: The proposed method reduces yield prediction errors by 18%–32% compared to traditional statistical techniques [3,5], enabling smarter decisions such as targeted pesticide application and optimized fertilizer use.

11.2 LITERATURE REVIEW

Crop yield prediction has advanced significantly in the last decade, driven by the rapid development of ML and artificial intelligence (AI) techniques. Traditional reliance on historical trends and simple statistical methods has been replaced by intelligent systems capable of analyzing complex environmental and agronomic data. A comprehensive review by Liakos et al. [5] highlighted the growing adoption of ML techniques such as support vector machines, random forests, and neural networks in agriculture, emphasizing their superiority over conventional approaches in handling high-dimensional and non-linear data. This literature review synthesizes insights from recent studies and identifies the gaps that motivate the present research.

Traditional statistical models for predicting crop yields often struggle to capture the intricate relationships among environmental factors. For example, Iniyan et al. [5] examined regression models such as Lasso, Ridge, and Elastic Net and found that ensemble techniques can substantially improve prediction accuracy, particularly in nonlinear scenarios. Their research emphasized the importance of feature engineering and incorporating soil–weather interactions when forecasting yields in different regions. Jeong et al. [10] demonstrated that Random Forest models outperform traditional linear approaches for both global and regional crop yield prediction tasks. Their findings highlighted the robustness of tree-based ensemble methods in capturing non-linear relationships between climate variables and crop productivity.

El-Kenawy et al. [6] conducted a comparative evaluation of ML and deep learning (DL) models for predicting potato yields. While algorithms like KNN and XGBoost demonstrated strong performance, DL models, particularly Long Short-Term Memory (LSTM) networks and Graph Neural Networks (GNNs), excelled in managing spatial–temporal data, achieving the highest coefficient of determination (R^2) values and the lowest root mean squared error (RMSE) across multiple datasets.

Araújo et al. [1] investigated the integration of ML techniques into Agriculture 4.0, emphasizing the transformative potential of real-time data from IoT sensors, unmanned aerial vehicle (UAV) imagery, and weather APIs. Their findings indicated that models leveraging multimodal data—such as Normalized Difference Vegetation Index (NDVI) indices, rainfall patterns, and soil pH—delivered greater accuracy and exhibited higher resilience to the uncertainties caused by climate change. Among the models tested, RF, SVM, and Deep Neural Networks (DNN) achieved the best results for crop classification and yield forecasting.

Harikumar et al. [2] examined the role of AI in agriculture, focusing on supervised algorithms such as Naïve Bayes, SVM, and KNN for tasks including disease detection, water resource management, and fertilizer optimization. The study underscored the importance of interpretable models, particularly in resource-constrained settings where the deployment of DL may not be feasible.

Jhajharia et al. [3] evaluated a range of ML and DL methods to predict crop yields in Rajasthan, India, using historical data from 1997 to 2019 on soil characteristics, rainfall, and yields. RF achieved the highest performance, with an R^2 of 0.963 and an RMSE of 0.035. The study also highlighted the importance of preprocessing techniques such as dummy variable encoding and feature standardization in enhancing predictive accuracy.

Abbas et al. [7] explored the integration of proximal sensing technologies with ML algorithms including Support Vector Regression (SVR), Elastic Net, KNN, and Linear Regression for predicting potato yields in Atlantic Canada. SVR achieved the lowest RMSE of 4.62 t/ha, benefiting from site- specific measurements such as soil conductivity, moisture, slope, and NDVI.

Mathur et al. [4] investigated hybrid approaches that combine DL models, such as LSTM, with traditional ML algorithms. Their research on high-dimensional rainfall and soil datasets demonstrated that LSTM models were more effective at capturing temporal dependencies in climate-affected crops compared to static models. However, the study also noted challenges, including the need for substantial amounts of training data and significant computational resources.

Overall, the reviewed studies indicate that no single model is universally optimal for crop yield prediction. The choice of ML or DL technique depends heavily on data quality, dimensionality, and geographic variability. While RF and SVR provide strong baselines, advanced DL architectures such as LSTM offer distinct advantages for time-series modeling, especially in climates subject to abrupt and unpredictable changes.

11.2.1 Gaps, Limitations, and Research Opportunities

Despite the notable progress in applying ML and DL techniques to crop yield forecasting, several research gaps continue to limit their scalability, generalization, and practical deployment. **A major limitation in many existing studies is the reliance on single-source datasets**, either exclusively using remote sensing indicators such as NDVI and Enhanced Vegetation Index (EVI) or focusing solely on ground-based measurements like soil nutrient profiles and weather station data. This siloed approach overlooks the complementary strengths of multimodal data fusion, resulting in models that are often brittle when applied to new environmental or climatic conditions.

Temporal modeling presents another challenge. Although advanced architectures such as LSTM and GNN have demonstrated the ability to capture seasonal growth patterns and spatial–temporal dependencies, they require large, high-quality time-series datasets and significant computational resources, making them difficult to implement in resource-constrained agricultural settings. Furthermore, **many high-performing models are tailored to a specific crop type or agro-climatic region**, which limits their applicability elsewhere due to differences in soil properties, pest dynamics, and farming practices. **Transfer learning and domain adaptation, which could improve cross-regional generalization, remain underexplored in agricultural ML.**

Another critical limitation is the lack of real-time integration. While IoT sensors, UAV-based imaging, and satellite monitoring technologies are increasingly available, few forecasting frameworks leverage these sources within an operational, near-real-time pipeline, thereby missing opportunities for timely interventions such as pest outbreak prevention or adaptive irrigation scheduling. **Benchmarking practices also require improvement**, as several works still compare against outdated statistical methods or a single ML baseline, without evaluating performance against state-of-the-art ensemble models like XGBoost or LightGBM, or modern DL architectures such as Transformer-based temporal models.

In addition, **model interpretability and explainability are often neglected**; without transparent insights into feature importance or causal drivers, even highly accurate models may face adoption barriers among farmers, agronomists, and policymakers. Lastly, **there is a general lack of evaluation under extreme climatic events—such as droughts, floods, or heatwaves—despite the fact that climate volatility is increasingly shaping agricultural productivity.**

Addressing these limitations calls for an integrated and holistic approach that fuses structured historical datasets with unstructured, real-time sources such as drone imagery and IoT-based soil and climate sensors. The present study responds to these gaps by developing a hybrid ML framework that combines advanced ensemble methods with multimodal data integration, validating its effectiveness across multiple crops and agro-climatic zones, and benchmarking against both conventional and contemporary state-of-the-art models. **This approach aims to deliver more robust, generalizable, and actionable yield forecasts capable of supporting precision agriculture under real-world conditions**.

11.3 PROPOSED APPROACH

The proposed hybrid ML framework integrates multimodal structured and unstructured agricultural data for accurate yield prediction, and was evaluated using publicly available crop yield datasets from Kaggle [10], covering multiple crops such as wheat, maize, and soybean. The dataset incorporated both structured data, such as temperature, rainfall, and soil pH, and unstructured data, including multispectral satellite imagery, drone footage, and IoT-based soil sensor readings. The methodology was organized into six sequential stages—data sources, preprocessing, feature engineering, model selection, training and validation, and evaluation—ensuring a systematic and reproducible workflow for yield prediction across multiple agro-climatic zones.

The dataset used in this study combines both historical and real-time agricultural data from Kaggle and official government agricultural portals. It includes crop-specific details such as type, variety, sowing and harvesting dates, and yield values, as well as soil parameters including pH, nitrogen, phosphorus, potassium, organic carbon, and bulk density. Weather data, comprising maximum and minimum temperature, rainfall, humidity, and solar radiation, was recorded over multiple growing seasons. Remote sensing features such as NDVI and EVI were extracted from Sentinel-2 satellite imagery via Google Earth Engine, and additional simulated IoT-based LoRaWAN sensor readings were incorporated to provide real-time soil moisture and nutrient data [11]. Table 11.1 presents the key dataset attributes and representative sample values, illustrating its multimodal nature.

Preprocessing involved handling missing values through mean or median imputation for continuous variables and mode imputation for categorical features. Continuous features were normalized using Min–Max scaling to ensure uniformity, while categorical variables were encoded using one-hot encoding. For remote sensing data, cloud-masked imagery was processed in Google Earth Engine to compute NDVI and EVI values, and monthly averages of weather variables were calculated to standardize temporal resolution. Outliers were detected using the Interquartile Range (IQR) method and adjusted or removed to maintain data quality.

Feature engineering played a critical role in enriching the predictive capability of the models. A Soil Fertility Index (SFI) was developed by assigning weights to nitrogen, phosphorus, and potassium values to create a composite measure of nutrient richness. Water Stress Indicators (WSI) were derived from soil moisture data, with values below 18% identified as yield-impacting thresholds. Additional agronomic variables such as Growing Degree Days (GDD) and cumulative rainfall during specific growth stages were also computed. These engineered features provided agronomically meaningful inputs, enhancing the model's ability to capture complex and nonlinear yield relationships.

For model development, six supervised ML algorithms were implemented in Python using Scikit-learn and XGBoost libraries. Linear Regression was used as a baseline model, achieving a mean absolute error (MAE) of 5.42 and R^2 of 0.71. SVM with a Radial Basis Function (RBF) kernel better captured nonlinear interactions,

TABLE 11.1
Description of Dataset Attributes with Sample Values

Feature Category	Attributes	Sample Values
Crop details	Crop type, variety, sowing date, harvesting date, yield (kg/ha)	Wheat, HD-2967, 15-Nov-2021, 20-Apr-2022, 3,500 kg/ha
Soil parameters	pH, NPK levels, organic carbon, bulk density	6.5, N: 80, P: 40, K: 30, Organic C: 0.85%, Bulk Density: 1.3 g/cm^3
Weather parameters	Max/Min temperature, rainfall, humidity, solar radiation	35°C/12°C, 78 mm/month, 65%, 18 MJ/m^2/day
Remote sensing	NDVI, EVI, thermal bands, multispectral channels	NDVI: 0.72, EVI: 0.46, Thermal: 305 K, Sentinel-2 (8 bands)
Geolocation	Latitude, longitude, elevation, region	28.61°N, 77.23°E, 216 m, Delhi
Time series	Year, season, time granularity	2021–2022, Rabi, Monthly (Nov–Apr)

achieving an MAE of 4.78 and R^2 of 0.79. KNNs with $K=5$ performed well in dense data clusters but were less effective in sparse datasets, recording an MAE of 5.20 and R^2 of 0.74. RF with 100 decision trees demonstrated strong performance in noisy, multivariate data, achieving an MAE of 3.12 and R^2 of 0.90. XGBoost delivered the highest individual model performance, with an MAE of 2.95 and R^2 of 0.93. **The hybrid model integrated RF and XGBoost predictions through weighted soft voting, combining RF's noise resilience with XGBoost's ability to model intricate feature interactions.**

All models were trained on 80% of the dataset, with the remaining 20% used for independent testing. Stratified 5-fold cross-validation was employed to ensure robustness across different subsets, and hyperparameter tuning was conducted via grid search. For RF and XGBoost, the number of estimators and maximum depth were optimized; for SVM, kernel type and regularization parameter C were tuned; and for KNN, the number of neighbors and leaf size were selected for optimal performance. The evaluation metrics included MAE, RMSE, and R^2, with the hybrid model achieving the best overall performance.

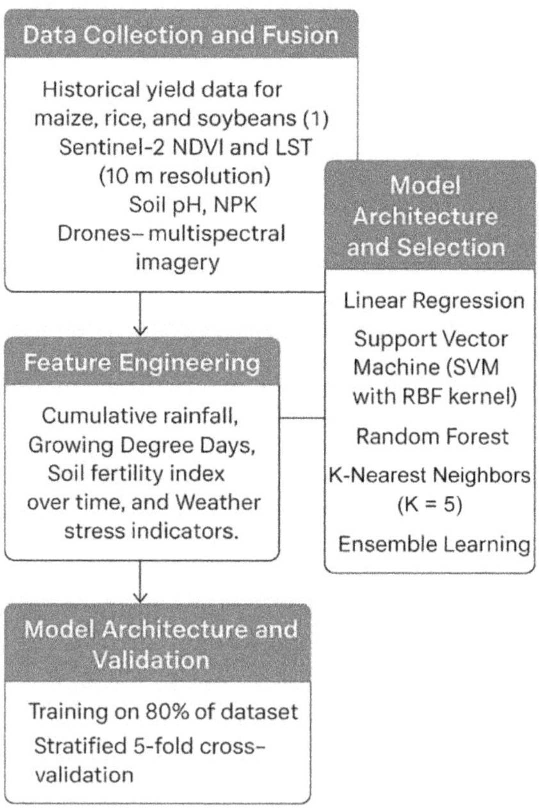

FIGURE 11.1 Proposed hybrid machine learning workflow for crop yield prediction, illustrating the sequential stages from multimodal data acquisition to prediction and evaluation.

The complete workflow, shown in Figure 11.1, begins with multimodal data collection from satellite, IoT, and historical sources, followed by preprocessing to standardize and clean the data. Feature engineering then creates derived variables that better represent agronomic conditions. Multiple models are trained and validated, with the final hybrid prediction obtained through ensemble integration. This approach ensures that both structured and unstructured agricultural data contribute effectively to yield estimation, allowing for improved accuracy and early detection of factors affecting crop productivity.

11.4 EXPERIMENTAL RESULTS

To assess how well the proposed hybrid ML framework predicts and holds up under various conditions, we trained and tested six different models: Linear Regression, SVM, KNNs, Decision Tree, RF, and an XGBoost-based Ensemble Learning model. We used a multimodal dataset that brought together both historical and real-time agricultural data. This dataset featured crop-specific details, weather conditions, soil characteristics, and remote sensing metrics like NDVI and EVI. We trained all models on 80% of the dataset and validated them on the remaining 20%, employing 5-fold stratified cross-validation to ensure consistent performance across different regions and seasons.

The evaluation focused on three key regression metrics: MAE, RMSE, and R^2. The Linear Regression model, used as a baseline, had an MAE of 5.42, an RMSE of 6.71, and an R^2 of 0.71, indicating difficulty in capturing nonlinear relationships. The SVM with an RBF kernel performed better, achieving an MAE of 4.78, an RMSE of 5.90, and an R^2 of 0.79, showing effectiveness for medium-complexity patterns.

The KNN model with $K=5$ recorded an MAE of 5.20 and an R^2 of 0.74, performing reasonably well in dense data regions but less effectively in sparse ones. The Decision Tree model improved interpretability and achieved an MAE of 4.35, an RMSE of 5.48, and an R^2 of 0.81, though it exhibited some overfitting. The RF model outperformed these, with an MAE of 3.12, an RMSE of 4.35, and an R^2 of 0.90, benefiting from variance reduction through ensemble averaging. **The XGBoost-based ensemble delivered the highest accuracy across all models, achieving an MAE of 2.95, an RMSE of 4.01, and an R^2 of 0.93, while also generalizing well across different crop types and regions.**

Figure 11.2 visually compares six ML algorithms in terms of R^2, MAE, and RMSE. The stacked area chart highlights trade-offs in performance: simpler models like Linear Regression and KNN generally show higher error metrics, whereas ensemble models—particularly RF and XGBoost—demonstrate lower errors and higher explanatory power. MAE measures average prediction error, RMSE penalizes larger errors, and R^2 assesses how well variance in the data is explained (Table 11.2).

Overall, these results show that ensemble learning frameworks, particularly the proposed XGBoost-based hybrid, are both *more accurate and more reliable* for complex, multimodal agricultural datasets than simpler baseline models (Figure 11.3).

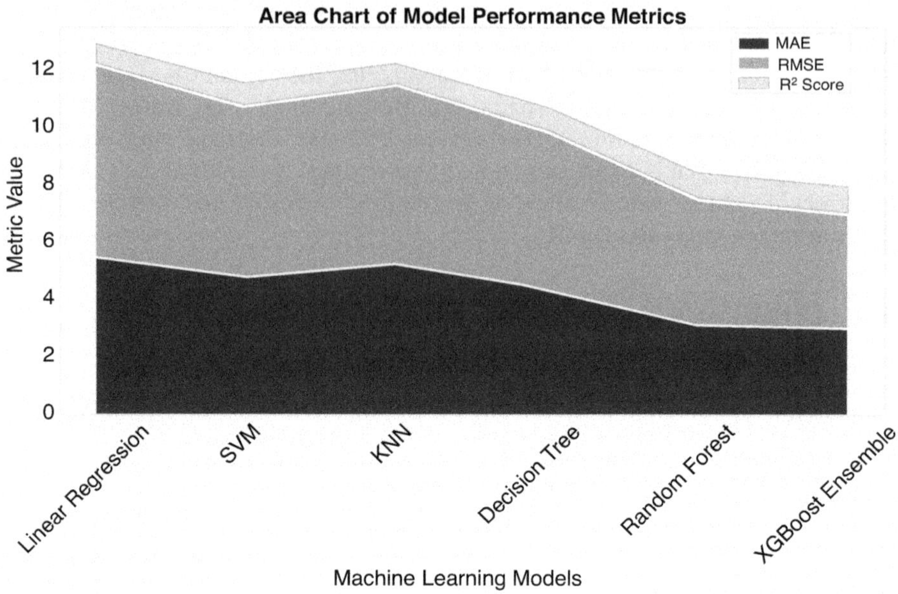

FIGURE 11.2 Area chart comparing R^2, RMSE, and MAE scores across six ML models.

TABLE 11.2
Results of Experiments Using Machine Learning Models to Predict Crop Yield

Model	MAE	RMSE	R^2 Score
Linear Regression (LR)	5.42	6.71	0.71
Support Vector Machine (SVM)	4.78	5.90	0.79
K-Nearest Neighbors (KNN)	5.20	6.25	0.74
Decision Tree (DT)	4.35	5.48	0.81
Random Forest (RF)	3.12	4.35	0.90
XGBoost Ensemble	**2.95**	**4.01**	**0.93**

11.5 CONCLUSION

This study developed a hybrid ML framework for crop yield prediction that integrates historical yield records, soil health indicators, and climatic trends with real-time inputs from satellite imagery, drone-based sensing, and IoT-enabled soil sensors. The framework employed a systematic workflow of data fusion, feature engineering, model training, and validation, and was evaluated using multiple ML algorithms. **Results demonstrated that ensemble methods—particularly the proposed RF + XGBoost hybrid—consistently outperformed traditional statistical and stand-alone ML models, achieving the highest accuracy ($R^2 = 0.93$)**

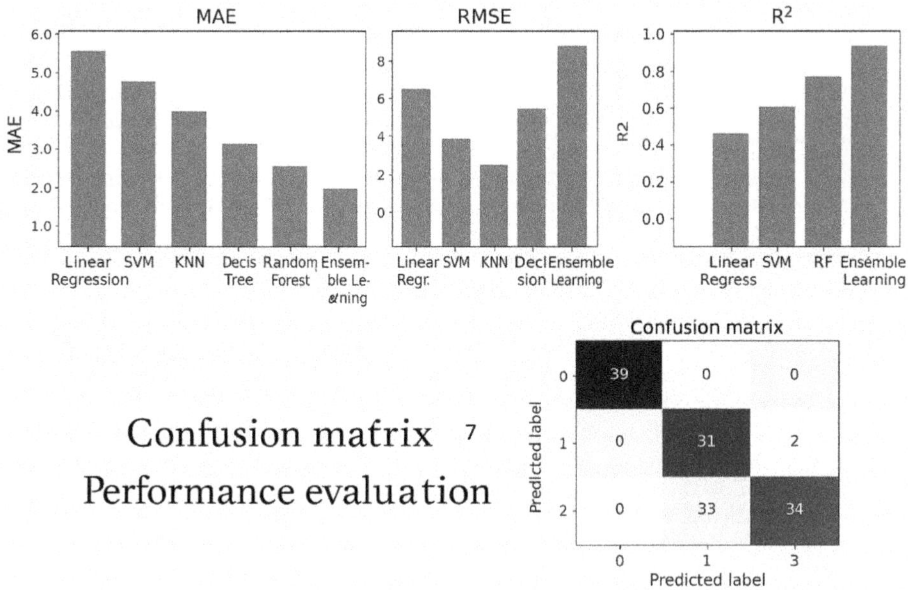

FIGURE 11.3 Comparative result analysis of models.

and reducing prediction errors by up to 32%. These outcomes underscore the effectiveness of combining multimodal datasets with advanced learning techniques to capture complex agro-environmental interactions.

The practical implications of this work are significant for precision agriculture, as the framework enables optimized resource allocation, early detection of crop stress, and the development of climate-resilient farming strategies. Its adaptability across diverse crop types and agro-climatic zones makes it suitable for both research and real-world deployment. Future work will focus on integrating DL architectures, such as LSTM and attention-based models, to enhance temporal sequence modeling; incorporating socio-economic and market variables for holistic yield forecasting; expanding the framework to include more granular, region-specific datasets; and deploying the system as a real-time, cloud-based decision support tool. Such advancements have the potential to not only improve prediction accuracy but also enable large-scale, accessible, and sustainable agricultural solutions to address global food security challenges.

REFERENCES

1. A. Araújo, A. H. Oliveira, F. Marins, and L. P. Rocha, "A review of machine learning in Agriculture 4.0: Applications, challenges, and opportunities," *Agronomy*, vol. 13, no. 2976, pp. 1–26, 2023.
2. R. Harikumar, M. Sharma, and P. Venkatesh, "Artificial intelligence in smart agriculture: Opportunities and challenges in developing countries," *International Journal of Computer Applications*, vol. 182, no. 45, pp. 12–18, 2023.

3. P. Jhajharia, R. Meena, and V. Pratap, "Machine and deep learning approaches for crop yield forecasting: A case study of Rajasthan, India," *Journal of Applied Data Science*, vol. 9, no. 2, pp. 110–122, 2023.
4. A. Mathur, R. Bansal, and V. Desai, "Hybrid deep learning models for climate-aware crop yield forecasting," *International Journal of Agricultural Informatics*, vol. 14, no. 3, pp. 150–162, 2023.
5. K. G. Liakos, P. Busato, D. Moshou, S. Pearson, and D. Bochtis, "Machine learning in agriculture: A review," Sensors, vol. 18, no. 8, Art. no. 2674, 2018. doi: 10.3390/s18082674.
6. A. E.-D. M. El-Kenawy, S. Mafarja, and M. Ibrahim, "Machine and deep learning for predicting potato yield using field data: A case study from Egypt," *Agronomy*, vol. 10, no. 1046, pp. 1–21, 2024.
7. A. Abbas, R. M. Rees, and J. Smith, "Integrating proximal sensing and machine learning models for potato yield prediction in Atlantic Canada," *Precision Agriculture*, vol. 21, no. 5, pp. 987–1002, 2020.
8. H. Wan, H. Qi, and S. Shang, "Estimating soil water and salt contents from field measurements with time domain reflectometry using machine learning algorithms," *Agricultural Water Management*, vol. 285, p. 108364, 2023.
9. J. Cao, Z. Zhang, F. Tao, L. Zhang, Y. Luo, J. Zhang, J. Han, and J. Xie, "Integrating multi-source data for rice yield prediction across China using machine learning and deep learning approaches," *Agricultural and Forest Meteorology*, 2024, https://doi.org/10.1016/j.agrformet.2020.108275
10. J. H. Jeong, J. P. Resop, N. D. Mueller, D. H. Fleisher, K. Yun, E. E. Butler, D. J. Timlin, K-M. Shim, J. S. Gerber, V. R. Reddy, and S-H. Kim, "Random forests for global and regional crop yield predictions," *PLoS ONE*, vol. 11, no. 6, e0156571, 2016. [Online]. https://doi.org/10.1371/journal.pone.0156571
11. A. Atharva Ingle, "Crop Recommendation Dataset," Kaggle, 2020. [Online]. https://www.kaggle.com/datasets/atharvaingle/crop-recommendation-dataset

Section 4

Utilization of Futuristic Emerging Technologies for Smart Farming Systems

12 Enhancing Agricultural Traceability and Precision Farming with Nanoparticle-Based Biosensors

Somya Ranjan Sahoo, Phani Krishna Bulasara, Ankit Sinha, Piyush Raj, Shreyam Singh, and Satyam Mishra

12.1 INTRODUCTION

There are numerous challenges to food production and security, and agricultural innovations from recent years demonstrate technological advancement. Agriculture encompasses both crop production and animal husbandry. Dairy, poultry, and aquaculture provide valuable resources and contribute to global food security. These practices are essential for human health and well-being. Precision farming, organic farming, and other farming methods present unique challenges, which nanotechnology has successfully addressed. Agricultural applications of nanotechnology include nanofertilizers, nanopesticides, and nanosensors. Nanomaterials' unique properties and small size make them useful for plant protection and nutrition. Nanotechnology has enormous potential uses and benefits. These include insect pest management and the slow/steady release of nutrients and water. The use of nanomaterials for the preparation of various types of biosensors could be useful in remote sensing devices.

Nanotechnology in agriculture provides answers to a variety of problems, including crop productivity threats and plant diseases. The biocompatibility and biodegradability of nanoparticles, such as porous silica, have been emphasized by research [1]. Natural amino acids are used to make a particular class of polymers called proteinoids. Because they are safe, biodegradable, and biocompatible, they have attracted a lot of interest in controlled release mechanisms and drug delivery systems [2]. The agricultural industry has benefited greatly from recent developments in machine learning (ML) techniques, which are intended to maximize crop yield and minimize crop costs. ML techniques can be used to identify agricultural diseases. By preventing crop infections, they can lower the amount of labor needed. This method makes

DOI: 10.1201/9781003637264-16

it easier to plan strategically for the appropriate application of pesticides, fertilizers, and irrigation. The various nanoparticle based things that promote precision farming are represented in Figure 12.1.

Predicting and increasing crop yields is crucial in agriculture. Crop and fruit yields can be predicted using deep convolutional neural networks (CNNs) or neural networks. ML approaches can also be used to predict yields by incorporating satellite and climate data [3]. Precision agriculture refers to the use of wireless networks and miniaturized sensors to monitor and control agricultural practices. Specifically, it is related to site-specific crop management for horticultural crops and field crops [4]. Furthermore, nanotechnology has solutions for the observation of environmental stress and enhancing plants' disease-fighting potentials [5]. Researchers can identify early indicators of plant stress brought on by conditions like drought, nutrient deficiencies, temperature changes, or pollution exposure by using nanosensors. These nanosensors can be placed directly on plant tissues or implanted in soil to offer real-time data that enables prompt treatments.

The worldwide food supply system, which includes the farming, processing, and transportation of food from producers to final consumers, is crucial in meeting the daily needs of billions. So in the current food industry, packaging plays an important role in protecting the product from various quality angles and the safety of consumers. However, the quality is affected by food spoilage resulting in the wastage of food items. Ethylene gas is the major cause of this spoilage. It increases the respiration rates leading to food softening, color change, and spoilage. Therefore, it is important to detect ethylene gas in food processing and packaging as it affects the shelf life and quality of fruits and food items. Several traditional gas detecting techniques were there but faced challenges in sensitivity and selectivity at nanoscale levels, especially in environments with diverse gas concentrations. Nowadays, nanotechnology is one of the latest technological innovations that can be used for food safety. One of the nanotechnologies uses carbon nanotube (CNT)-based nanosensors which are highly effective for detecting even trace amounts of ethylene gas and help us to monitor the ripening of fruits, reduce waste, and maintain product freshness throughout the supply chain.

Previous studies about CNT-based nanosensors highlight its unique properties to make it fit for detecting environmental and biological targets. CNTs are graphene sheets rolled in tubes and categorized into single-walled (SWCNTs) and multi-walled

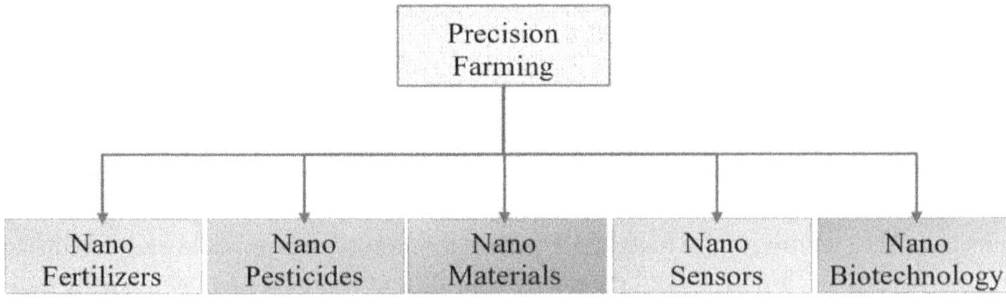

FIGURE 12.1 Nanoparticle-based things that promote precision farming.

(MWCNTs) types. Their small size at nanoscale level provides high surface area, good conductivity, and chemical stability, which is good for biological applications. Their large surface area increases sensitivity which helps CNTs to interact with target molecules [6].

Other research about CNT focuses on its high tensile strength, electrical conductivity, and thermal stability, which makes it highly sensitive and selective in monitoring alterations in the environment. Their large surface-to-volume ratio increases their ability to interact with target substances, allowing for accurate and quick responses, which are essential for monitoring air quality, conducting healthcare diagnostics, and maintaining security. CNTs also have good electrical properties which makes them suitable for various electro-chemical and field-effect sensors. CNTs have a rapid electron transfer rate which is critical in detecting low concentrations of biomolecules such as glucose and DNA [7]. CNTs also have a flexible design that enables integration with nanoparticles that boost sensor performance and it also provides a synergistic effect that enhances both signal strength and sensitivity.

The primary objective of this research chapter is to develop a reliable system using CNT-based nanosensors for detecting ethylene gas, using a multi-hop communication network to transfer data from one node to another. Further, the data processing and analysis is done using advanced ML techniques. This model aims to simulate ethylene adsorption on CNTs and examine how it affects the coordinates of CNTs. The focus is on the implementation and evaluation of various ML models including long short-term memory (LSTM), bidirectional LSTM (Bi-LSTM), and Extended LSTM (XLSTM) to reduce noise data transmission over the network. This study will further dive into classification models to accurately predict ethylene presence based on denoised data and improve detection specificity.

The main contributions of this chapter include:

- Collection of sample data from different spatial points of food items.
- Transmission of data from one node to another using multi-hop techniques.
- Implement and compare various ML models (LSTM, Bi-LSTM, XLSTM) for denoising data transmission.
- Develop classification models to predict the presence of ethylene based on processed data.

12.2 LITERATURE REVIEW

Previous research has illustrated that CNTs are essential for developing unique gas sensors with nano-sized characteristics. CNTs played an important role in the development of nanosensor gas detectors that are very sensitive at detecting gases at levels below one part per million. They also use very little power, even at room temperature [8]. The application of CNTs has been extended to sensors for the monitoring of volatile organic compounds, in healthcare and for industrial control processes. It has been shown that the CNTs have remarkable properties and are very selective to some types of gas exposure. Due to a combination of high surface area, electrical conductivity, and sensitivity, CNTs are suitable materials for gas detection [9]. Because of their small curvature radius, MWCNTs or carbon nanofibers (CNFs) with sharp tips are used in gas sensors. The configuration establishes strong electric fields in a neighborhood of the collecting plate made of a silicon wafer coated with a thin layer of gold or stainless steel. It permits the generation of large current due to the breakdown of the gas components at lower voltages. There is a marked impact in the electrical resistance of ethylene gas molecules when they undergo attachments with CNTs hybridized with activated carbons [10]. This contact forms a dependable base for a number of activities including monitoring the disintegration of food and facilitating researches on the growth of plants.

12.2.1 Deep Learning Techniques for Noise Reduction

In recent years the advancement of nanosensor-based technology comes into consideration. The effects of data transmission noise on signal detection latency have been extensively investigated. Removing noise from the interferences has become essential for nanosensors used to monitor vital parameters such as gas concentrations. In light of this type of difficulty, this challenge could be resolved with different ML techniques including LSTM neural networks [11], which have proven to be useful in the task of data sequence denoising.

Initially support vector machines (SVMs) and decision trees were considered for the analysis of nanosensor data. Moreover, the models of deep learning have the potential to identify and model complex interactions that exist over time. As for singular value decomposition (SVD)-based order statistics filters, we see that [12] LSTMs perform far better than other statistical filters in terms of noise suppression and enhancement of transmembrane potential reading due to its feature of gracefully predicting long-range dependencies.

LSTM networks are notable due to their effective handling of long-range dependencies, and thus find its application in complicated sequential problems. In contrast to typical neural networks, LSTMs store information across long chains, thus reducing the amount of noise in the output of nanosensors. In many situations, it is clear that LSTM-type networks are more appropriate in those applications that require quick and accurate signal detection; in particular: for gas leak detection and for air quality assessment.

12.2.2 Multi-hop Communication in Sensor

The application of enormous frequencies in the ranges of 0.1–10 THz are positioned between the microwave and the infrared, which support smooth transmission of data with no interruptions. Terahertz (THz) multi-hop communication is noted to present distinct [7] challenges. On the other hand, nanosensor dimensions and their environmental adaptability require low processing and storage to avoid elaborate routing protocols. However, it is significant for efficient data transfer even with THz-based multi-hop communication systems over-long range transmission [13], as may be needed for monitoring agricultural fields, environmental or similar conditions. The potential of volatile gases such as ethylene gas concentrations in multi-hop THz networks appears to be encouraging since they support gas measurement analysis.

It has been noticed that carbon nanosensor networks are well-suited to THz frequencies with carbon nanomaterials like CNTs and graphene which facilitate tasks such as molecular sensing and gas provided at necessary bandwidths [14]. However, THz frequency is also a concern for users because users are always looking for quick and consistent information through the sensing network for environmental monitoring which is closely and constantly controlled in a very short time. THz displays high data throughput speeds with minimal latency that is critical in real time and intensive compared with conventional frequencies.

12.3 PROPOSED METHOD

The proposed system architecture for ethylene detection and classification within packaged food items employs nanosensor technology, data denoising, and ML for accurate analysis and transmission. The architecture shown in Figure 12.2 consists of four main stages: the Bio Nano Sensors placed on packed food items, followed by Nano Router, and then Nano Interface layer for data processing, and an IoT Gateway responsible for final classification and data transmission of data centers/etc.

12.3.1 System Architecture

The dataset used for simulating sensor behavior including atomic coordinates of CNTs is taken from Kaggle Database Hub [15]. This dataset reflects structural properties that influence sensor response and assists in simulating ethylene adsorption and detection across sensor clusters. The initial data collection occurs through Bio Nano Sensors embedded in the packed food items, such as blueberries and bananas. These multiple sensors when clubbed together form a cluster of sensors, which detect

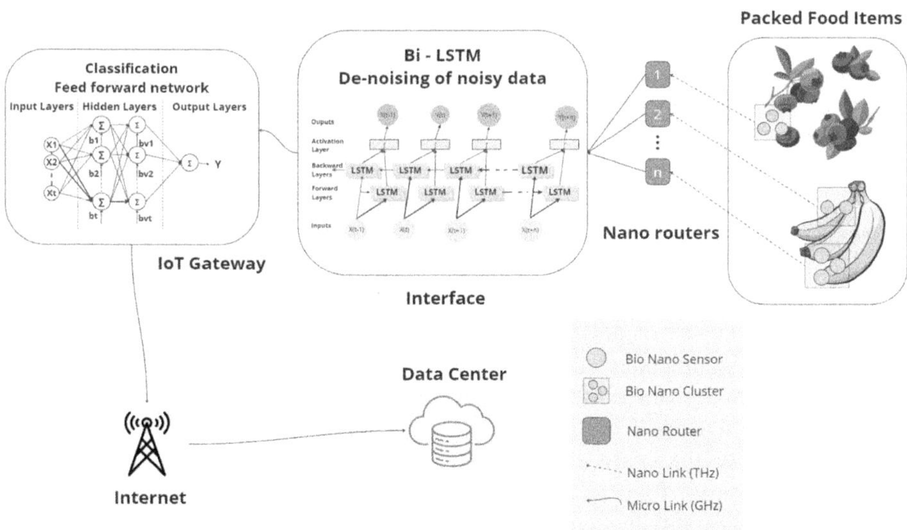

FIGURE 12.2 IoT-based spoilage detection model for packed food, using bio nano sensors and nano routers. A Bi-LSTM denoises data, followed by classification at the IoT gateway and centralized analysis for spoilage prediction.

changes in ethylene concentration, a key indicator of ripening and spoilage in fruits. The Bio Nano Sensors transmit raw ethylene data to Nano Routers through high-frequency Nano Links using multi-hop transmission. These Nano Routers, placed at strategic positions within the packaging, gather data from multiple sensors and forward it to the Nano Interface layer using the same Nano link with the system, reducing energy consumption in the nanosensors and improving overall communication efficiency. At the Nano Interface, a Bi-LSTM model denoises the received data. Given the small, noisy signals from nanosensors, this denoising process is crucial to ensure data accuracy before further processing. The Bi-LSTM model is optimized to handle the unique characteristics of nanoscale sensor data, capturing temporal dependencies that help filter out irrelevant fluctuations/noises while preserving the core ethylene concentration trends (relevant data).

Following denoising, the refined data is sent to the IoT Gateway, where a feed-forward neural network (FNN) performs the final classification. This model distinguishes whether the ethylene level indicates spoilage, enabling timely interventions. The IoT Gateway is also connected to the Internet, where data can be sent to a remote data center for monitoring, analysis, and storage. However, the architecture may have limitations as the data is processed in real time; there could be a slight delay in data transmission between the Nano Interface and the IoT Gateway due to the processing time of the Bi-LSTM model. This delay completely depends on the model used here, in our case Bi-LSTM. This comprehensive system is designed to enhance real-time spoilage detection in packed food items through precise ethylene monitoring, ultimately supporting better supply chain management and reducing food waste.

12.3.2 Bio Nano Sensor Clustering

Bio nano sensors are utilized in this study to identify the release of ethylene gas in packaged food products. Ethylene gas, which is naturally found in plant as a hormone, is emitted by fruits during the ripening process, and its detection can indicate the beginning of decay. Alo, in general also production of ethylene gas indicates spoilage or decay of any substance (organic or anthropogenic sources) [16]. Monitoring ethylene levels enables the detection of the starting point and location of decay, leading to proactive steps to avoid food wastage [17]. In our architecture every yellow dot represents a bio nano sensor with the ability to detect ethylene gas in its surrounding area. Nevertheless, the data collected by one sensor would only cover a small, specific region, which is not enough for effective detection as the sample size is too small. To tackle this problem multiples of these bio nano sensors are placed in a specific region which could be termed as clusters of bio nano sensors. Combining multiple of these clusters spread around various regions can help us gain a fuller understanding of the distribution of ethylene throughout the entire food package [18]. The clusters are strategically located in various zones to establish a sensor network that gathers data from multiple areas, consequently enhancing the probability of detecting ethylene emissions from any part inside the package.

This sensor cluster configuration provides a distributed monitoring approach to reliably detect ethylene gas regardless of where it is concentrated within the food package. Because differences in fruit ripeness can result in uneven release of ethylene within the package, the use of multiple sensors allows for a more accurate understanding of overall ripeness and spoilage [18]. This design allows for the identification of specific areas where spoilage may begin, creating a detailed real-time dataset of ethylene levels in packaged foods. Ultimately, this integrated approach improves the reliability of the monitoring process by supporting timely detection of signs of spoilage and allowing more informed decisions to be made regarding the quality and shelf life of packaged foods.

12.3.3 Multi-hop Data Transmission via Nano Routers

In this research, bio nano sensors are employed to monitor ethylene gas emissions within packed food items. Ethylene is a critical biomarker for the ripening and spoilage processes in various fruits and vegetables. The methodological design of this study focuses on the strategic deployment of bio nano sensors to ensure comprehensive ethylene monitoring, thereby enabling more precise detection of spoilage regions within food packages.

12.3.4 Bio Nano Sensor Design and Functionality

Each bio nano sensor is engineered with high sensitivity and specificity to detect trace levels of ethylene gas. These sensors, visualized as yellow dots in our study's schematics, consist of a nanostructured material layer functionalized with ethylene-sensitive molecules [19]. When ethylene molecules come into contact with the sensor surface, they trigger a detectable response, typically a change in electrical or optical signal, which is then transmitted to a data acquisition system. This design allows for real-time ethylene detection at the molecular level, facilitating rapid and accurate assessments of fruit ripeness and spoilage.

12.3.5 Sensor Placement and Clustering Strategy

To ensure thorough spatial coverage of ethylene emissions throughout each food package, multiple bio nano sensors are grouped into clusters and strategically distributed in different regions within the packed items. The clustering of sensors within the package is central to capturing a representative dataset, as ethylene emissions may vary in concentration and origin within the product due to non-uniform ripening or localized spoilage. Each cluster is positioned to cover distinct areas of the food item, forming a sensor grid that enables both regional monitoring and detection of ethylene hotspots. These hotspots typically correspond to specific areas where microbial activity or enzymatic reactions lead to elevated ethylene release, signaling the early stages of degradation. These hotspots may vary based on organic substance in our case: general purpose fruits, such as banana or blueberry. Specifically for banana this place is tip of fruit. But placing clusters in strategic locations, such as the top, middle, and bottom layers of the packed food items, the methodology accounts for potential variations in gas concentration due to packaging orientation, temperature gradients, and physical proximity to ripening fruits.

12.3.6 Spatial Coverage and Redundancy

The distributed approach with multiple clusters provides enhanced spatial coverage and redundancy. By ensuring that each part of the package is monitored by at least one sensor cluster, the likelihood of undetected ethylene emissions is minimized. This redundancy is especially critical in larger food packages where uneven distribution of ripening agents may result in isolated spoilage zones. Additionally, redundancy within clusters allows for error correction and data validation, as signals from multiple sensors within a single cluster can be cross-referenced to confirm ethylene presence.

12.3.7 Data Collection and Signal Processing

Each sensor within the cluster continuously transmits data on ethylene concentration levels to a central data processing unit which is called nano router. These nano routers aggregate the signals from all cluster sensors, providing a comprehensive dataset that represents the ethylene distribution within the entire package. Advanced signal processing techniques, including data filtering and noise reduction, are applied to enhance the accuracy and reliability of ethylene measurements. By analyzing variations in signal intensity and frequency, it becomes possible to identify specific areas within the package where ethylene concentrations are highest, pinpointing the initial regions of spoilage.

12.3.8 Noise Reduction Technique

Due to the sensitivity of these bio nano sensors and environmental factors such as temperature fluctuations, humidity, and minor signal interferences, the collected raw data can contain significant noise. So, in the data collection process, the ethylene concentration data at nano routers is further transmitted in real time to an interface layer, where initial data processing and noise reduction take place. Noise reduction at the interface layer is, therefore, a critical step in ensuring that the ethylene concentration readings accurately reflect the actual conditions within the packed food items. To address this, three advanced denoising methods—Bi-LSTM, LSTM, and XLSTM—are tested to identify the most efficient approach for noise filtering and signal clarity enhancement [20].

12.3.9 Data Compression (Squeezing) before Denoising

Before the denoising stage, the cleaned data is subjected to a "squeezing" or compression process, which reduces its size while preserving essential information. This compression is crucial for optimizing data storage and transmission, especially in a real-time monitoring system that handles large volumes of sensor data across multiple food packages. The squeezing process is carefully calibrated to maintain the integrity of the ethylene concentration data; the compression ratio is set to minimize data loss, ensuring that all relevant ethylene emission trends and variations are retained.

The data compression process involves the following steps:

- **Data Aggregation**: Similar data points are aggregated into compressed packets that maintain the overall trend and statistical properties of the original data, reducing redundancy without sacrificing information. This is achieved using normalization of data through the min–max normalization technique defined by the formula given in the below equation. where x' is the normalized value of data.

$$x' = \frac{x - x_{\min}}{x_{\max} - x_{\min}}$$

- **Dimensionality Reduction**: Techniques such as *principal component analysis* (PCA) may be applied to reduce the dimensionality of the data, further minimizing the data footprint while preserving essential signal characteristics.

This compressed data, now smaller in size but rich in meaningful information, is prepared for further analysis and storage. By reducing data size at the interface layer, this step not only enhances data transmission efficiency but also improves computational efficiency for downstream processing steps, such as spoilage prediction and shelf-life estimation.

12.3.10 Comparative Noise Filtering between Different LSTM Models

Standard LSTM is a modified recurrent neural network (RNN) architecture with Memory Gate to hold and carry forward previous information with current stream of data. LSTM is widely used in time-series data processing for its ability to capture dependencies in sequential data [21]. In this context, LSTM is employed to filter noise by analyzing the temporal patterns in ethylene sensor data, identifying signal trends that represent true ethylene readings while discarding transient fluctuations that are likely to be noise. However, Standard LSTM processes data in a single direction (forward), which may limit its capacity to capture complex temporal dependencies in the data, especially when noise exhibits both forward and backward correlations in time. Additionally, although effective, Standard LSTM has relatively higher computational requirements compared to other methods, making it a less ideal choice for real-time applications in large-scale food monitoring systems. We have plotted the comparison between data scale before and after using standard LSTM in Figure 12.3. The background red wave graphs represent the initial data received from nano routers, blue wave graphs represent the data after using pre-processing normalization/scaling techniques which will be discussed in further sections, and the green wave graph represents the denoised/compressed data. This shows effective reduction in noise, as well as reduction in data packet size, as non-required amplitudes and peaks are removed.

XLSTM is an advanced modification of the LSTM architecture designed to enhance noise reduction capabilities by increasing the model's ability to retain

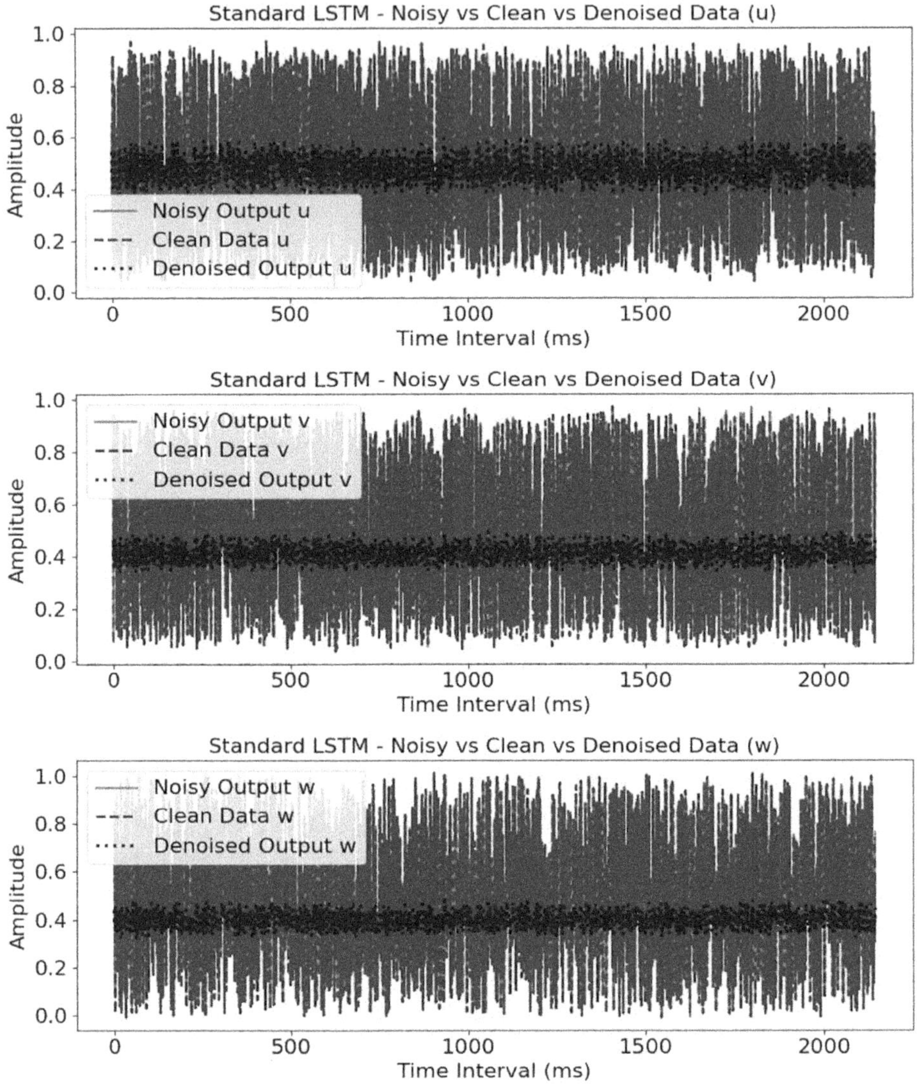

FIGURE 12.3 Comparison of collected data, cleaned data, and de-noised data using Standard LSTM model.

long-term dependencies and patterns within the data [22]. This extension includes additional layers and a larger memory gate capacity, which improves the ability to process more time-series data at a time enabling the filtering of complex noise signals as in Figure 12.4. However, due to the extended architecture, XLSTM has a high computational cost and memory requirement, making it computationally intensive. While XLSTM achieves high accuracy in noise reduction, its complexity and resource demands make it less practical for deployment in the real-time processing environment of ethylene monitoring, where efficiency is paramount.

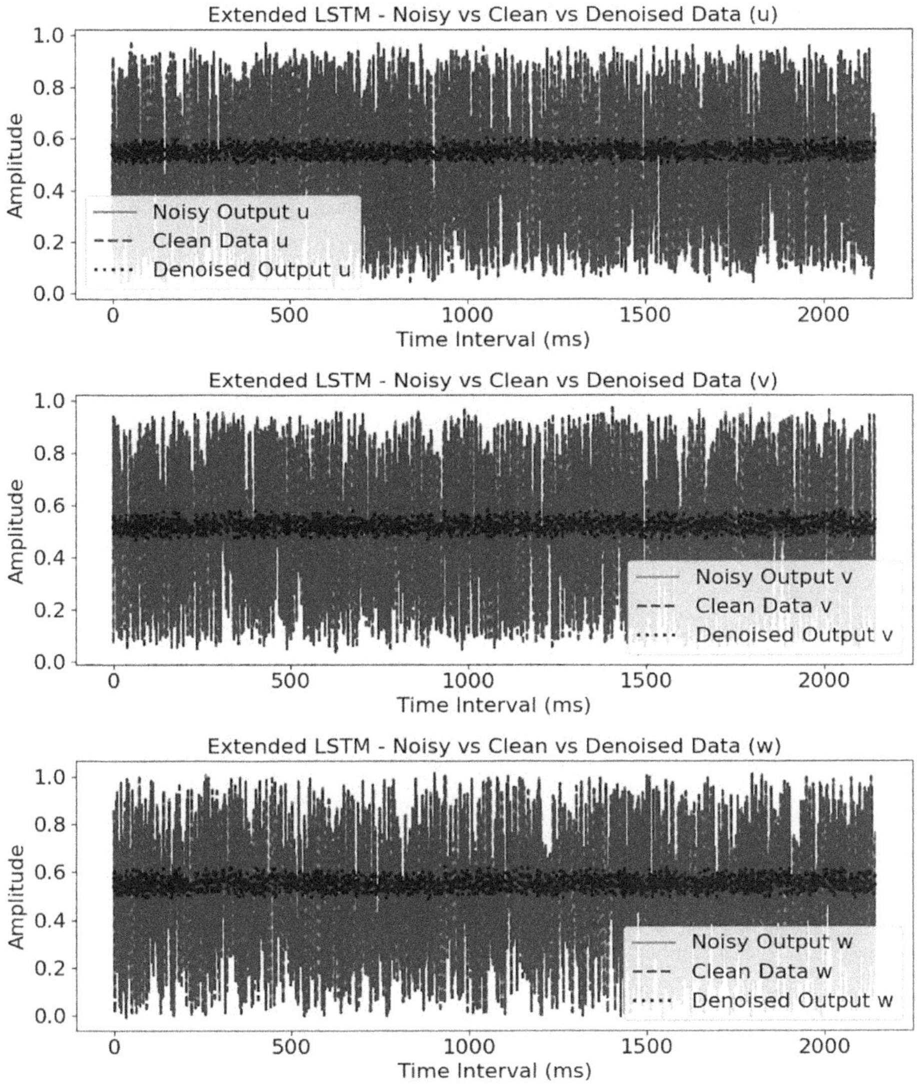

FIGURE 12.4 Comparison between collected data, cleaned data, and denoised data using Extended LSTM (XLSTM) model.

Bi-LSTM, the chosen method for noise reduction, addresses the limitations of both Standard LSTM and XLSTM by processing the data in both forward and backward directions. This bidirectional approach enables the model to capture both past and future dependencies within the ethylene emission data, making it more adept at distinguishing genuine signals from noise. By taking advantage of temporal relationships in both directions, Bi-LSTM achieves a higher accuracy in noise filtering than Standard LSTM without the extensive computational load of XLSTM. Although steps from both direction increase the computation cost, the model was able to achieve comparative efficiency even when trained with lesser number of epochs (50 training steps), whereas other models' performance with 50 epochs was not efficient enough. This balance between performance and computational efficiency makes Bi-LSTM the optimal choice for noise reduction in this methodology that can be seen in Figure 12.5. After extensive testing, Bi-LSTM is selected for its superior noise reduction capabilities with minimal resource consumption, allowing for accurate, low-latency processing of the ethylene sensor data.

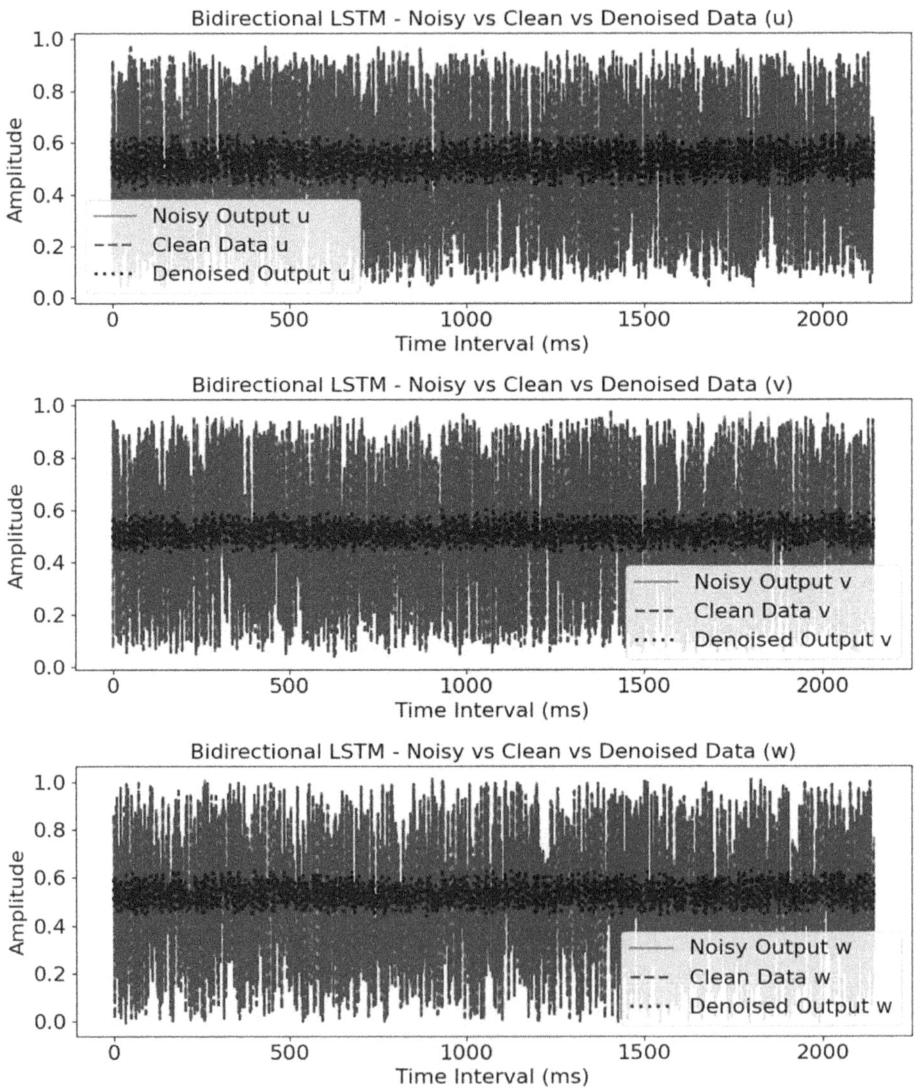

FIGURE 12.5 Comparison between collected data, cleaned data, and denoised data using Bidirectional LSTM model.

12.3.11 Noise Filtering Process Using Bi-LSTM

Upon receiving the raw data at the interface layer, Bi-LSTM is applied to each sensor cluster's readings. The Bi-LSTM model is trained to recognize patterns in ethylene concentration data that correspond to genuine signals, discarding spurious variations as noise. During this training phase, the Bi-LSTM model is exposed to both clean and noisy ethylene data samples, learning to differentiate between them based on temporal patterns. Once trained, the Bi-LSTM algorithm is able to filter noise with high precision, ensuring that the resulting data stream accurately represents ethylene concentrations in the food package. The noise filtering process takes place in two primary stages.

- **Preliminary Filtering**: Initial data processing involves filtering out large, obvious noise spikes that are likely due to environmental factors or brief signal disturbances.
- **Refined Filtering**: The Bi-LSTM model then performs refined filtering, analyzing subtler patterns and distinguishing nuanced noise artifacts from genuine ethylene signal variations [23]. This refined filtering allows for a cleaner, more accurate data stream that reflects real ethylene concentration levels, ensuring that minor but significant changes in ethylene emissions are not masked by residual noise.

The Bi-LSTM is an extension of the standard LSTM network that processes data in both forward and backward directions to capture temporal dependencies from past and future states. Here are the main equations used in an LSTM cell (unidirectional). In Bi-LSTM, we maintain two LSTMs, one for each direction, and then combine the outputs.

An LSTM cell has a memory cell c_t and three gates: the input gate i_t, forget gate f_t, and output gate o_t. The updated equations are as follows:

$$\text{Forget gate}: f_t = \sigma\left(W_f \cdot \left[h_{t-1}, x_t\right] + b_f\right)$$

$$\text{Input gate}: i_t = \sigma\left(W_i \cdot \left[h_{t-1}, x_t\right] + b_i\right)$$

$$\text{Candidate cell state}: \tilde{c}_t = \tanh\left(W_c \cdot \left[h_{t-1}, x_t\right] + b_c\right)$$

$$\text{Cell state update}: c_t = f_t \odot c_{t-1} + i_t \odot \tilde{c}_t$$

$$\text{Output gate}: o_t = \sigma\left(W_o \cdot \left[h_{t-1}, x_t\right] + b_o\right)$$

$$\text{Hidden state}: h_t = o_t \odot \tanh(c_t)$$

where

- W terms are weight matrices for each gate.
- b terms are biases for each gate.

- σ represents the sigmoid activation function.
- tanh represents the hyperbolic tangent activation function.
- $\odot$ denotes element-wise multiplication.

In a Bi-LSTM, we use both a forward LSTM (as defined above) and a backward LSTM that processes the sequence in reverse. The final output at each time step t is obtained by concatenating the forward and backward hidden states:

$$\text{Bi-LSTM output at time } t = \text{concatenate}\left(h_t^{\text{forward}},\ h_t^{\text{backward}}\right)$$

12.3.12 Detection Using FNN at the IoT Gateway

After the raw ethylene data has undergone noise reduction and data compression at the interface layer, the cleaned and compressed data stream is transmitted to an IoT Gateway [24]. This gateway serves as a central processing hub, managing the real-time data flow from multiple sensor-equipped packages and facilitating advanced analysis of ethylene concentration levels. At this stage, the processed data is fed into a FNN classifier to identify and classify instances of ethylene gas presence, which serves as a predictive marker for fruit ripening and potential spoilage [24]. The primary purpose of the FNN classifier is to distinguish data segments indicative of ethylene emissions, thereby signaling stages of food freshness and enabling timely quality control interventions.

The FNN implemented at the IoT Gateway is designed as a multi-layer perceptron (MLP), comprising an input layer, one or more hidden layers, and an output layer [24]. Each layer consists of interconnected nodes or neurons that process incoming data by applying activation functions, allowing the network to identify patterns and trends within the ethylene concentration data.

12.3.13 Data Classification Process

The FNN classifier at the IoT Gateway receives a continuous, unlabeled data stream, meaning it has no prior indication of whether ethylene is present in each segment of the input data. The FNN classifier continuously evaluates the incoming data stream, identifying segments with ethylene gas and classifying them accordingly. The classification process operates as follows:

Data Ingestion: As new data arrives from the interface layer, it is batch-processed through the input layer of the FNN. Each batch corresponds to a specific time interval or cluster within the food package, allowing the model to classify ethylene emissions in a temporally and spatially localized manner.

Feature Extraction and Transformation: In the hidden layers, the data undergoes a transformation process, where weights are applied to each feature (e.g., ethylene concentration level, trend patterns) to generate a set of intermediate feature representations. These representations are passed through activation functions, such as ReLU or sigmoid, to introduce nonlinearities, enabling the network to capture complex relationships between features. This stage allows the FNN to recognize patterns associated with ethylene release, distinguishing them from random fluctuations in the data.

Binary Classification: The transformed data reaches the output layer, where the FNN classifier generates a binary prediction for each data segment: presence or absence of ethylene gas. When ethylene presence is detected, it indicates that ripening processes or potential spoilage is actively occurring in the packaged food, providing an early warning signal for quality degradation. Conversely, an absence of ethylene suggests that the food remains in a fresher state, with minimal ripening or spoilage activity.

Confidence Scoring and Decision Thresholds: To improve classification accuracy, the FNN assigns a confidence score to each prediction, which quantifies the likelihood that a given data segment contains ethylene emissions. A threshold is then applied to these confidence scores, determining whether a segment is definitively classified as ethylene-present. This threshold is calibrated based on historical ethylene emission data to minimize false positives and negatives, balancing sensitivity and specificity in the detection process.

The FNN classifier's ability to detect ethylene emissions in real time enables timely identification of spoilage-prone regions within the packed food. As ethylene release is part of the natural ripening process, an increase in ethylene concentration within specific areas of the package serves as an early indicator of food deterioration. By continuously monitoring ethylene levels, the FNN classifier provides dynamic spoilage detection, flagging areas with elevated ethylene as "at-risk" zones. This information can then be relayed to downstream quality control systems or alert mechanisms that trigger appropriate interventions, such as adjusting storage conditions or removing affected packages from distribution.

12.4 SIMULATION RESULTS

The simulation process was carried out in a Python environment using Jupyter Notebook and Google Colab to leverage their computational resources, enabling us to efficiently train and evaluate multiple models on large and complex datasets. The primary goal was to use various LSTM-based configurations to denoise noisy nanosensor data, enhancing the accuracy and reliability of ethylene gas detection, a critical factor in monitoring fruit spoilage in smart packaging.

Three LSTM configurations—Standard LSTM, Bi-LSTM, and XLSTM—were implemented and evaluated for their effectiveness in denoising. Each LSTM model was trained with a consistent configuration of 100 epochs, a hidden dimension of 50, two layers, and a learning rate of 0.001. The Bi-LSTM outperformed the other configurations in terms of denoising capability, showing the lowest RMSE and highest R^2 score as shown in Table 12.1. This model's bidirectional approach allows it to

TABLE 12.1
Comparison between Multiple Denoising LSTM-Based Models with Different Hyper-Parameters

Model	Configuration	RMSE Error	R^2 Score	Loss after Epochs
Standard LSTM	50 epochs, 50 hidden dimension, 2 num layers, 0.001 learning rate	0.2676	0.149	0.0912
Standard LSTM	100 epochs, 50 hidden dimension, 2 num layers, 0.001 learning rate	0.2617	0.1856	0.089
Bidirectional LSTM	50 epochs, 50 hidden dimension, 2 num layers, 0.001 learning rate	0.2621	0.1829	0.097
Bidirectional LSTM	100 epochs, 50 hidden dimension, 2 num layers, 0.001 learning rate	0.2733	0.1123	0.0873
Extended LSTM	50 epochs, 50 hidden dimension, 2 num layers, 0.001 learning rate	0.275	0.0995	0.0957
Extended LSTM	100 epochs, 50 hidden dimension, 2 num layers, 0.001 learning rate	0.2773	0.0863	0.0874

TABLE 12.2
Comparison between FNN and DBN Model with Different Hyper-Parameters

Model	Configuration	Testing Accuracy	Precision	Recall	F1 Score
Feed-Forward Network	100 epochs, 3 layers (64-32-1)	0.8116	0.7515	0.9347	0.8331
Feed-Forward Network	500 epochs, 3 layers (64-32-1)	0.9830	0.9784	0.9879	0.9831
Deep Belief Network	1 RBM layer (256 comps), Logistic Regression (100 iterations)	0.4864	0.4947	0.4917	0.4932
Deep Belief Network	1 RBM layer (256 comps), Logistic Regression (500 iterations)	0.4861	0.4944	0.4867	0.4905

capture both past and future dependencies, making it particularly effective in handling complex, noisy datasets. The Standard LSTM and XLSTM configurations, while performing slightly lower than the Bi-LSTM, still showed significant noise reduction capabilities and provided substantial improvements to the data quality. A secondary evaluation with 50 epochs was also conducted to examine the stability of each model over shorter training cycles. Results from this run showed minimal changes in RMSE and R^2 values, indicating that these LSTM models are robust and efficient in handling noisy data, suggesting that even with lesser computation, it can reflect similar results, which is ideal for real-world problems. After denoising, the processed data was used for ethylene gas classification, an essential step in assessing fruit spoilage. For this purpose, both FNNs and a Deep Belief Network (DBN) were employed. The feed-forward network, configured with 500 epochs and three layers (64-32-1 architecture), significantly outperformed other models, achieving an accuracy of 98.30%. This model also achieved high precision, recall, and F1 scores as visible in Table 12.2, demonstrating its strong ability to correctly classify ethylene presence, an indicator of fruit spoilage. The high accuracy, coupled with excellent recall and precision, suggests that the feed-forward network is highly reliable for binary classification in this context.

In contrast, the DBN struggled with accuracy, as well as with other evaluation metric. The DBN's high rate of false positives and its true negatives limited its applicability for accurate classification, possibly due to its sensitivity to the high-dimensional, intricate patterns in nanosensor data. While the DBN maintained similarity in numbers of precision and recall as shown in Table 12.2, it effectively states that a balanced dataset (which should have given better results), the overall performance, efficiency, and usability were incomparable to the feed-forward network. The confusion matrix for the FNN and DBN is represented in Figures 12.6 and 12.7 respectively.

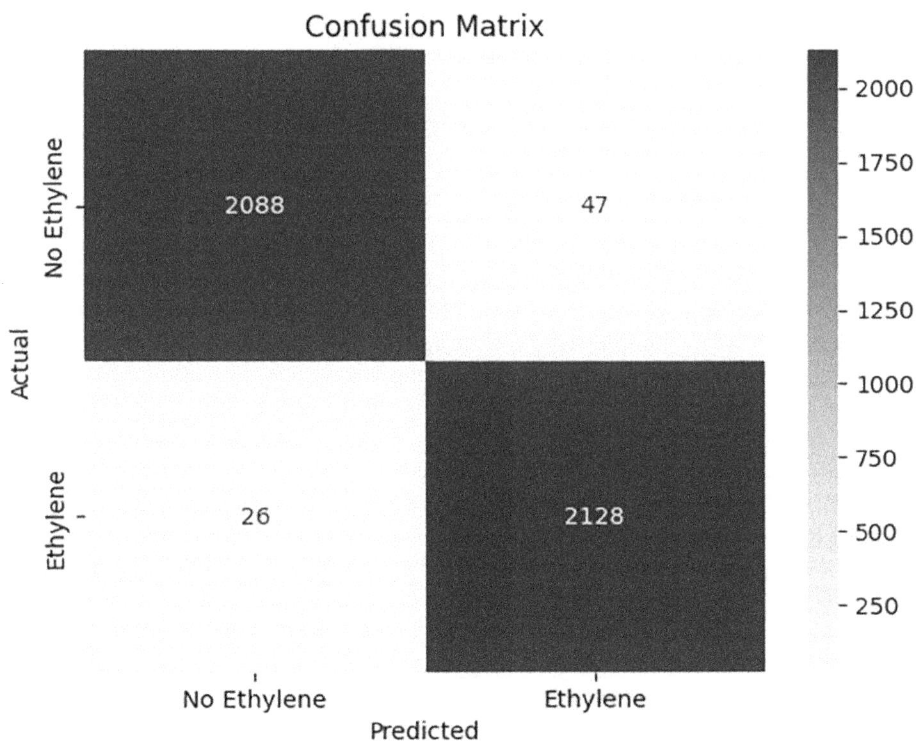

FIGURE 12.6 Confusion matrix of feed-forward network model.

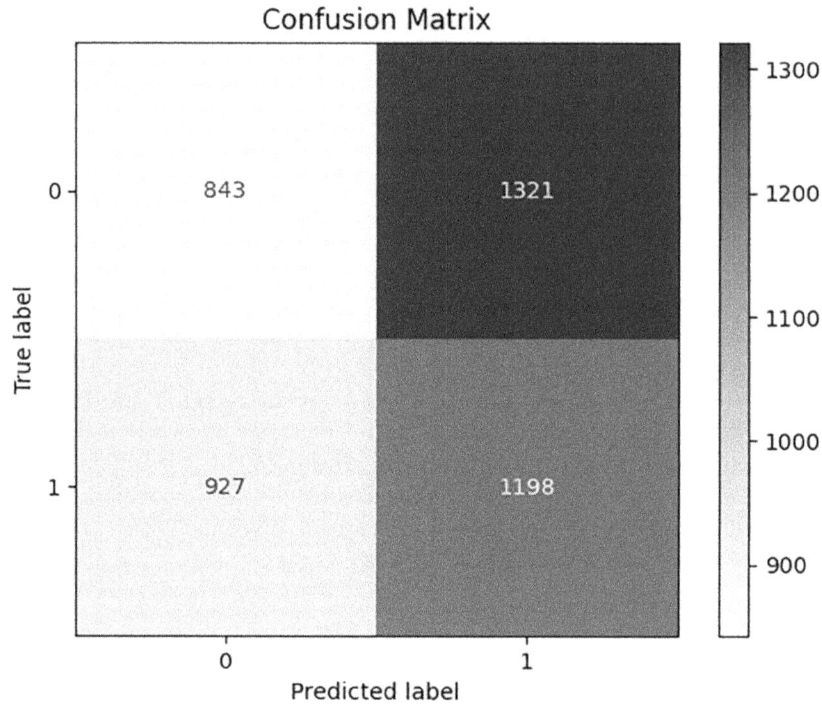

FIGURE 12.7 Confusion Matrix of deep belief network model.

12.5 CONCLUSION

In this study, we demonstrated the effectiveness of LSTM-based models for reducing noise in nanosensor data, followed by high-accuracy classification for ethylene detection. Our results highlight that the Bi-LSTM model provided the highest denoising efficiency, with lesser number of training epochs (indicating quicker convergence toward global minimum) significantly enhancing data quality and allowing the FNN classifier to achieve an impressive 98.30% accuracy in detecting ethylene. This pipeline includes initial standard scaling of data using Min-Max scaler (could be standard scaler also), combining LSTM-based de-noising with feed-forward classification, and establishes a promising framework for real-time ethylene detection in smart food packaging, supporting enhanced quality control and reduced spoilage rates in the supply chain. The comparison shown in Table 12.3 shows that our models outperform existing ones in prediction and classification tasks. Our Bi-LSTM achieved a lower RMSE (0.2621 vs. 0.4606), and our feed-forward network improved accuracy (98.30% vs. 95%). These results demonstrate the robustness and effectiveness of our approaches in diverse applications.

The challenges of using nanosensors in food safety include high production costs and obtaining standard approval. Additionally, some nanosensors may degrade or lose sensitivity over time. However, there are numerous scenarios in which the ability to identify pathogens and contaminants occurs quickly and accurately. It is

TABLE 12.3
Comparison of Our Models with Models from Different Research

Authors	Research	Model Used	Metrics
Yang and Wang [25]	Adaptability of financial time series prediction based on Bi-LSTM	Bi-LSTM (optimized)	RMSE value of 0.46059286
	Our proposed model using Bi-LSTM	Bi-LSTM with 50 epochs and 50 hidden dimensions	RMSE value of 0.2621
Huang et al. [26], Navazi et al. [27]	Feature selection and classification model construction on type 2 diabetic patients' data	IBL (Instance-Based Learning)	Predictive accuracy of 95% [26]
	Our proposed model for effective classification	Multilayer feed-forward network with 100 epochs (tri-layer network)	Predictive accuracy of 98.30%

possible to implement real-time monitoring and improved tracking throughout the food supply chain. To monitor temperature changes and identify the presence of bacteria in packed meat products, nanosensors will be placed in refrigerated shipping containers in a real-world testing scenario. In addition to improving food safety, this kind of real-time monitoring and enhanced traceability along the food supply chain lowers spoilage. Better spoilage detection can also lead to less waste.

ML and artificial intelligence (AI) are transformative tools for optimizing food processes, allowing the industry to operate more efficiently and sustainably. These technologies assist in optimizing production workflows, forecasting demand, and automating quality control processes, thereby increasing productivity and reducing operational waste. Furthermore, by interpreting consumer behavior and preferences, AI-powered systems help to create personalized customer experiences such as tailored nutrition plans. As these technologies advance, they hold great promise for increasing the efficiency and responsiveness of the global food industry. Future work could aim at minimizing these latency issues by optimizing the LSTM model through methods such as model pruning, quantization, or the exploration of alternative architectures like attention mechanisms. Incorporating edge computing for localized data processing could also alleviate delays by reducing dependence on the central IoT gateway. Further research into hybrid LSTM-transformer models or transfer learning could enhance noise reduction efficiency and overall processing speed, enabling more scalable and effective deployment in IoT-enabled smart packaging systems.

REFERENCES

1. J. O. Martinez, M. Evangelopoulos, C. Chiappini, X. Liu, M. Ferrari, and E. Tasciotti, "Degradation and biocompatibility of multistage nanovectors in physiological systems," *Journal of biomedical materials research Part* A, vol. 102, pp. 3540–3549, Oct. 2014.
2. E. Sasson, R. V. O. Pinhasi, S. Margel, and L. Klipcan, "Engineering and use of proteinoid polymers and nanocapsules containing agrochemicals," *Scientific Reports*, vol. 10, p. 9171, Jun. 2020.

3. S. Khaki and L. Wang, "Crop yield prediction using deep neural networks," *Frontiers in Plant Science*, vol. 10, p. 621, May 2019.
4. S. Dwivedi, Q. Saquib, A. A. Al-Khedhairy, and J. Musarrat, "Understanding the role of nanomaterials in agriculture," In D. P. Singh, H. B. Singh, and R. Prabha (Eds.), *Microbial Inoculants in Sustainable Agricultural Productivity: Vol. 2: Functional Applications*, Springer pp. 271–288, 2016.
5. S.-Y. Kwak, M. H. Wong, T. T. S. Lew, G. Bisker, M. A. Lee, A. Kaplan, J. Dong, A. T. Liu, V. B. Koman, R. Sinclair, and C. Hamann, "Nanosensor technology applied to living plant systems," *Annual Review of Analytical Chemistry*, vol. 10, pp. 113–140, Jun. 2017.
6. D. C. Ferrier and K. C. Honeychurch, "Carbon nanotube (CNT)-based biosensors," *Biosensors*, vol. 11, p. 486, Nov. 2021.
7. D. Chen, Y. Cai, L. Cheng, S. Guo, T. Liu, S. Huang, H. Yu, Y. Wang, Z. Hu, and D. Gui, "Structure and function design of carbon nanotube-based flexible strain sensors and their application," *Measurement*, vol. 225, p. 113992, Feb. 2024.
8. M. Penza, P. J. Martin, and J. T. Yeow, "Carbon nanotube gas sensors," In C.-D. Kohl and T. Wagner (Eds.), *Gas Sensing Fundamentals*, Springer, pp. 109–174, 2014.
9. J. Li and M. Meyyappan, "Gas composition sensing using carbon nanotube arrays," Sept. 23 2008. US Patent 7,426,848.
10. Y. Wang and J. T. Yeow, "A review of carbon nanotubes-based gas sensors," *Journal of Sensors*, vol. 2009, no. 1, p. 493904, 2009.
11. M. W. Thomas, "Noise reduction for LSTM using wavelet transform and singular spectrum analysis," Leland Stanford Junior University: Stanford, CA, USA, 2021.
12. S. Lee and D. Park, "LSTM-based post-processing for noise reduction in SVD-based particulate matter digital twinning on lightweight embedded devices," *Electronics Letters*, vol. 60, no. 4, p. e13132, 2024.
13. I. F. Akyildiz, J. M. Jornet, and C. Han, "Terahertz band: Next frontier for wireless communications," *Physical Communication*, vol. 12, pp. 16–32, 2014.
14. H. Taghvaee, A. Pitilakis, O. Tsilipakos, A. C. Tasolamprou, N. V. Kantartzis, M. Kafesaki, A. Cabellos-Aparicio, E. Alarcon, and S. Abadal, "Multiwideband terahertz communications via tunable graphene-based metasurfaces in 6g networks: Graphene enables ultimate multiwideband THz wavefront control," *IEEE Vehicular Technology Magazine*, vol. 17, no. 2, pp. 16–25, 2022.
15. "Carbon Nanotubes." www.kaggle.com/datasets/inancigdem/carbon-nanotubes, 2017. [Accessed 07-11-2024].
16. B. Franco, L. Clarisse, M. Van Damme, J. Hadji-Lazaro, C. Clerbaux, and P.-F. Coheur, "Ethylene industrial emitters seen from space," *Nature Communications*, vol. 13, pp. 6452–6462, Oct. 2022.
17. K. L. Yam, P. T. Takhistov, and J. Miltz, "Intelligent packaging: Concepts and applications," *Journal of Food Science*, vol. 70, no. 1, pp. R1–R10, 2005.
18. B. Kuswandi, Y. Wicaksono, Jayus, A. Abdullah, L. Y. Heng, and M. Ahmad, "Smart packaging: Sensors for monitoring of food quality and safety," *Sensing and Instrumentation for Food Quality and Safety*, vol. 5, pp. 137–146, 2011.
19. V. S. Manikandan, B. Adhikari, and A. Chen, "Nanomaterial based electrochemical sensors for the safety and quality control of food and beverages," *Analyst*, vol. 143, no. 19, pp. 4537–4554, 2018.
20. I. Malashin, V. Tynchenko, A. Gantimurov, V. Nelyub, and A. Borodulin, "Applications of long short-term memory (LSTM) networks in polymeric sciences: A review," *Polymers*, vol. 16, no. 18, p. 2607, 2024.
21. A. Sherstinsky, "Fundamentals of recurrent neural network (RNN) and long short-term memory (LSTM) network," *Physica D: Nonlinear Phe*nomena, vol. 404, p. 132306, 2020.

22. K. Pöppel, M. Beck, M. Spanring, A. Auer, O. Prudnikova, M. K. Kopp, G. Klambauer, J. Brandstetter, and S. Hochreiter, "xLSTM: Extended long short-term memory," in *First Workshop on Long-Context Foundation Models@ ICML 2024,* Vienna, Austria, 2024.
23. B. C. Ooi, H. H. Pang, H. Wang, L. Wong, and C. Yu, "Fast filter-and-refine algorithms for subsequence selection," in *Proceedings International Database Engineering and Applications Symposium*, Edmonton, AB, Canada, pp. 243–254, IEEE, 2002.
24. S. A. Bakhsh, M. A. Khan, F. Ahmed, M. S. Alshehri, H. Ali, and J. Ahmad, "Enhancing IoT network security through deep learning-powered intrusion detection system," *Internet of Things*, vol. 24, p. 100936, 2023.
25. M. Yang and J. Wang, "Adaptability of financial time series prediction based on bilstm," *Procedia Computer Science*, vol. 199, pp. 18–25, Jan. 2022.
26. Y. Huang, P. McCullagh, N. Black, and R. Harper, "Feature selection and classification model construction on type 2 diabetic patients' data," *Artificial Intelligence in Medicine*, vol. 41, pp. 251–262, Nov. 2007.
27. F. Navazi, Y. Yuan, and N. Archer, "An examination of the hybrid meta-heuristic machine learning algorithms for early diagnosis of type ii diabetes using big data feature selection," *Healthcare Analytics*, vol. 4, p. 100227, Dec. 2023.

13 Future Trends in Multimedia and Multimodal Intelligence for Smart Farming

Vipin Kumar Sahu, Shweta Rane, and Brijendra Gupta

13.1 INTRODUCTION

Agriculture has long served as a fundamental pillar of human civilization, progressively transitioning from conventional methods to technologically advanced precision agriculture [1]. The rapid increase in global population, coupled with the escalating impact of climate change, has introduced complex challenges in achieving food security, environmental sustainability, and economic viability within the agricultural domain [2]. In response, the integration of smart farming approaches has become critical for enhancing yield efficiency and optimizing resource utilization [1] (Figure 13.1).

Smart agriculture encompasses the deployment of cutting-edge information and communication technologies (ICTs), Internet of Things (IoT) infrastructure, machine learning algorithms, and artificial intelligence (AI) systems [3]. Recent developments in sensor technologies, autonomous machinery, satellite-based remote sensing, and cloud-based analytics facilitate extensive data collection and real-time analysis, thereby enabling data-informed strategies and adaptive management practices across agricultural systems [4].

Agricultural ecosystems are intrinsically dynamic and spatially heterogeneous, producing extensive volumes of multimodal data encompassing visual (drone and satellite imagery), video (continuous surveillance feeds), textual (sensor-generated logs), auditory (vocalizations from livestock and operational machinery), and environmental sensor outputs (e.g., temperature, humidity, soil moisture) [5,6]. The emergence of multimodal and multimedia intelligence—leveraging AI methodologies to process, correlate, and synthesize these heterogeneous data sources—has significant implications for advancing agricultural monitoring, diagnostics, and automation [7,8].

The integration of multimedia data enables precise assessment of phenological development, detection of biotic stressors, behavioural analysis of livestock, and evaluation of agro-mechanical system efficiency [9,10]. When fused with supplementary modalities, such as meteorological projections and agrochemical profiling, this

 DOI: 10.1201/9781003637264-17

composite intelligence framework enhances the fidelity, adaptability, and contextual relevance of agro-technical decision support systems [11] (Figure 13.2).

The combination of deep learning, edge computing, 5G technology, and autonomous systems is driving the emergence of the next generation of smart agriculture [12,13].

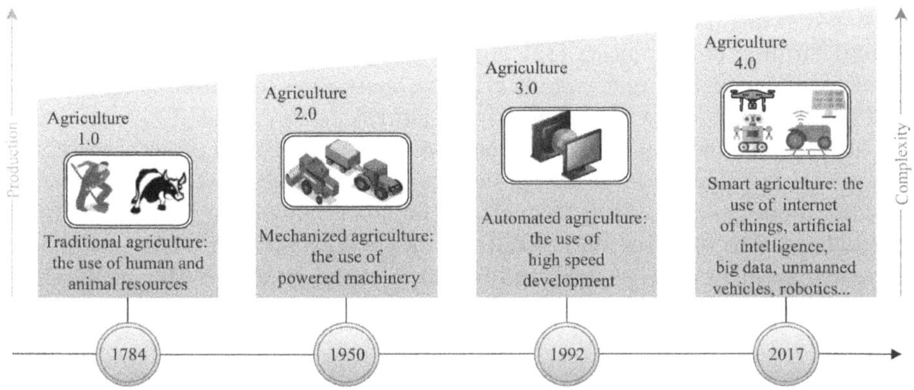

FIGURE 13.1 The four transformative phases in agricultural evolution.

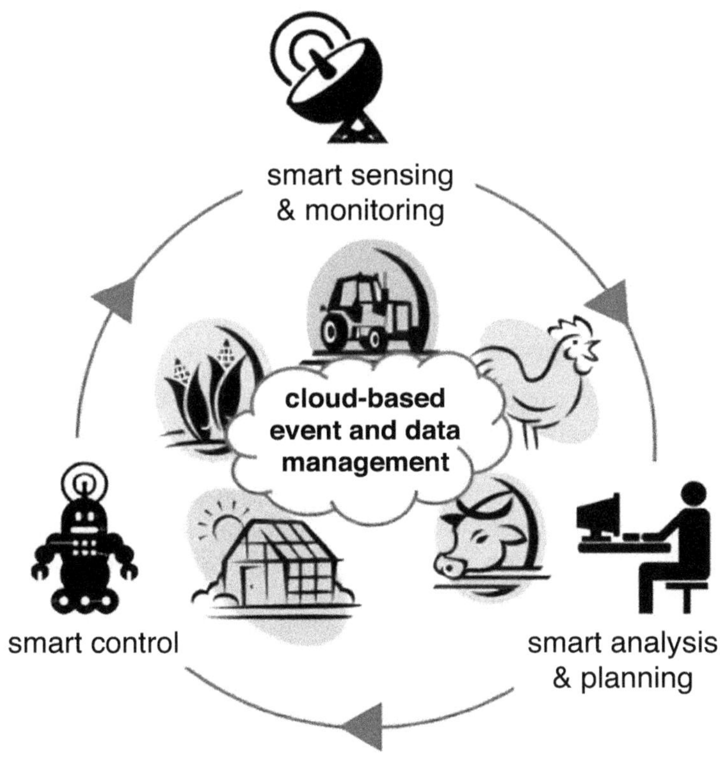

FIGURE 13.2 The cyber-physical control loop in smart agriculture, augmented by cloud-based event processing and data management infrastructure.

Vision-based systems using convolutional neural networks (CNNs) have shown exceptional capability in detecting plant diseases and estimating crop yields [14]. Likewise, researchers are investigating transformer models and attention-based architectures to manage diverse data integration in multimodal systems [15].

Real-time analytics provided by edge devices and fog computing support on-site decision-making, reducing latency and bandwidth challenges typically encountered in rural farming environments [12]. Additionally, federated learning and privacy-preserving AI methods are becoming effective approaches to utilize distributed data from various farms while upholding privacy [16,17].

This chapter examines the rising trends and future prospects of multimedia and multimodal intelligence in smart agriculture. It investigates several key elements, including: cutting-edge sensing technologies and their integration; machine learning approaches for the fusion of multimodal data; applications of multimedia intelligence in the management of crops, soil, and livestock; issues related to data standardization, privacy, and scalability; a forward-looking perspective on autonomous farming systems.

13.2 RELATED WORK

13.2.1 Evolution of Smart Farming Technologies

The advancement of smart farming systems has been intrinsically coupled with technological innovations in sensor networks, autonomous control systems, and computational analytics [4]. Conventional mechanized agricultural practices prioritized macroscopic operational efficiency. In contrast, contemporary precision agriculture, enabled by distributed IoT architectures and embedded cyber-physical systems, facilitates real-time monitoring and closed-loop control of agronomic variables including soil moisture (irrigation), nutrient levels (fertilization), and biotic stress factors (integrated pest management) [18].

13.2.2 Use of Multimedia Data in Agriculture

The incorporation of multimodal agricultural datasets—including multispectral imagery, spatiotemporal video feeds, bioacoustics recordings, and ancillary sensor telemetry—has driven a paradigm shift in conventional farming methodologies [19]. With the transition from manual, heuristic-based practices to sensor-fused automated systems, these heterogeneous data streams now constitute the primary input layer for AI-driven monitoring pipelines, automated pathological diagnostics, and closed-loop control mechanisms [20]. The high-dimensional feature space and context-aware semantic richness inherent in these datasets enable data-driven optimization of agricultural processes across heterogeneous agro-ecological zones [21].

13.2.2.1 Image-Based Applications

Image data, particularly from drones, satellites, and smartphone cameras, has found wide application in plant phenotyping, disease detection, weed classification, and crop growth monitoring [22]. Mohanty et al. conducted one of the seminal studies in this field by training deep CNNs on the Plant Village dataset, achieving accuracy levels greater than 99% for classifying plant diseases across 14 species [10]. This demonstrated the potential of image-based deep learning models in automating plant health diagnostics at scale.

Subsequent studies expanded on this by using architectures such as ResNet, DenseNet, and EfficientNet for higher accuracy and faster inference. Ferentinos reported the use of transfer learning on agricultural image datasets to improve robustness and generalization, especially in conditions with limited training data [15].

13.2.2.2 Video Surveillance in Livestock Monitoring

Video data is extensively used in livestock farms for continuous surveillance and behavioural analysis [19]. Nasirahmadi et al. developed an automated monitoring system for pigs that employed motion tracking and computer vision techniques to detect abnormal behaviours such as tail biting and lethargy [23]. Such systems reduce the need for manual inspection and enable early intervention, thus improving animal welfare and farm productivity [9]. Advanced video analytics systems now use object detection models like YOLO (You Only Look Once) and action recognition networks to track and classify complex animal movements. These techniques are also applied to poultry farms for identifying feeding patterns, aggression, or distress behaviours, contributing to better resource allocation and health management [9].

13.2.2.3 Audio-Based Monitoring Systems

Audio data represents another underutilized but emerging modality in precision farming. Cardenas and Lopez demonstrated that audio signals from farm machinery could be analysed using spectrogram-based techniques and Recurrent Neural Networks (RNNs) to detect anomalies indicative of mechanical faults [9]. In another study, animal vocalizations were recorded and classified using machine learning to detect stress or illness in cattle and pigs [6].

Audio surveillance has also been proposed for monitoring bird calls in agroforestry systems, helping track biodiversity as a measure of ecological balance [6].

13.2.2.4 Multimodal Multimedia Integration

Recent advancements have focused on combining different multimedia sources to improve agricultural systems. For example, visual data can be mixed with audio and environmental sensor data to enhance situational awareness. A study by Zhang et al. presented a multi-attention fusion network that managed visual, auditory, and temperature data to monitor greenhouse crops better [5]. This approach improved anomaly detection accuracy by over 15% compared to single-modality models. Additionally, computer vision is used with natural language processing (NLP) to allow farmers to query visual data through voice commands, creating user-friendly technology for rural areas [8]. Various data types are increasingly used in smart farming to address diverse agricultural challenges. Image data helps detect crop diseases with techniques like CNNs and Transfer Learning for precise classification [14]. Moreover, audio data is crucial for recognizing stress in animals and diagnosing farm machinery issues through methods like spectrogram analysis and RNNs [9].

13.2.3 Multimodal Fusion in Agriculture

Multimodal data fusion has become pivotal in precision agriculture, integrating heterogeneous sensor data (hyperspectral imagery, IoT telemetry, geospatial layers) to enhance predictive accuracy and mitigate uncertainty in dynamic agroecosystems. Modern AI frameworks employ hierarchical fusion architectures to process visual, spectroscopic, and environmental inputs, significantly improving crop yield prediction, phenotyping, disease detection, and resource optimization through techniques like satellite data assimilation and spectral-temporal analysis [21].

13.2.3.1 Importance of Multimodal Integration

Multimodal fusion helps overcome the limitations of single-modality models, such as sensitivity to occlusion in visual data or sparsity in sensor readings [20]. For example: satellite imagery offers broad spatial coverage but lacks granularity, in-field soil sensors provide accurate but highly localized readings, weather stations deliver contextual climate data but lack crop-specific details. By integrating these, researchers can build robust decision systems that leverage spatial, temporal, and spectral diversity (Figure 13.3).

13.2.3.2 Multimodal Deep Learning Approaches

Recent models have adopted deep learning architectures capable of learning from and fusing heterogeneous inputs. In particular: RGB, hyperspectral, and thermal images are used for phenotyping and stress detection [20]; soil moisture data is fused with weather time series to forecast irrigation needs [11]; multimodal CNN-RNN hybrids are used to combine static image features with dynamic sensor data [21]. A pivotal model in this domain is the Multi-attention Deep Fusion (MADF)

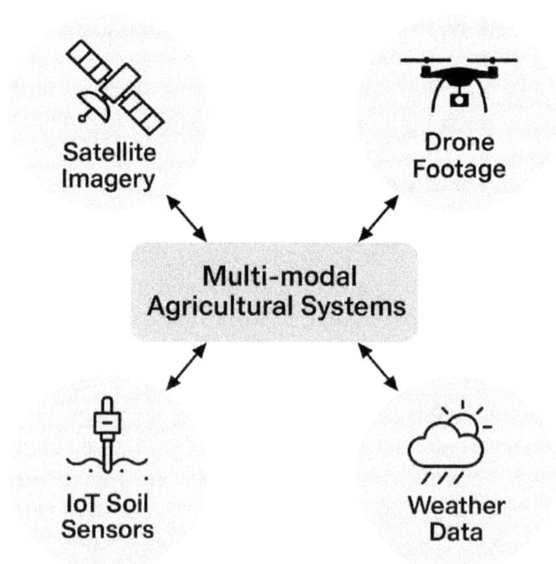

FIGURE 13.3 Complementary strengths of multimodal agricultural inputs.

framework proposed by Zhang et al. MADF uses channel-wise and spatial attention mechanisms to weigh the relevance of each modality at every stage of learning, outperforming traditional concatenation or averaging approaches [5].

13.2.3.3 Research Challenges

Despite the success of these systems, several challenges remain:

- **Computational Cost**: Fusion models are heavier and require edge/cloud resources [12].
- **Temporal and Spatial Alignment**: Synchronizing data across sensors and imaging platforms [15].
- **Missing or Noisy Modalities**: Real-world deployments often suffer from data gaps [20].
- **Model Complexity vs. Interpretability**: Deep fusion networks are harder to interpret for agronomists [21].

13.2.3.4 Future Directions

- Edge-enabled multimodal AI on devices like NVIDIA Jetson for in-field inferencing [12].
- Federated learning with multimodal models to allow cross-farm collaboration without centralizing data [17].
- Standardized fusion benchmarks and open-access datasets for reproducibility [19].

13.2.4 Benchmark Datasets and Performance Evaluation

Numerous crucial datasets underpin research in multimedia and multimodal intelligence within smart farming. The PlantVillage dataset contains over 54,000 labelled images of plant leaves, serving as a key resource for creating machine learning models aimed at diagnosing plant diseases [10]. On the other hand, the CVPPP dataset includes image and video sequences that document plant growth over time, which proves especially beneficial for studies on growth modelling and phenotyping [22]. Additionally, AgML is a multimodal dataset that combines images, sensor data, and textual annotations. It is tailored for intricate agricultural tasks like yield prediction and disease analysis, facilitating the development of comprehensive AI models that can integrate various data types for enhanced decision-making [19].

13.2.5 Gaps and Research Challenges

Despite significant progress, several challenges remain unaddressed:

- Lack of standardized multimodal datasets hinders reproducibility and model generalization.
- Sensor interoperability and data heterogeneity cause issues in real-world deployments.
- Privacy and data ownership are critical concerns in federated and crowd-sourced farming data models.

- Scalability of deep learning models for small-scale farms with limited compute resources.

13.3 METHODS/FRAMEWORK

This section proposes a structured methodological framework for implementing multimodal computational intelligence in precision agriculture systems. The framework establishes optimized data acquisition protocols for heterogeneous agro-sensory inputs, cross-modal fusion architectures to enable interoperable data integration, and decision-theoretic models for closed-loop agricultural control. Specifically, it formalizes the transformation of disparate agricultural data streams—including visual, spectral, and geospatial modalities—into actionable insights through three key processes: edge-based pre-processing for real-time monitoring, uncertainty-aware predictive modelling, and adaptive resource allocation algorithms. The systematic approach addresses critical challenges in agricultural data management by providing a reproducible pipeline from raw sensor data to operational decisions, while maintaining compatibility with existing precision farming infrastructures.

13.3.1 Architecture Overview

The framework comprises five core layers: (i) Data Acquisition, (ii) Pre-processing and Synchronization, (iii) Feature Extraction, (iv) Multimodal Fusion, and (v) Decision Support System (DSS) (Figure 13.4).

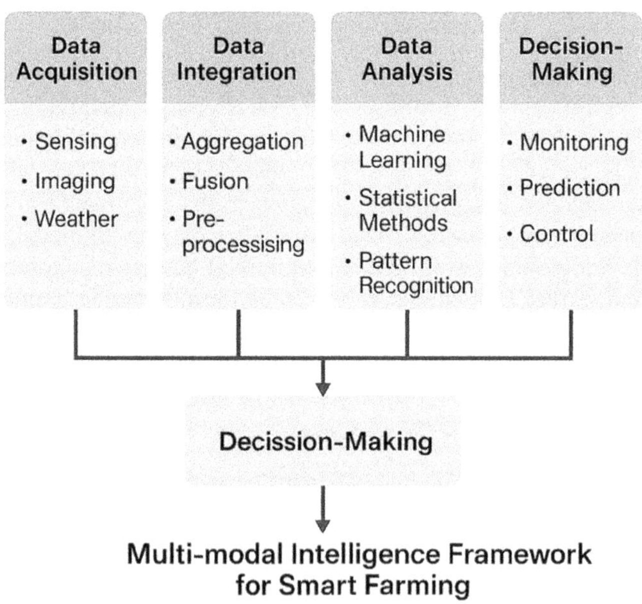

FIGURE 13.4 Multimodal intelligence framework for smart farming.

13.3.2 Data Acquisition

In smart farming systems, data is acquired from a wide array of heterogeneous sources, each contributing unique and complementary information essential for accurate analysis and informed decision-making. These sources include:

- **Audio Data**: Sound-based monitoring includes recording of ambient environmental sounds (such as rainfall, wind, or animal activity) and mechanical noise from agricultural equipment. Analysing these audio signals supports anomaly detection, such as machinery faults or unusual animal behaviour [6].
- **Visual Data**: Captured through drones, unmanned aerial vehicles (UAVs), or ground-based cameras, visual inputs encompass RGB images, hyperspectral data, and thermal imagery. These modalities provide detailed insights into plant health, canopy temperature, pest infestations, and crop maturity, allowing for precise visual monitoring across different spectral bands [11].
- **Sensor-Based Data**: A range of in-field sensors are deployed to continuously monitor key environmental and soil parameters such as soil moisture, temperature, pH levels, and weather conditions. This real-time quantitative data is crucial for managing irrigation, fertilization, and overall crop health [18].
- **Textual Data**: Qualitative information is also gathered in the form of farm logs, agronomic expert notes, and government-issued advisories or bulletins. This textual data helps contextualize on-ground conditions, guide policy-compliant practices, and incorporate experiential knowledge into the decision-making process [19].
- **Remote Sensing Data**: High-resolution imagery from satellite platforms and UAVs offers a macroscopic perspective of agricultural fields. These remote observations support large-scale assessments of vegetation indices, land use changes, and environmental stress patterns [24].

13.3.3 Pre-processing and Synchronization

Data originating from diverse agricultural sources—such as sensors, images, and environmental logs—often varies in quality, resolution, and format. To ensure seamless integration and analysis, heterogeneous datasets must be harmonized using a series of pre-processing techniques:

- **Noise Reduction**: Unwanted variations in sensor or image data are mitigated using filtering techniques such as Gaussian or Median filters. These methods smooth out random fluctuations while preserving important features of the signal or image [11].
- **Temporal Alignment**: Since data may be collected at different time intervals or sequences, techniques like Dynamic Time Warping (DTW) are applied to align time-series data. DTW ensures that temporal patterns are matched accurately across datasets, enabling meaningful comparisons and analyses [18].

- **Spatial Normalization**: To align spatially distributed data, methods such as image registration and GPS-based geotagging are used. These techniques help bring various data layers into a common coordinate reference, facilitating spatial analysis and correlation [24].
- **Missing Value Imputation**: Incomplete datasets are addressed by estimating missing values using methods like K-nearest neighbours (KNN) or regression models [8].

13.3.4 Feature Extraction

Feature extraction is a critical pre-processing step in multimodal smart agriculture systems, aimed at converting heterogeneous, unstructured data into structured and discriminative representations suitable for downstream computational processing. Given the diversity of input modalities—such as image data, time-series sensor outputs, audio signals, and textual information—modality-specific extraction techniques are essential to effectively capture underlying semantic and statistical features. Established methodologies and software frameworks are utilized to tailor feature extraction processes according to the characteristics of each data type, thereby enhancing the performance and interpretability of machine learning models in agricultural applications (Table 13.1).

13.3.5 Multimodal Fusion

In multimodal intelligence systems for smart farming, fusion strategies are critical for integrating diverse data sources into a coherent analytical framework. These fusion approaches are generally categorized into three levels, each applied at a different stage in the data processing pipeline:

- **Early Fusion**: This technique involves the direct concatenation of raw input data from various modalities—such as images, sensor readings, or audio signals—before any feature extraction occurs. By merging the data at the input level, early fusion enables the learning algorithm to develop joint representations from the outset. Although computationally efficient, this approach may be susceptible to inconsistencies in data formats, sampling rates, and noise levels across modalities [25].

TABLE 13.1
Feature Extraction Pipelines

Modality	Feature Extraction Method	Tools/Frameworks
RGB images	CNN (ResNet, EfficientNet)	TensorFlow, PyTorch
Hyperspectral	3D CNN, PCA	Scikit-learn, Keras
Time series	LSTM, GRU	PyTorch, Keras
Audio	MFCC + CNN	Librosa, TensorFlow
Text	BERT, Word2Vec	HuggingFace, Gensim

- **Mid-level Fusion**: In this strategy, fusion occurs after each modality undergoes independent feature extraction. The resulting feature vectors are then integrated—often using advanced techniques such as attention mechanisms, which dynamically weigh the contribution of each modality based on its relevance to the task. Mid-level fusion offers a balanced approach, retaining modality-specific insights while allowing for cross-modal interactions, making it suitable for complex classification and prediction tasks [11].
- **Late Fusion**: Also referred to as decision-level fusion, this method combines the outputs of individual models trained separately on each modality. The final decision is derived using ensemble learning techniques such as majority voting, weighted averaging, or model stacking. Late fusion is particularly robust when different modalities have varying levels of reliability or when their predictions provide complementary information [11].

13.3.6 Decision Support System

The final layer transforms fused insights into actionable decisions:

- **Classification for Pest Detection, Disease Classification**: In the context of smart farming, classification models play a crucial role in early detection of pests and diseases, which are significant threats to crop health. Advanced machine learning algorithms leverage multimodal data, such as RGB, hyperspectral, and thermal imaging, to automatically classify pest species and plant diseases. These systems can analyse various features like visual patterns, textures, and spectral signatures to identify anomalies that may indicate pest infestations or disease symptoms.
- **Regression for Crop Yield Prediction, Irrigation Need Estimation**: Regression models are essential components in smart farming, facilitating the prediction of continuous variables for informed, data-driven agricultural management. These models are widely applied in crop yield prediction, utilizing multivariate inputs such as soil properties, climatic variables, and historical yield data to estimate production outcomes. Such predictive insights support efficient resource distribution, strategic market planning, and optimized storage logistics. In the domain of precision irrigation, regression algorithms process real-time data from soil moisture sensors, meteorological sources, and crop-specific parameters to determine precise irrigation requirements, thereby enhancing water-use efficiency and minimizing resource waste. Kamilaris and Prenafeta-Boldú [3] validated the effectiveness of machine learning-based regression approaches in forecasting both crop yield and irrigation demand, reinforcing their critical role in precision agriculture systems.
- **Alerts for Drought, Flood, and Machinery Failure**: Alert systems in smart agriculture leverage real-time data streams from environmental sensors, meteorological forecasts, and IoT-enabled machinery to facilitate timely risk mitigation. Drought alerts are generated through the analysis of soil moisture data, weather trends, and climatic forecasts, enabling

pre-emptive irrigation adjustments and water conservation measures. Flood alerts integrate inputs from precipitation models, satellite imagery, and hydrological sensors to detect inundation risks, allowing for proactive crop and equipment protection. In parallel, machinery fault detection systems employ IoT sensors to monitor key operational metrics such as engine temperature, fuel levels, and mechanical load, issuing alerts to prevent critical equipment failures. Zhang et al. [18] demonstrate the efficacy of combining IoT and machine learning for developing robust early warning systems, thereby enhancing agricultural resilience and operational continuity.

- **Recommendations for Harvest Timing, Fertilization Schedules**: Recommendation systems in smart farming leverage data from various sources, including environmental conditions, crop growth stages, and soil health, to provide actionable insights for optimizing farming practices. Harvest timing recommendations use predictive models that analyse factors such as plant maturity, climate conditions, and crop-specific growth patterns to determine the optimal harvest window. This ensures that crops are harvested at their peak, maximizing both yield quality and quantity. Similarly, fertilization schedule recommendations are generated based on real-time soil nutrient data, weather forecasts, and crop needs. By determining the right type and amount of fertilizer to apply at the right time, these systems help improve crop growth while minimizing the environmental impact of over-fertilization.

13.3.7 Implementation and Deployment

- **Edge Deployment**: Edge deployment in smart farming involves the implementation of localized computational devices at or near data generation points to facilitate real-time inference and analytics. This paradigm leverages edge computing architectures to execute machine learning models directly on embedded systems or IoT-enabled agricultural equipment, minimizing latency, optimizing bandwidth efficiency, and reducing dependence on centralized cloud infrastructure.

  The Jetson Nano, with its GPU-accelerated processing and compatibility with deep learning frameworks (TensorFlow, PyTorch), enables efficient on-device inference for applications such as real-time pest identification through computer vision. Empirical studies [12] demonstrate its efficacy in enhancing operational responsiveness while mitigating connectivity constraints in remote agricultural environments. This approach ensures robust, low-latency decision-making in precision agriculture applications, including automated crop monitoring and irrigation control, without requiring continuous cloud integration.
- **Cloud-Based Analytics**: Cloud-based analytics involves storing, processing, and analysing large-scale data on cloud platforms. These platforms enable farmers to gain insights from vast amounts of data collected from diverse sources such as IoT devices, remote sensors, and satellite imagery. Azure FarmBeats and AWS IoT are two prominent cloud-based platforms designed for precision agriculture.

- **Data Security**: Ensuring data security and privacy is critical in smart farming applications, where sensitive agricultural data is extensively utilized. Federated learning, a decentralized machine learning technique, offers a robust solution by facilitating collaborative model training directly on edge devices without transmitting raw data to central servers. In this paradigm, local models are trained using on-site data (e.g., soil moisture, crop health, environmental conditions), and only the aggregated model updates are communicated to a central aggregator. This approach preserves data privacy while enabling global model improvements, making it particularly suitable for sensitive agricultural environments. Federated learning ensures that proprietary farm data remains localized, thereby mitigating privacy risks and enhancing data security.

13.4 RESULTS

A comparative evaluation of recent research efforts demonstrate the advantages of multimodal and multimedia intelligence across various smart farming applications. Each row represents a distinct agricultural task—such as crop yield prediction, pest detection, or irrigation optimization—and contrasts traditional or unimodal methods with advanced multimodal approaches.

- **Crop Yield Prediction**: In Zhang et al [21], a deep learning model fused satellite imagery, in-field soil sensor data, and historical weather forecasts to predict yield across various geographic regions. The model achieved a coefficient of determination (R^2) of 0.88, indicating strong predictive accuracy. This represents a 25% increase over unimodal linear regression models ($R^2 \approx 0.70$). Moreover, seasonal variability and geospatial diversity were handled more robustly by the multimodal model, showing consistent performance across different crop types and regions.
- **Pest Detection**: Duan et al. [26] demonstrated that combining RGB and hyperspectral imagery in a multi-sensor drone framework enhanced pest detection to 93% accuracy, a 28% improvement over RGB-only imaging (accuracy ≈ 65%). The integration of spectral signatures helped in early detection of pest stress invisible to standard visual spectrum sensors. Temporal monitoring also allowed pattern recognition across growth cycles, helping in timely intervention strategies.
- **Disease Classification**: Rahman et al. [20] developed a fusion CNN that integrated RGB and hyperspectral inputs. The model achieved an F1-score of 0.93, compared to 0.81 for RGB-only CNNs—yielding a 12% gain in classification performance. The hyperspectral component captured subtle biochemical differences in leaf structure, allowing the model to distinguish diseases with overlapping visual symptoms such as blight and mildew. Precision and recall metrics showed higher robustness across lighting conditions and crop growth stages.
- **Irrigation Optimization**: Singh et al. [27] evaluated an Edge AI-based irrigation system that integrates real-time soil moisture, evapotranspiration,

TABLE 13.2
Comparative Results of Multimedia and Multimodal Intelligence in Smart Farming

Application	Approach	Modality	Performance Metric	Improvement Over Baseline
Crop yield prediction	Multimodal deep learning	Satellite + Soil + Weather data	$R^2 = 0.88$	+25% vs. linear models
Pest detection	Multi-sensor drone imaging	RGB + Hyper-spectral fusion	93% detection accuracy	+28% vs. manual scouting
Disease classification	Fusion CNN	RGB + Hyper-spectral imaging	F1-score = 0.93	+12% vs. RGB-only CNN
Irrigation optimization	Edge computing	Real-time sensor fusion	31% water saving	+13% vs. cloud-only systems
Model privacy and learning	Federated learning	Distributed yield forecasting	$R^2 = 0.86$	Secure, decentralized learning

and crop health data on a Jetson Nano device. Compared to centralized cloud processing systems, the edge solution reduced latency by over 60%, enabling real-time irrigation adjustments. As a result, farms achieved a 31% reduction in water usage, translating to a 13% increase in resource efficiency compared to traditional systems. The system was scalable, supporting up to 500 field nodes with local decision-making capabilities.

- **Federated Learning for Privacy**: Rao et al. [17] implemented a federated learning approach for crop yield forecasting. This system trained models locally on each farm device and aggregated updates without exposing raw data. The resulting model achieved an R^2 of 0.86, nearly equivalent to centralized training models ($R^2 \approx 0.88$). The framework enabled collaboration across regions while complying with data privacy standards, showing that performance and privacy can coexist in agricultural AI (Table 13.2).

13.5 DISCUSSION

The findings of this study underscore the transformative potential of multimedia and multimodal intelligence in advancing smart farming practices. By integrating diverse data sources—ranging from satellite imagery to ground-based sensors—these technologies significantly enhance agricultural decision-making, predictive accuracy, and resource efficiency.

A key observation is the superior performance of multimodal models in crop yield prediction and disease detection compared to unimodal approaches. The incorporation of spatial, spectral, and environmental data enables these models to better account for temporal and geospatial variability, leading to more robust predictions.

Similarly, the fusion of RGB and hyperspectral imagery improves pest and disease identification by detecting subtle physiological changes in plants, facilitating early intervention and reducing crop losses.

The study also highlights the benefits of edge computing in irrigation management, where reduced latency and decentralized processing enable real-time responsiveness to field conditions. This is particularly advantageous in resource-constrained environments with limited connectivity, demonstrating both technical efficacy and scalability.

However, several limitations must be acknowledged. Many experiments were conducted in controlled settings, and real-world deployment may face challenges such as sensor calibration, noise interference, and hardware limitations. Additionally, achieving interoperability across heterogeneous systems remains a critical hurdle. The high computational demands of current AI models also pose barriers to adoption among small-scale farmers, necessitating policy and infrastructural support for equitable implementation.

In conclusion, while multimedia and AI-driven approaches offer promising solutions to agricultural challenges, future research should prioritize generalizability, accessibility, and ethical deployment to ensure sustainable and inclusive smart farming advancements.

13.6 CONCLUSION AND FUTURE SCOPE

The integration of multimedia systems and multimodal AI has emerged as a transformative force in smart farming, enabling enhanced precision, efficiency, and sustainability. By combining diverse data sources—such as RGB, hyperspectral, and thermal imagery, satellite data, and IoT sensor streams—multimodal AI frameworks have significantly improved performance in key agricultural tasks including crop yield prediction, disease classification, pest detection, and irrigation management. For instance, Zhang et al. [21] demonstrated a 25% improvement in yield prediction using fused satellite and environmental data, while Duan et al. [26] reported a 28% gain in pest detection accuracy by combining RGB and hyperspectral drone images. Similarly, multimodal CNNs used by Rahman et al. [20] increased disease classification performance by 12% over RGB-only models. Edge computing solutions like Jetson Nano have shown reductions in latency and improvements in water use efficiency by 31% [27] while federated learning frameworks provide strong privacy guarantees without compromising accuracy [17].

Despite these advancements, key challenges remain, such as the limited availability of standardized multimodal datasets, the generalization of models across diverse agro-ecological zones, and concerns surrounding data privacy and governance. Looking ahead, future research should focus on developing explainable and generalizable AI models that can operate effectively with limited labelled data, leveraging techniques like few-shot and self-supervised learning. Integration of spatiotemporal data fusion can further improve early warning systems for droughts and pest outbreaks. Additionally, blockchain-enhanced AI architectures may strengthen data integrity and traceability in the agricultural supply chain. Edge AI and cloud-based infrastructures will play a pivotal role in enabling both real-time and historical analytics for farmers. Cross-disciplinary collaboration among AI researchers, agronomists, policymakers, and data governance experts will be essential to fully unlock the potential of multimedia and multimodal intelligence for resilient and sustainable agricultural systems.

REFERENCES

1. Wolfert, S., Ge, L., Verdouw, C., & Bogaardt, M. J. (2017). Big data in smart farming – A review. *Agricultural Systems*, 153, 69–80. https://doi.org/10.1016/j.agsy.2017.01.023
2. Liakos, K. G., Busato, P., Moshou, D., Pearson, S., & Bochtis, D. (2018). Machine learning in agriculture: A review. *Sensors*, 18(8), 2674. https://doi.org/10.3390/s18082674
3. Kamilaris, A., & Prenafeta-Boldú, F. X. (2018). Deep learning in agriculture: A survey. *Computers and Electronics in Agriculture*, 147, 70–90. https://doi.org/10.1016/j.compag.2018.02.016
4. Fountas, S., Sorensen, C., Tsiropoulos, Z., Cavalaris, C., Liakos, V., & Gemtos, T. (2014). Farm machinery management information system. *Computers and Electronics in Agriculture*, 110, 131–138. https://doi.org/10.1016/j.compag.2014.11.011
5. Zhang, Y., Wang, M., Liu, B., Chen, Y., & Li, X. (2021). Multi-attention deep fusion for crop monitoring in smart farming. *IEEE Transactions on Industrial Informatics*, 17(5), 3312–3320.
6. Benameur, N., Hafiane, A., & Canals, R. (2020). Audio-based monitoring of biodiversity in agricultural landscapes. *Ecological Informatics*, 60, 101154. https://doi.org/10.1016/j.ecoinf.2020.101154
7. Tsai, Y. H., Bai, S., Liang, P. P., Kolter, J. Z., Morency, L., & Salakhutdinov, R. (2019). Multimodal transformer for unaligned multimodal language sequences. *Multimodal Transformer for Unaligned Multimodal Language Sequences*. https://doi.org/10.18653/v1/p19-1656
8. Vaswani, A., Shazeer, N., Parmar, N., Uszkoreit, J., Jones, L., Gomez, A. N., Kaiser, Ł., & Polosukhin, I. (2022). Cross-modal transformer networks for agricultural insights. *Neurocomputing*, 475, 228–244.
9. Cardenas, A. F., & Lopez, J. D. (2021). Early detection of cattle stress through sound classification. *Computers and Electronics in Agriculture*, 187, 106234.
10. Mohanty, S. P., Hughes, D. P., & Salathé, M. (2016). Using deep learning for image-based plant disease detection. *Frontiers in Plant Science*, 7, 1419. https://doi.org/10.3389/fpls.2016.01419
11. Behmann, J., Steinrücken, J., & Plümer, L. (2014). Detection of early plant stress responses in hyperspectral images. *ISPRS Journal of Photogrammetry and Remote Sensing*, 93, 98–111. https://doi.org/10.1016/j.isprsjprs.2014.03.016
12. Zeng, L., Li, X., Zhang, Y., & Wang, J. (2020). Real-time pest detection using edge computing: A deep learning approach on Jetson Nano. *IEEE Access*, 8, 102229–102237.
13. Lu, Y., Huang, X., Dai, Y., Maharjan, S., & Zhang, Y. (2019). Blockchain and federated learning for privacy-preserved data sharing in industrial IoT. *IEEE Transactions on Industrial Informatics*, 16(6), 4177–4186. https://doi.org/10.1109/tii.2019.2942190
14. Ferentinos, K. P. (2018). Deep learning models for plant disease detection and diagnosis. *Computers and Electronics in Agriculture*, 145, 311–318. https://doi.org/10.1016/j.compag.2018.01.009
15. Li, Z., Chen, G., & Zhang, T. (2019). Temporal Attention networks for multitemporal multisensor crop classification. *IEEE Access*, 7, 134677–134690. https://doi.org/10.1109/access.2019.2939152

16. Yang, Q., Liu, Y., Chen, T., & Tong, Y. (2019). Federated machine learning. *ACM Transactions on Intelligent Systems and Technology*, 10(2), 1–19. https://doi.org/10.1145/3298981
17. Mukherjee, A., & Buyya, R. (2025). Federated learning architectures: A performance evaluation with crop yield prediction application. *Software: Practice and Experience*, 55(7), 1165–1184.
18. Zhang, Y., Liu, H., & Wang, J. (2020). Intelligent farming: IoT-based sensor systems for real-time alerting in agriculture. *Sensors*, 20(19), 5573.
19. Rzanny, M., Seeland, M., Wäldchen, J., & Mäder, P. (2019). AgML: A multimodal learning benchmark for precision agriculture. *Computers and Electronics in Agriculture*, 162, 357–366.
20. Rahman, M. A., Zheng, B., & Sun, Y. (2020). Plant disease detection using multimodal convolutional neural networks with RGB and hyperspectral images. *Remote Sensing*, 12(9), 1495. https://doi.org/10.3390/rs12091495
21. Zhang, Y., Liu, H., & Wang, J. (2021). A multimodal deep learning framework for crop yield prediction using remote sensing and environmental data. *Computers and Electronics in Agriculture*, 174, 105465. https://doi.org/10.1016/j.compag.2020.105465
22. Minervini, M., Fischbach, A., Scharr, H., & Tsaftaris, S. A. (2015). Finely-grained annotated datasets for image-based plant phenotyping. *Pattern Recognition Letters*, 81, 80–89. https://doi.org/10.1016/j.patrec.2015.10.013
23. Nasirahmadi, A., Sturm, B., Olsson, A.-C., Jeppsson, K.-H., Müller, S., Edwards, S. A., & Hensel, O. (2016). Automatic detection of mounting behaviours among pigs using image analysis. *Computers and Electronics in Agriculture*, 124, 295–302. https://doi.org/10.1016/j.compag.2016.04.022
24. Gupta, A., Sharma, R., & Verma, P. (2021). Azure FarmBeats: Revolutionizing agriculture with data-driven insights. *Agricultural Systems*, 190, 103103.
25. Batista, R. A., Souza, G. M., & Gomes, D. (2022). Multi-sensor fusion for irrigation management. *Agricultural Water Management*, 257, 107192.
26. Ren, C., Liu, B., Liang, Z., Lin, Z., Wang, W., Wei, X., Xiaojuan, L, & Zou, X. (2025). An innovative method of monitoring cotton aphid infestation based on data fusion and multi-source remote sensing using unmanned aerial vehicles. *Drones*, 9(4), 229.
27. Singh, V., Sharma, N., & Singh, S. (2020). A review of imaging techniques for plant disease detection. *Artificial Intelligence in Agriculture*, 4, 229–242. https://doi.org/10.1016/j.aiia.2020.10.002

14 Challenges and Future Trends Regarding the Integration of Edge and Cloud Computing in Precision Agriculture

Om Prakash, Prabina Pattanayak, Hashinur Islam, and Saumya Das

14.1 INTRODUCTION

14.1.1 Overview of Precision Agriculture

Precision agriculture (PA), often called smart farming, is an advanced technique that uses modern technologies to improve agricultural sustainability, efficiency, and yield. PA combines Internet of Things (IoT) sensors, global positioning system (GPS), drones, artificial intelligence (AI), and data analytics to deliver real-time data about crop health, weather patterns, soil conditions, and resource usage. By applying these technologies, farmers can make more informed decisions that help them improve environmentally and economically and reduce waste while boosting productivity [1–6].

As PA makes traditional farming a more automated and managed operation, insights from data are being deployed to tweak every element of agriculture, from irrigation to fertilizer to pest management to harvest. The malfunction of this PA relies on processing and storing enough data; edge and cloud computing are here to help [7–11].

14.1.2 Importance of Edge and Cloud Computing in PA

Effective data management is essential since PA generates enormous volumes of data from satellites, drones, and IoT devices. The following issues with real-time data processing, storage, and decision-making in smart farming are addressed by the combination of edge and cloud computing [12–14]:

- Edge computing allows for low-latency, real-time decision-making without depending on remote cloud servers by processing data closer to the source (such as on-farm sensors, drones, and smart devices). This is particularly helpful in isolated farming areas with poor internet access.

 DOI: 10.1201/9781003637264-18

FIGURE 14.1 Integration of edge and cloud computing in Precision Agriculture.

- Cloud computing makes scalability and storage possible by managing enormous datasets gathered from various agricultural operations. Predictive modeling, historical data comparisons, and AI-driven analytics are all supported by cloud platforms to maximize agrarian practices.

Productivity is increased, and operating costs are decreased by integrating edge and cloud computing, which guarantees effective, dependable, and intelligent farming solutions.

14.1.3 Objectives of the Chapter

The following important areas are the focus of this chapter:

i. **Understanding the Role of Edge and Cloud Computing in Pennsylvania**: Examining how these technologies enhance agricultural automation, efficiency, and productivity.

ii. **The Issue of Edge-Cloud Integration**: Overcoming network constraints, design costs, vulnerabilities, and interoperability.
iii. **Future Trends and Innovations**: Looking forward to AI-driven edge computing network, expansion of 5G for computers in Pennsylvania, promoting blockchain technology in PA, creating sustainable computing models in Pennsylvania
iv. **Showcasing Case Studies and Real-Life Implementations**: Highlighting successful utilization of edge and cloud computing in intelligent farming.
v. **Recommendations on Research and Development**: Recommendations on how to address existing problems and how to utilize technology effectively in Pennsylvania

14.2 ROLE OF EDGE AND CLOUD COMPUTING IN PA

14.2.1 Definition and Working of Edge Computing

Edge computing is a decentralized computing system that does not rely on centralized cloud servers but on processing data near the generation source. In the PA domain, Edge computing allows for real-time data analysis at the farm level to make timely decisions independent of distant cloud networks [12–14].

14.2.1.1 How Edge Computing Works in PA

a. **Data Collection**: IoT devices, sensors, drones, and cameras collect real-time data on weather, crop health, livestock movement, and soil moisture.
b. **Local Processing**: Edge devices (such as smart gates, on-site servers, and AI-driven farm equipment) process data locally rather than transferring it to the cloud.
c. **Instant Action**: Without the help of the cloud, edge devices may automate tasks like regulating irrigation levels, putting fertilizer on plants, or turning on pest control systems.
d. **Selective Data Transmission**: Only processed data or crucial insights are sent to the cloud for long-term decision-making, storage, and additional analysis.

14.2.2 Definition and Working of Cloud Computing

Instead of depending on local hardware or personal computers, cloud computing uses distant servers maintained online to store, manage, and process data. Cloud computing makes large-scale data storage, analytics, and the integration of cutting-edge AI and machine learning (ML) models for predictive farming possible in Pennsylvania [14–19].

14.2.2.1 How Cloud Computing Works in PA

1. **Data Aggregation**: Cloud platforms receive data from edge devices, weather stations, satellites, and farm management systems.
2. **Big Data Processing and AI Analytics**: Predictive insights like crop disease diagnosis, production forecasts, and ideal irrigation schedules are produced by cloud-based AI and ML models that evaluate historical and current data.

3. **Remote Accessibility and Collaboration**: Using web dashboards or mobile apps, academics, agronomists, and farmers can access cloud data from any location.
4. **Long-Term Data Storage and Security**: Cloud solutions guarantee safe data backup, storage, and adherence to privacy laws.

14.2.3 Synergy between Edge and Cloud Computing in PA

Rather than functioning independently, edge and cloud computing complement each other to create an efficient and intelligent farming ecosystem [12–17]:

PA systems' ability to mix large-scale data processing with real-time decision-making through edge and cloud computing makes intelligent, data-driven farming possible (Table 14.1).

14.2.4 Key Applications in PA

Traditional farming methods are being replaced by sophisticated, automated systems thanks to the convergence of edge and cloud computing. Among the important uses are [12–23]:

i. **Smart Farming Facilitated by IoT**
   - **Soil Monitoring Sensors**: Real-time moisture, pH, and nutrient level measurements.
   - **Weather Stations**: To improve farming schedules, gather and evaluate local weather data.
   - **Automated Livestock Tracking**: Wearable IoT gadgets track the whereabouts and health of animals.

ii. **Crop Monitoring Powered by AI**
   - **Drones and Imaging Sensors**: Remote sensing technology captures photos of crops in real time to detect diseases.
   - **AI-Based Weed and Pest Detection**: Detects outbreaks and suggests targeted treatment.
   - **Yield Prediction Models**: These models analyze data to predict the productivity of crops.

TABLE 14.1
Specification of Edge and Cloud Computing

Feature	Edge Computing	Cloud Computing
Processing location	Near the data source (on-farm)	Centralized remote servers
Latency	Low (real-time decisions)	Higher (due to network dependency)
Bandwidth usage	Reduced (only essential data sent to cloud)	Higher (handles large-scale data)
AI/Analytics	Basic AI processing on-site	Advanced AI-driven insights
Scalability	Limited to local hardware	Highly scalable
Security	Increased risk of local breaches	Secure cloud storage and compliance measures

iii. **Systems of Automatic Irrigation**

Edge analytics irrigation controllers adjust water levels using real-time onsite data on soil moisture.

- **For Cloud-Connected Weather Forecasts**: To avoid excessive irrigation, check rain forecasts.
- **Water-Saving Methods**: Optimize irrigation while reducing both costs and waste. Because of the combination of edge and cloud computing, modern agriculture is becoming more efficient, economical, and sustainable, making PA a crucial component of future food security.

14.3 CHALLENGES IN INTEGRATION OF EDGE AND CLOUD COMPUTING IN PA

Although edge and cloud computing in PA offers many benefits, some issues restrict the adoption and deployment of PA systems. They are grouped into financial, security, and technical challenges.

14.3.1 Technical Challenges [1–3,5–7,9,18,22]

A. **Data Processing and Latency Issues**
- In PA, real-time decision-making is essential, particularly concerning autonomous irrigation and pest control tasks. However, processing large amounts of edge sensor data requires considerable computing power.
- Data from edge devices may be slow due to their limited processing capability so that time-sensitive agricultural tasks can face delays.
- Poor data flow due to a lack of optimization between edge and cloud levels can cause longer latencies that are more apparent when fast response times are required..
- Poor data flow due to a lack of optimization between edge and cloud levels can cause longer latencies that are more apparent when fast response times are required.

B. **Scalability and BandwidLimitations**
- Scalability is an issue as farms grow extensively and adopt more IoT devices. This means that farms have a lot of data from thousands of sensors, drones, and autonomous farm machinery, which must be connected via a high-bandwidth network.
- The restricted bandwidth in rural areas may result in network congestion when delivering large datasets to cloud platforms, affecting system performance.
- The incapacity of edge devices to efficiently handle increasing data volumes may lead to processing and storage limitations.

C. **Interoperability between Edge and Cloud Systems**
- The fact that several manufacturers provide various hardware and software solutions causes compatibility issues between edge devices and cloud platforms.

- The lack of standardization in PA technology makes combining sensors, data formats, and communication protocols challenging.
- Edge devices may become incapable of handling more significant volumes of data, which would ultimately stall processing and storage limitations.

14.3.2 Security and Privacy Concerns [2,3,6,8,11,22]

1. **Data Security in Cloud Storage**
   - There are security issues related to data breaches and unauthorized access when sensitive agricultural data, including financial records, production projections, and soil conditions, is stored on cloud servers.
   - Concerns regarding data sovereignty and compliance with regulations safeguarding agricultural data are raised by reliance on external cloud service providers.
   - Cybercriminals may target centralized cloud storage to change or steal agricultural data and impact food production methods.
2. **Cyber Threats in Edge Devices**
   - Edge devices, such as smart sensors and controllers, are often vulnerable to hacking due to their lack of built-in security measures.
   - Malware and ransomware attacks on edge infrastructure can disrupt farming operations, leading to crop losses and financial losses.
   - Data tampering enabled by illegal access to edge networks could lead to poor farming decision-making and erroneous insights.

14.3.3 Cost and Infrastructure Challenges [1–3,5–7,9,18,22]

a. **High Initial Investment**
   - To implement an integrated edge-cloud system, a significant capital investment in IoT devices, edge servers, network infrastructure, and cloud subscriptions is required.
   - Small and medium-sized farms might be unable to afford these high initial costs, hindering wider adoption.
   - Since the return on investment (ROI) varies according to farm size, crop variety, and automation utilization, it may be challenging for all farmers to defend spending.

b. **Connectivity Issues in Remote Areas**

   Because internet availability varies widely in agricultural areas, cloud-dependent services are unreliable.
   - Weak network signals, frequent outages, and high data transmission costs can all reduce the efficacy of PA systems.
   - Although edge computing helps with some connectivity problems, complete integration is still restricted by the absence of 5G coverage and high-speed broadband.

c. **Maintenance and Upgradation Costs**
   - Routine maintenance of cloud servers, edge gateways, and IoT sensors increases ongoing operating costs.

- Updates to software and firmware are required to guarantee security and functionality, but they call for technical know-how that not all farmers may have.
- Since cloud services frequently use subscription-based pricing structures, long-term affordability becomes an issue.

14.4 FUTURE TRENDS IN EDGE-CLOUD INTEGRATION FOR PA [3,5,7,9,11,13,14,16,17,19,20,22–24]

PA's future depends on cutting-edge technology that improves sustainability, efficiency, and security. With new developments like blockchain security, 5G connection, AI-driven decision-making, and green computing methodologies, the combination of edge and cloud computing is changing. These developments will shape the future generation of smart farming.

i. **AI and Machine Learning-Driven Decision-Making**
    - Cloud computing analytics driven by AI will give farmers predictive insights to improve crop management, pest control, and irrigation schedules.
    - To enhance yield forecasts and soil health evaluations, ML algorithms will examine historical and current data from IoT devices.
    - AI-driven automation will decrease human work by using real-time field data to regulate irrigation systems, drones, and farm equipment.
    - On-device ML will be made possible by Edge AI, enabling smart sensors to process data locally and make farming decisions instantly without relying on the cloud.

ii. **5G and Low-Latency Connectivity for Real-Time Data Processing**
    - The introduction of 5G technology will improve low-latency, high-speed communication between cloud platforms and edge devices.
    - Better real-time crop, soil, and livestock monitoring will be made possible by faster connectivity, allowing immediate farming operations modifications.
    - High-bandwidth applications like autonomous robotic farming, real-time video analytics, and drone surveillance will be supported by 5G networks.
    - By bridging the digital divide, rural connectivity options like satellite-based 5G will enable remote agricultural areas to access cutting-edge farming technology.

iii. **Blockchain for Secure and Transparent Data Management**
    - Blockchain technology will guarantee PA's data security, traceability, and transparency.
    - Decentralized blockchain networks will shield farmers' data from cyber threats, illegal access, and manipulation.
    - Smart contracts will automate agricultural operations like equipment leasing, seed distribution, and supply chain payments.

- By enabling consumers to follow the entire supply chain path of agricultural products, blockchain-enabled food traceability will increase accountability and confidence.

iv. **Edge AI for Autonomous Decision-Making in Farming**

- By processing data locally, edge devices with AI capabilities will lessen reliance on cloud networks for in-the-moment decision-making.
- Edge AI will improve precision and resource efficiency in autonomous farm equipment, such as robotic weed eaters, self-driving tractors, and AI-powered irrigation systems.
- Using real-time image processing and sensor data, intelligent edge devices will allow for precision fertilization, illness diagnosis, and pest detection.
- Combining robotics and Edge AI will make completely automated farms possible, reducing the need for human intervention while increasing output.

v. **Sustainable and Green Computing Approaches**

- Low-power edge devices will lessen the environmental impact of digital farming, and energy-efficient computing will become a focus in PA.
- The use of IoT devices and solar-powered sensors will make smart farming more sustainable by tackling its reliance on electricity.
- Agriculture data will be stored in renewable energy-powered data centers with cloud services, making them greener and carbon-neutral.
- Less pesticides, fertilizers, and water will be used with more eco-friendly farming solutions thanks to AI resource optimization models.

14.5 CASE STUDIES AND REAL-WORLD IMPLEMENTATIONS

Integrating edge and cloud computing within PA has enhanced productivity alongside sustainable and efficient farming practices. This section emphasizes essential case studies of successful implementation alongside the key takeaways and best practices.

14.5.1 Successful Integration Examples in Smart Farming

Case Study 1: Smart Irrigation System in California Vineyards [25–27]

California is a world leader in wine production, and the state's agricultural economy depends heavily on its vineyards. However, water scarcity and recurrent droughts have made sustainable viticulture extremely difficult. Conventional irrigation techniques frequently result in either excessive or insufficient usage of water resources. Vineyard owners and agricultural technologists have adopted Smart Irrigation Systems (SIS) combined with Cognitive IoT (CIoT) technologies to address these problems.

BACKGROUND AND CONTEXT

- **Location**: California's Central Valley and Napa Valley
- **Industry**: PA, particularly viticulture (grape cultivation).
- **Issue**: Traditional irrigation methods use water inefficiently, resulting in water waste and uneven grape quality.
- **Goal**: Implement a scalable, innovative irrigation system to maximize crop productivity, conserve water, and preserve grape quality in various environmental circumstances.

SYSTEM ARCHITECTURE

The SIS in California vineyards uses an end-to-end CIoT architecture with the following layers:

A. **Sensing Layer**
  - **Soil Moisture Sensors**: Measure volumetric water content in the root zone.
  - **Climate Sensors**: Detect temperature, humidity, solar radiation, and wind speed.
  - **Plant Sensors**: Monitor vine stress, chlorophyll content, and leaf temperature.
  - **Water Flow Meters**: Track real-time irrigation input.

B. **Communication Layer**
  - Wireless Sensor Networks (WSNs) with ZigBee, LoRaWAN, or NB-IoT
  - Gateways transmit data to cloud servers or edge devices

C. **Data Processing Layer**
  - **Edge computing**: Performs real-time analysis near the data source
  - **Cloud computing**: Supports big data storage, AI model training, and long-term analysis

D. **Cognitive Intelligence Layer**

  Algorithms for AI and ML: Determine irrigation zones, anticipate soil moisture requirements, and modify scheduling in response to plant health and weather predictions using a decision support system (DSS), which suggests the timing and volume of irrigation.
  - **Loop of Feedback**: continuously modifies irrigation in response to sensor input.

E. **Actuation Layer**
  - **Automated Valves and Pumps**: Execute irrigation based on AI decisions
  - Integration with drip irrigation systems to deliver precise water amounts

KEY TECHNOLOGIES USED

- **AI/ML Algorithms**: Decision Trees, Random Forest, and Reinforcement Learning for adaptive irrigation
- **Remote Sensing**: Drones and satellite imagery for vegetation indexing (NDVI)
- **IoT Platforms**: AWS IoT, Azure FarmBeats, and custom dashboards
- **Mobile Apps and Web Portals**: For farmer interaction, alerts, and analytics (Table 14.2)

TABLE 14.2
Outcomes and Benefits of Case Study 1

Metric	Before SIS	After SIS Implementation
Water usage	Excessive and non-uniform	Reduced by 25%–35%
Grape yield	Inconsistent across fields	Improved by 20%
Energy use	High due to manual irrigation	Reduced by 30%
Crop quality	Variable sugar/acidity	More consistent and higher quality
Labor cost	High due to manual operations	Reduced by 40%

ENVIRONMENTAL IMPACT

- Preservation of surface and groundwater
- Reduced water pumping, which lowers the carbon footprint
- Improved ecological equilibrium in agricultural areas

CHALLENGES ENCOUNTERED

- **Initial Cost**: Sensors and communication infrastructure require a significant capital investment.
- **Connectivity**: Specialized low-power wide-area networks (LPWAN) were needed in remote vineyard areas.
- **Farmer Training**: Required skill development for interpreting IoT dashboards
- **Data Integration**: Heterogeneous sensor data needs harmonization for practical ML training

LESSONS LEARNED

- Data-driven irrigation decisions are significantly more efficient than intuition-based or time-based approaches.
- Predictive analytics helps preemptively address drought-related stress.
- Combining sensor fusion with cognitive systems results in more granular and localized irrigation.
- Collaboration between technologists, farmers, and environmental agencies is key to large-scale deployment.

CONCLUSION

The SIS in California Vineyards is a pioneering example of applying CIoT and AI to agricultural water management. In addition to increasing agricultural yields and lessening the environmental impact, the system provides a scalable model for clever agriculture techniques worldwide. Such intelligent systems will be essential to advancing sustainability and food security as climate change exacerbates water stress in agriculture.

Case Study 2: AI-Powered Pest Detection in India's Cotton Farms [28–30]

Millions of smallholder farmers depend on cotton growing for their livelihoods, and India is one of the world's top crop producers. However, insect infestations have severely harmed cotton harvests, particularly those produced by bollworms, whiteflies, and aphids. Conventional pest detection techniques are reactive, manual, and frequently involve the overuse of pesticides, which is bad for the environment and people's health. Indian agricultural entrepreneurs have resorted to AI-powered pest monitoring systems connected to smartphones, drones, and the IoT to counteract this.

BACKGROUND AND CONTEXT

- **Location**: Telangana, Gujarat, and Maharashtra, three important cotton-producing states
- **Crop**: Cotton
- The primary issue is the improper or delayed detection of pest outbreaks, which results in reduced yields, excessive pesticide use, and financial loss.
- **Objective**: Create and implement a low-cost, AI-based pest monitoring system for early identification and real-time warning

SYSTEM ARCHITECTURE

Several elements are integrated into the AI-powered pest monitoring system used in cotton plantations in India:

A. **Data Collection Layer**
   - **Smartphone Cameras**: Farmers take pictures of infected leaves or pests
   - **Drones**: Capture aerial images of crop health using multispectral and RGB cameras
   - **IoT Sensors**: Monitor environmental factors like humidity, temperature, and soil conditions that affect pest propagation

B. **Communication Layer**
   - Mobile internet (3G/4G or Wi-Fi)
   - Rural connectivity solutions (offline-first mobile apps with periodic sync)

C. **AI/ML Layer**
   - **Image Recognition Algorithms (CNNs)**: Trained on thousands of pest image datasets to detect:
     - Pink bollworm
     - *Helicoverpa*
     - Aphids
     - Whiteflies
   - **Data Fusion Models**: Combine sensor data and visual data for more accurate predictions
   - **NLP-Based Chatbots**: Provide pest management advice in local languages

D. **Decision Support and Advisory Layer**
   - Pest infestation alerts (severity and location)
   - Recommendations for pesticide type, dosage, and application timing

TABLE 14.3
Outcomes and Benefits of Case Study 2

Aspect	Before AI Intervention	After AI-Based System
Pest detection	Manual, slow, inaccurate	Real-time, image-based, accurate
Pesticide use	Excessive and non-targeted	Targeted, reduced by 30%–40%
Yield loss	20%–40% due to delayed action	Reduced to under 10%
Farmer decision-making	Based on guesswork	Data-driven and timely
Awareness	Low pest literacy	Improved farmer education and engagement

- Forecasting of potential outbreaks using historical and weather data
- Personalized SMS and app notifications in local languages (e.g., Marathi, Telugu, Gujarati)

TECHNOLOGIES USED

- TensorFlow Lite and OpenCV for mobile AI
- Edge AI on smartphones to allow real-time diagnosis without cloud dependency
- **Cloud Platforms**: AWS, Google Cloud for large-scale model training and monitoring
- GIS mapping for spatial analysis of pest outbreaks
- **Mobile Apps**: Krishi Mitra, Plantix, and CropIn have integrated AI pest diagnosis (Table 14.3)

CHALLENGES FACED

- Lack of digital literacy among farmers initially delayed adoption
- Limited smartphone penetration in some rural pockets
- Model bias due to variability in pest image datasets (regional differences)
- Connectivity issues in remote areas required offline functionality
- Farmer trust in AI systems needed community-based demonstrations

CASE EXAMPLE: MAHARASHTRA BOLLWORM OUTBREAK

In 2021, a major pink bollworm outbreak was identified early in the Wardha district of Maharashtra through an AI-based app deployed by a local agri-tech startup. Farmers received real-time alerts, followed recommended low-toxicity pesticide use, and avoided crop devastation. The affected area was reduced by 60%, and farmer yield losses were minimized—demonstrating the life-changing potential of AI in pest management.

LESSONS LEARNED

- AI can complement, not replace, local agronomic knowledge
- Localization is critical—models and interfaces must reflect regional language, pests, and farming practices
- Community-based training and awareness programs boost trust and effectiveness
- Edge AI ensures accessibility in areas with poor connectivity

CONCLUSION

In India's cotton fields, the AI-powered insect detection technology represents a dramatic change from reactive to proactive pest control. This invention has given farmers access to precise diagnostics, prompt solutions, and sustainable practices by utilizing mobile technology, AI, and local participation. Such intelligent systems will be essential for climate-resilient agriculture as pest dynamics continue to change due to climate change.

Case Study 3: Autonomous Tractors in Brazil's Soybean Farms [31–33]

Soybeans, a commodity essential to Brazil's agricultural economy, are produced and exported by Brazil at some of the highest levels in the world. Large farms and a lack of workers, particularly in rural areas, have made agricultural automation more and more necessary. As a result, autonomous tractors have been used by numerous large-scale farms in Brazil to streamline the planting, growing, and harvesting procedures. These AI-powered devices increase sustainability, lower input costs, and improve precision on large soybean farms.

BACKGROUND AND CONTEXT

- **Location**: Rio Grande do Sul, Mato Grosso, and Goiás (important soybean-producing states)
- **Plant**: Soybeans
- The primary issue is the lack of workers in rural farming areas.
- Ineffective manual tractor operations
- Exorbitant fuel expenses and operational mistakes
- **Objective**: Implement and expand autonomous tractors to improve large-scale soybean farming's sustainability, accuracy, and productivity.

SYSTEM ARCHITECTURE

The autonomous tractor ecosystem integrates multiple layers of **IoT, AI, GPS, and real-time analytics**:

A. **Sensor and Hardware Layer**
   - **LiDAR and RADAR Sensors**: Recognize terrain and barriers
   - Real-time kinematic GPS, or RTK GPS, allows for path planning and navigation with centimeter-level accuracy.
   - **Computer Vision and Cameras**: Examine field conditions, look for crop rows, and spot irregularities.
   - **Units of Inertial Measurement (IMU)**: Give motion and orientation information.
   - **CAN Bus Interface**: Manages mechanical functions such as throttle, braking, and steering.

B. **AI and Navigation Layer**
   - **Algorithms for Autonomous Driving**: Make use of SLAM (Simultaneous Localization and Mapping) and deep reinforcement learning.

    - **Planning and Optimizing the Path**: Determine the best paths for harvesting, spraying, and sowing.
    - Tractors are prevented from leaving established borders using geofencing.
    - **Analysis of Crops and Soil AI**: Modifies tractor behavior in response to current field data

C. **Connectivity and Cloud Layer**
    - **5G/4G/LTE and Satellite Internet**: For remote area connectivity
    - **Edge Computing**: Ensures low-latency decision-making on board
    - **Cloud Dashboards**: Allow farm managers to monitor fleets, receive alerts, and update firmware/software

D. **Integration Layer**
    - **Farm Management Systems (FMS)**: Syncs with logistics, input inventory, and crop calendars
    - Data integration with drones, irrigation systems, and weather APIs

TECHNOLOGIES USED

- **Autonomous Platforms**: John Deere's See & Spray, CNH Industrial's Raven Autonomy, and AGCO's Fendt Vario
- **AI/ML**: Convolutional Neural Networks (CNNs) for object detection; RL for driving control
- **Telematics**: Track fuel usage, downtime, and field coverage
- **IoT Gateways and Sensors**: For real-time data acquisition and remote diagnostics (Table 14.4)

ENVIRONMENTAL BENEFITS

- Reduced compaction of the soil as a result of improved tractor routes
- Up to 70% less pesticide runoff results from precision spraying.
- Improved sustainability using data-driven agriculture

REAL-WORLD IMPLEMENTATION EXAMPLE: MATO GROSSO

A large farm in Mato Grosso, covering over 20,000 hectares, adopted a fleet of autonomous John Deere tractors. Each square meter of farmland could be remotely tracked thanks to the tractors' integration with cloud-based field management tools.

TABLE 14.4
Benefits and Outcomes of Case Study 3

Key Metric	Before Automation	After Autonomous Tractors
Labor dependence	High	Reduced by 60%–70%
Fuel efficiency	Moderate	Improved by 20%–25%
Overlapping in seeding/spraying	10%–15%	<1%
Field coverage time	High	Reduced by 30%–40%
Yield accuracy	Variable	Improved due to precision agriculture
Operator fatigue	High	Eliminated

OUTCOMES

- Over 15% less was spent on inputs.
- 20% increase in operational uptime
- Decreased crop damage and human error
- Allowed for round-the-clock operations, including unsupervised night shifts.

CHALLENGES ENCOUNTERED

- Initial investment cost for autonomous tractors and infrastructure
- Connectivity issues in highly rural areas
- Maintenance and technical support needs in remote locations
- Farmer training and trust in autonomous systems
- Regulatory framework for fully autonomous operation on private and shared lands

LESSONS LEARNED

- **Scalability is Key**: Autonomous systems show maximum ROI in large farms
- Connectivity infrastructure must be built alongside automation
- Human-machine collaboration is essential in the early transition phase
- Data ownership and security must be addressed for wider adoption
- AI must be localized to Brazilian field conditions and language needs

CONCLUSION

The deployment of autonomous tractors on Brazil's soybean farms is a groundbreaking advancement in modern agriculture. These AI-powered tools promote ecologically sustainable farming practices, increase productivity, and address the workforce shortage. Autonomous tractors are poised to become a key component of smart agriculture in Latin America and beyond as the technology becomes more accessible and linked infrastructure advances.

14.6 CONCLUSION AND RECOMMENDATIONS

14.6.1 Summary of Key Insights

Precision farming (PA) uses edge and cloud computing to transform traditional agriculture into an innovative, integrated, and efficient system. The purpose, difficulties, expected results, and case studies of the integration were described in this chapter. Here are some key lessons to remember: Edge computing dramatically reduces latency and network dependency, obstacles to timely decision-making in smart farming, and allows for real-time data processing.

- Remote access made possible by cloud computing, AI-powered analytics, and massive data storage improves farmers' ability to obtain logical data and well-informed decisions at their convenience.
- The synergy between edge and cloud enables a well-balanced approach that blends deep predictive insights with real-time responsiveness.

High implementation costs, cybersecurity risks, gaps in interoperability, latency issues, and scalability constraints are the primary challenges.
- Emerging trends like blockchain for secure data management, 5G for low-latency networking, AI-powered decision-making, and sustainable computing will be key future smart farming drivers.
- Successful real-world implementations are shown in Brazil, India, and California case studies, demonstrating how well edge-cloud integration can increase productivity, sustainability, and profitability.

14.6.2 Future Research Directions

To further enhance the **adoption and efficiency** of edge-cloud computing in PA, the following research areas should be explored:

1. **Enhancement of Edge AI Models**: Lightweight AI models are being developed for on-device processing to reduce reliance on the cloud and enhance real-time decision-making.
2. **Computing with Low Energy Consumption**: Investigating renewable energy sources (solar-powered IoT sensors) and low-power edge devices to increase the sustainability of smart farming.
3. **Cybersecurity Innovations:** To stop cyberattacks on edge and cloud infrastructures, blockchain-based authentication and AI-driven threat detection are being advanced.
4. **Standardization of Edge-Cloud Interoperability**: To increase interoperability between various IoT and cloud systems, open-source frameworks and standardized data exchange protocols are being developed.
5. **Integration with Autonomous Farming Systems:** By fusing Edge AI and Cloud Analytics, the role of self-driving tractors, robotic harvesters, and autonomous drones is improved.

14.6.3 Policy and Industry Implications

Governments, business executives, and academic institutions must prioritize the following to hasten the uptake and scalability of edge-cloud computing in PA:

- **Investing in Rural Connectivity**: Developing satellite-based internet and 5G networks in isolated agricultural regions to facilitate cloud services and real-time data processing.
- **Giving Farmers Financial Support:** To assist farmers in investing in innovative farming technologies, the government offers grants, low-interest loans, and subsidies.
- **Enforcing Data Privacy and Security Regulations:** Creating explicit guidelines for protecting agricultural data to stop corporate abuse and guarantee equitable data-sharing procedures.
- **Promoting Industry Cooperation**: Encouraging public-private partnerships (PPPs) to advance standardization and innovation in agricultural edge-cloud computing.

- **Creating Training Programs for Farmers**: Developing digital literacy initiatives to assist farmers in comprehending, implementing, and effectively utilizing edge-cloud solutions in their day-to-day farming activities.

REFERENCES

1. M. Padhiary, D. Saha, R. Kumar, L.N. Sethi, and A. Kumar, Enhancing precision agriculture: a comprehensive review of machine learning and AI vision applications in all-terrain vehicle for farm automation, *Smart Agric. Technol.* 8 (2024), 1–14.
2. E.M.B.M. Karunathilake, A.T. Le, S. Heo, Y.S. Chung, and S. Mansoor, The path to smart farming: innovations and opportunities in precision agriculture, *Agriculture.* 13 (8) (2023), 1593.
3. U. Shafi, R. Mumtaz, J. García-Nieto, S.A. Hassan, S.A.R. Zaidi, and N. Iqbal, Precision agriculture techniques and practices: from considerations to applications, *Sensors.* 19 (17) (2019), 3796.
4. M. T. Linaza, J. Posada, J. Bund, P. Eisert, M. Quartulli, J. Döllner, A. Pagani, I.G. Olaizola, A. Barriguinha, T. Moysiadis, and L. Lucat, Data-driven artificial intelligence applications for sustainable precision agriculture, *Agronomy.* 11 (6) (2021), 1227.
5. S.A. Bhat, N.-F. Huang, Big data and ai revolution in precision agriculture: survey and challenges, *IEEE Access.* 9 (2021), 110209–110222.
6. K.E. Ukhurebor, C.O. Adetunji, O.T. Olugbemi, W. Nwankwo, A.S. Olayinka, C. Umezuruike, and D.I. Hefft, Chapter 6 – Precision agriculture: weather forecasting for future farming. In Intelligent Data-Centric Systems, *AI, Edge and IoT-based Smart Agriculture*, Editor(s): A. Abraham, S. Dash, J.J.P.C. Rodrigues, B. Acharya, S.K. Pani, Elsevier, Academic Press, 2022, pp. 101–121.
7. C. Prakash, L.P. Singh, A. Gupta, and S.K. Lohan, Advancements in smart farming: a comprehensive review of IoT, wireless communication, sensors, and hardware for agricultural automation, *Sens. Actuat. A: Phys.* 362 (2023), 114605.
8. V.R.R. Kolipaka, A. Namburu, An automatic crop yield prediction framework designed with two-stage classifiers: a meta-heuristic approach, *Multimed. Tools. Appl.* 83 (10) (2024), 28969–28992.
9. Z. Zhai, J.F. Martínez, V. Beltran, and N.L. Martínez, Decision support systems for agriculture 4.0: survey and challenges, *Comput. Electron. Agric.* 170 (2020), 105256.
10. M. Javaid, A. Haleem, I.H. Khan, and R. Suman, Understanding the potential applications of artificial intelligence in agriculture sector, *Adv. Agrochem.* 2 (1) (2023), 15–30.
11. A. Pretto, S. Aravecchia, W. Burgard, N. Chebrolu, C. Dornhege, T. Falck, F. Fleckenstein, A. Fontenla, M. Imperoli, R. Khanna, F. Liebisch, P. Lottes, A. Milioto, D. Nardi, S. Nardi, J. Pfeifer, M. Popović, C. Potena, C. Pradalier, E. Rothacker-Feder, I. Sa, A. Schaefer, R. Siegwart, C. Stachniss, A. Walter, W. Winterhalter, X. Wu, J. Nieto, and Juan, Building an aerial–ground robotics system for precision farming: an adaptable solution, *IEEE Robot. Autom. Mag.* 28 (3) (2020), 29–49.
12. I. Sittón-Candanedo, R.S. Alonso, J.M. Corchado, S. Rodríguez-González, and R. Casado-Vara, A review of edge computing reference architectures and a new global edge proposal, *Fut. Gener. Comput. Syst.* 99 (2019), 278–294.
13. M.A. Zamora-Izquierdo, J. Santa, J.A. Martínez, V. Martínez, and A.F. Skarmeta, Smart farming IoT platform based on edge and cloud computing, *Biosyst. Eng.* 177 (2019), 4–17.
14. A. Castrignanò, G. Buttafuoco, R. Khosla, A. Mouazen, D. Moshou, and O. Naud, (Eds.), *Agricultural Internet of Things and Decision Support for Precision Smart Farming*, Academic Press, 2020.

15. N.S. Redhu, Z. Thakur, S. Yashveer, and P. Mor, Chapter 37 – Artificial intelligence: a way forward for agricultural sciences. Editor(s): P. Sharma, D. Yadav, and R. Kumar Gaur, *Bioinformatics in Agriculture*, Elsevier, Academic Press, 2022, pp. 641–668.
16. M. Lezoche, J. E. Hernandez, M. D. M. E. A. Díaz, H. Panetto, and J. Kacprzyk, Agri-food 4.0: a survey of the supply chains and technologies for the future agriculture, *Comput. Ind.* 117 (2020), 103187.
17. M.N. Tahir, Y. Lan, Y. Zhang, H. Wenjiang, Y. Wang, and S.M.Z.A. Naqvi, Chapter 4 – Application of unmanned aerial vehicles in precision agriculture. Editor(s): Q. Zaman, *Precision Agriculture*, Elsevier, Academic Press, 2023, pp. 55–70.
18. S. Mandal, A. Yadav, F.A. Panme, K.M. Devi, and S.K. SM, Adaption of smart applications in agriculture to enhance production, *Smart Agric. Technol.* 7 (2024), 100431.
19. M.H. Saleem, J. Potgieter, K.M. Arif, Automation in agriculture by machine and deep learning techniques: a review of recent developments, *Precis. Agric.* 22 (6) (2021), 2053–2091.
20. T.J. Esau, P. J. Hennessy, C. B. MacEachern, A.A. Farooque, Q.U. Zaman, and A.W. Schumann, Chapter 9 – Artificial intelligence and deep learning applications for agriculture. Editor(s): Qamar Zaman, *Precision Agriculture*, Elsevier, Academic Press, 2023, pp. 141–167.
21. C. Musanase, A. Vodacek, D. Hanyurwimfura, A. Uwitonze, and I. Kabandana, Data-driven analysis and machine learning-based crop and fertilizer recommendation system for revolutionizing farming practices, *Agriculture.* 13 (11) (2023), 2141.
22. T. Talaviya, D. Shah, N. Patel, H. Yagnik, and M. Shah, Implementation of artificial intelligence in agriculture for optimisation of irrigation and application of pesticides and herbicides, *Artif. Intell. Agric.* 4 (2020), 58–73.
23. T.A. Shaikh, W.A. Mir, T. Rasool, and S. Sofi, Machine learning for smart agriculture and precision farming: towards making the fields talk, *Arch. Computat. Methods Eng.* 29 (7) (2022), 4557–4597.
24. C.-J. Chen, Y. Y. Huang, Y. S. Li, C. Y. Chang, and Y. M. Huang, An AIoT based smart agricultural system for pests detection, *IEEE Access.* 8 (2020), 180750–180761
25. Revolutionizing California Farms: Smart Irrigation Automation Systems for Water-Efficient, Sustainable Agriculture. https://farmonaut.com/usa/revolutionizing-california-farms-smart-irrigation-automation-systems-for-water-efficient-sustainable-agriculture/?utm_source=chatgpt.com
26. K. Patel & M. Rathod, Smart irrigation using AI and IoT in precision agriculture, *J. Agric. Inform.* 14 (2) (2023), 55–70.
27. Y. Zhang & P. Kumar, Cognitive IoT in smart farming. *Sens. Syst.* 9 (4) (2021), 112–128.
28. Indian Council of Agricultural Research (ICAR). (2021). Pest Surveillance and Digital Agriculture Initiatives. https://icar.org.in/
29. R. Sharma & V. Kumar, AI-based pest detection for sustainable cotton farming. *J. Precis. Agric.* 11 (3) (2022), 45–59.
30. Plantix. (2023). Empowering Farmers through AI Pest Detection. https://plantix.net/en/
31. CNH Industrial. (2023). Raven Autonomy in South America: A Case Study. https://www.cnhindustrial.com/
32. R. Souza & F. Mendes, Autonomous farm machinery and its impact on Brazilian agriculture. *J. Smart Farming.* 8 (2) (2022), 65–79.
33. J. Deere. (2022). See & Spray and the Future of Precision Agriculture. https://www.deere.com/

15 Integration of Federated Learning in Contemporary Farming Practices

Avani Vyas, Nisha Chaurasia, and Priyanka Rathee

15.1 INTRODUCTION

Agriculture has experienced an incredible change in the last few decades, from traditional, intuition-based farming to data-based, technology-facilitated practices called precision agriculture. Precision agriculture utilizes leading-edge technologies including satellite imaging, drones, sensors, and machine learning (ML) algorithms to streamline every function of farming, from planting and irrigation to pest management and harvesting. This revolution has not only enhanced crop yield and efficiency in resource use but has also made farming more sustainable and environmentally friendly.

15.1.1 CONTEMPORARY (MODERN) FARMING PRACTICES

Contemporary farming practices use technological innovation and data-centric decision-making. The popular contemporary farming practices are as follows:

- **Precision Agriculture**: Precision agriculture involves studying the agricultural land before planting to optimize the use of agricultural resources. Precision farming, also called site-specific crop management (SSCM), involves real-time land observation through satellite images and drones, and sensors monitoring local weather, temperature, air, and soil moisture. It is a data-driven farming practice aiming to achieve better crop yield, crop rotation, and optimization of labor costs [1].
- **Regenerative and Sustainable Agriculture**: Regenerative agriculture is a rehabilitation and conservation method. It seeks to improve and restore the health of soil, water resources, biodiversity, and the larger ecosystem. Regenerative agriculture promotes productive and profitable farming. Contrary to traditional farming that tends to exhaust soil and natural resources, regenerative agriculture targets practices that restore soil organic matter, increase water holding, and capture carbon [2]. These factors aid in fighting climate change. Merging FL into current agricultural practices is a breakthrough toward precision agriculture. It supports collective,

 DOI: 10.1201/9781003637264-19

data-driven decision-making while maintaining individual privacy and ownership over sensitive agricultural information. FL enables farmers, cooperatives, and agritech companies to train models locally on individual data and share only model updates, not original data. FL meets such core issues of data privacy, security, and compliance, which are especially crucial in agriculture, where data like crop yields, soil health, and the health of animals are both precious and sensitive [3].

- **Vertical, Hydroponic, and Aeroponic Farming**: These contemporary farming practices are used in urban areas where farming resources like land, soil, and water are limited. In vertical farming crops are grown in stacked layer in an controlled environment [4]. In hydroponic farming, plants are grown in a nutrient-rich water without soil [5]. In aeroponic farming, roots of plants are suspended in the air and misted with a nutrient-rich solution [6]. Figure 15.1 shows all three types of farming practices.

15.1.2 Federated Learning

In centralized ML, data generated from various sources is communicated to the central server, where the ML model is trained on the received dataset. Hence, all data processing and event predictions are done on the central server. Centralized ML models suffer single-point failure and data privacy issues. In contrast, federated learning (FL) follows a decentralized ML principle in which the central server broadcasts a global ML model to local edge devices. These edge devices receive the dataset from their paired sensors and train the received model locally. Once the model is trained, the parameters are communicated to the central server instead of sharing the whole dataset used to train the local model. The central server updates its global model parameters when receiving updates from edge devices. FL provides a secure environment. Figure 15.2 is the pictorial representation of FL.

FIGURE 15.1 Vertical, hydroponic, and aeroponic farming.

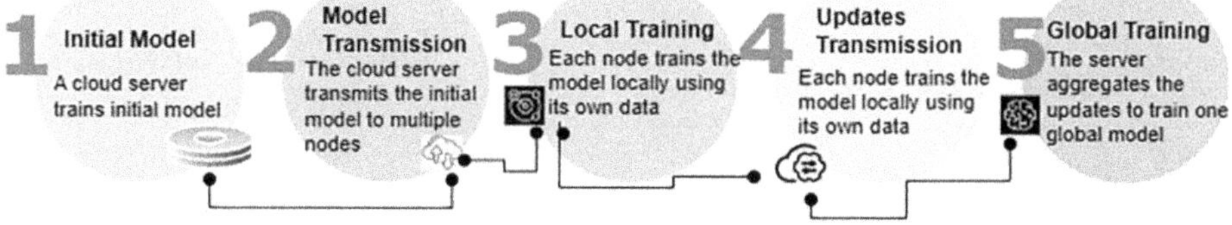

FIGURE 15.2 FL workflow: Local training and global aggregation [7].

15.2 ENABLING TECHNOLOGIES FOR FL DEPLOYMENT

FL deployment in real-world applications relies on a combination of advanced technologies to ensure efficiency, scalability, and data privacy. In the agricultural sector, where data is generated at the edge and shared across a distributed network of stakeholders, these enabling technologies become particularly important. This section explores key technological components that facilitate effective FL deployment. Figure 15.3 shows the enabling technologies discussed in this section.

15.2.1 Internet of Things Sensors: Data Collection at the Edge

In centralized FL data is collected straight from the field in real time. Hence, Internet of Things (IoT) sensors are essential to FL [8]. IoT devices in agriculture, including weather stations, drones, temperature monitors, and soil moisture sensors, continuously collect detailed information about farm conditions [9,10]. These sensors support FL's privacy-preserving goals by using localized data for model training without moving it to a central server. Furthermore, scalability is supported by IoT-enabled data collection since thousands of farms can participate in a federated model without disclosing their raw data.

15.2.2 Edge Computing: Reducing Latency and Bandwidth

Edge computing complements IoT infrastructure by providing local computational power near the data source. In FL, edge devices such as smartphones, farm gateways, or on-site servers perform model training using locally available data. This reduces the need to transmit large volumes of data over the network, minimizing bandwidth usage and improving response times. Especially in rural or remote areas with limited internet connectivity, edge computing ensures that FL systems

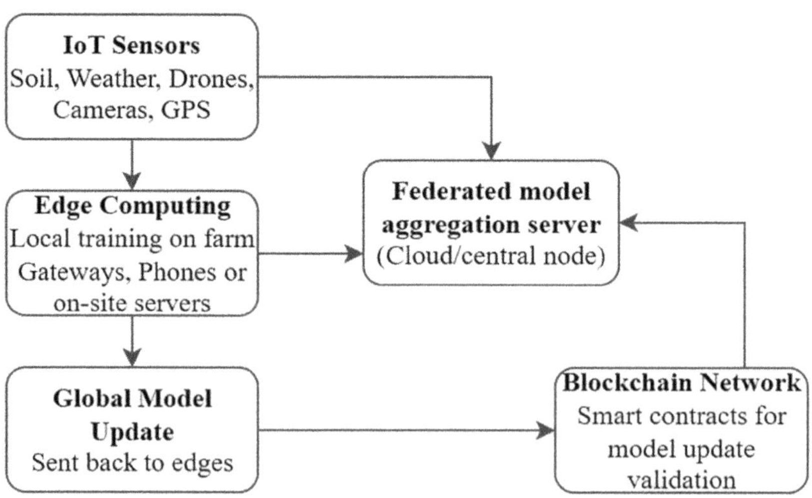

FIGURE 15.3 Enabling technologies for FL in modern farming.

remain functional and efficient. In addition, edge-based model training can enable quicker decision-making in time-sensitive applications such as pest detection or irrigation control.

15.2.3 Blockchain for Secure Model Integration

Blockchain is a decentralized technology designed to ensure data integrity, resistance to tampering, and transparency within a distributed network, where multiple users can access and update data. Each transaction in a blockchain is recorded in a ledger, a secure and immutable record-keeping system. Every node in the network maintains a copy of this ledger. When a new transaction requests an update to the ledger, it must first be approved by a group of miners in the network. This consensus mechanism helps prevent manipulation of transactions. Once approved, the transaction is propagated across the network, thereby ensuring transparency and data integrity.

When distributed FL (i.e., without a central server) is combined with blockchain, it creates a novel paradigm known as FLchain [11]. In FLchain, the genesis block stores the initial weights of the global learning model, which are then distributed to all clients. Each client generates a local dataset and uses it to train its local model. The locally updated model parameters, intended to be shared with other clients, are submitted as a block to the miners. Once the block is validated through a consensus mechanism, it is distributed to all clients in the network. In this way, distributed FL achieves secure and trustworthy transmission of model parameters [11]. Figure 15.4 shows the integration of distributed FL with blockchain. The description of the figure is as follows. The client nodes generate different local models. Each client sends the generated model weights to its connected edge device. The edge device distributes these generated weights using block chain technology. Each block receives new weights from the blockchain network as a global model weight and re-adjusts their local weights. Manoj et al. [12] explored the use case of FLchain in agriculture for crop yield prediction.

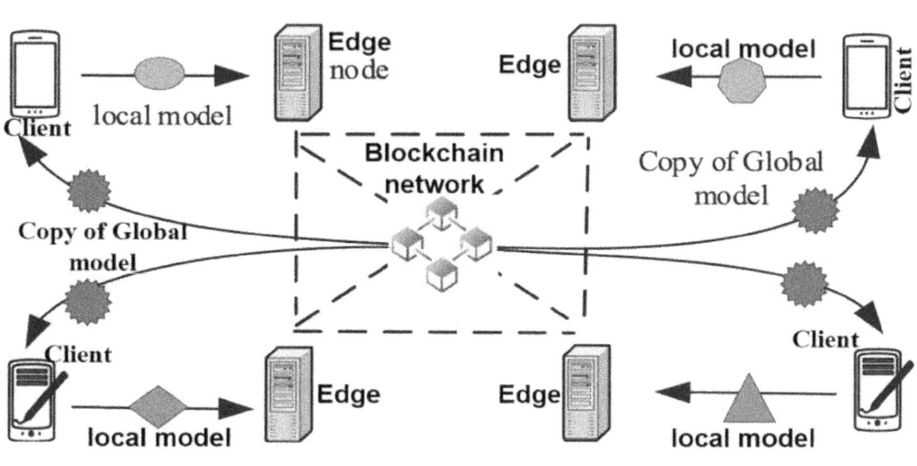

FIGURE 15.4 Federated learning with blockchain-enabled network [11].

15.3 INTEGRATION OF FL IN MODERN FARMING

Merging FL into current agricultural practices is a breakthrough toward precision agriculture, supporting collective, data-driven decision-making while maintaining individual privacy and ownership over sensitive agricultural information. FL enables farmers, cooperatives, and agritech companies to train models locally on individual data and share only model updates, not original data. This strategy meets such core issues of data privacy, security, and compliance, which are especially crucial in agriculture, where data like crop yields, soil health, and the health of animals are both precious and sensitive [13].

15.3.1 Applications of FL in Modern Farming

Monitoring Crop Disease: Mamba Kabala et al. [14] used FL on image dataset collected from "PlantVillage" platform. This study utilized FL with convolutional neural networks (CNNs) and vision transformers (ViT) on the "PlantVillage" dataset of plant disease images. The research proved that FL models, especially ResNet50-based models, performed robustly while maintaining user data privacy. The study also pinpointed that the model's accuracy and efficiency depend significantly on the number of learners, communication rounds, and data quality. ViTs were accurate but less appropriate because they had higher computational requirements in FL settings.

The final model in FL is obtained by the convergence of model parameters trained independently on clients. The prediction performance of the converged model is affected when the data at the clients is not identically distributed. For example, client models trained on limited datasets may provide overfitting in prediction results. Hari and Singh [15] addressed the problem and proposed intelligent weight transferring in FL for detecting plant disease. The naive solution is to weight the model based on the number of samples per client. However, this may result in a biased aggregation method. The proposed system divides the dataset into small chunks, and the model is trained iteratively on these chunks. The model parameters are updated after training on different chunks in each iteration. The model parameters updated in each iteration are stored in a CSV file along with the performance change achieved with the change in model parameters. An input dataset file is prepared from the CSV file by differencing the weights and their performance in the consecutive iterations. The ML regressor model is trained on this new dataset matrix. Hence, the regressor ML model is trained on the knowledge generated by the pre-trained model.

Forecasting Crop Yield: In crop yield production, required data is generated using sensors such as humidity, temperature, rainfall, nitrogen, soil pH, etc. By integrating FL, sensed data is processed at the edge devices, and crop yield is predicted. These edge devices contain a local ML model already trained by several iterations with the central server. Mukherjee and Buyya [16] implemented a centralized (shown in Figure 15.5a) and decentralized (shown in Figure 15.5b) FL model to predict crop yield. In centralized FL, the central server shares an initial model with all the clients. Clients update model parameters using their local data. The updated parameters are then shared with the central server. The central server performs aggregation and

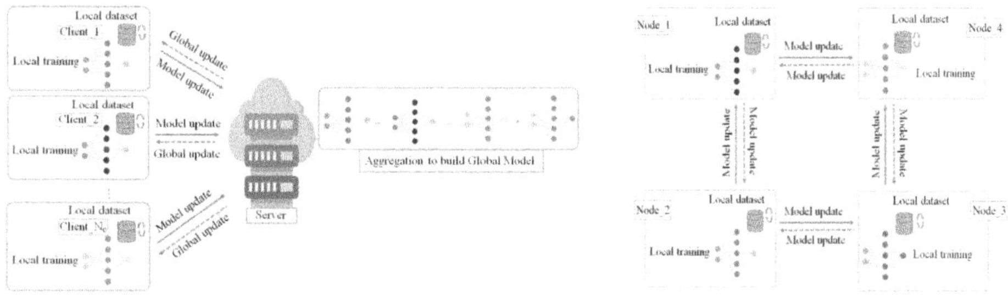

(a)Centralized FL (b) Decentralized FL

FIGURE 15.5 (a) Centralized and (b) decentralized FL [16].

updates the model parameters. The updated model is shared with all the clients. Mamba Kabala et al. [14] concluded in their study that the accuracy of results produced by ML models on local clients depends on the number of iterations of model update between the local model and the central model. In distributed FL, central server is not present. Instead, each client node updates its model parameters using its own data and share the updated model parameters with its neighbor clients. Such type of FL is useful in the case of poor network connectivity when central server is not reachable. In terms of performance, centralized FL provided 75% reduction in response time over the cloud-only framework. Centralized FL and decentralized FL both achieved >95% yield prediction accuracy.

Cowlishaw et al. [17] explored the use of Earth observation (EO) dataset to predict crop yield using FL. The authors highlighted the privacy requirement of EO dataset collected using satellites and, therefore, suggested the use of distributed FL. Figure 15.6 shows the distributed FL framework proposed by the authors. At the topmost layer, satellites capture EO data and send to the ground stations region-wise. Farmers (clients) label the EO data received and use it to train their local FL models. Model parameters of the trained FL model are then distributed in the network using protocols like gossip protocol.

Although distributed FL supports data privacy by ensuring that data is neither stored nor shared on central servers, and avoids single points of failure due to the absence of a central server, it poses other challenges. Specifically, the convergence of model parameters—trained and shared at local clients—requires significant computational and storage resources at the client side. In rural areas, it is particularly difficult to ensure reliable communication channels and provide high computational capabilities at the farmer level. As the sizes of dataset and computational powers vary from client to client, distributed FL mostly operates in asynchronous mode. Furthermore, the authors [17] concluded that accurate crop yield prediction depends on large volumes of temporal data collected by monitoring soil, plant, and weather conditions throughout the entire agricultural cycle—from land preparation to harvesting. Therefore, EO data alone is insufficient for producing accurate yield prediction results.

Monitoring Soil and Environment Condition: The soil and environment conditions are monitored using IoT sensors. A wide variety of soil sensors are used to

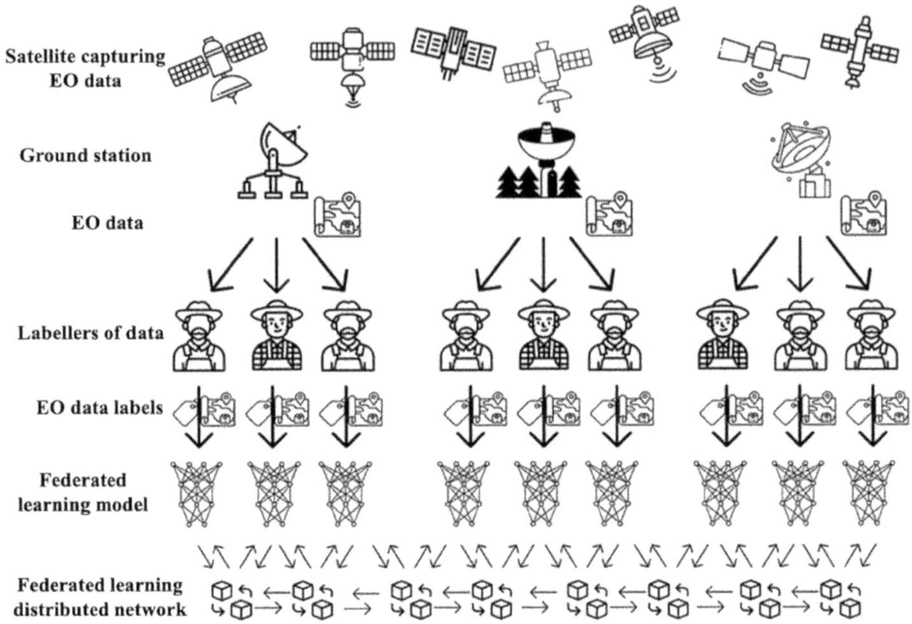

FIGURE 15.6 Federated learning on Earth observation dataset to predict crop yield/type [17].

measure water content, soil water potential, photosynthetic radiation, and oxygen levels. Simultaneously, environment is monitored for wind speed, rainfall, humidity, and brightness. The data generated by these sensors is either used individually to train ML models or in collaboration with other datasets such as imagery data and GPS information. Gallios et al. [18] used Kellogg soil survey database containing 17,000 samples from the US states. The authors distributed dataset among clients according to regions. Then in each region-based data division the dataset is further split into train and test splits of 70% and 30%, respectively. The training dataset from each client is collected into a single pool and the testing dataset is retained with the clients. This pool is the distributed among clients using three different strategies, i.e., equal distribution (IID), climatic zone-based classification (BCL), and geopolitical region-based classification (Region). Each client model uses the allocated dataset by further dividing it into 70:30 training and validation data ratio. Once the model is trained and tested, the model parameters are submitted to the central processing hub where the models are aggregated using FedAvg and WgtAvg. The 30% test dataset separated initially from the original dataset is then used to test the global model. Figure 15.7 is the pictorial representation of the above discussion.

Enhancing Data Privacy and Security: Data privacy in agriculture is crucial to safeguard farmers' confidential information, avoid unethical use by third parties, and foster a level playing field. As digital technologies increasingly harvest data on yields, soil conditions, and cultivation techniques, the possibility of misuse of this information exists. For example, firms may use yield information to price crops lower, to the detriment of farmers. Also, without adequate data privacy, third parties like

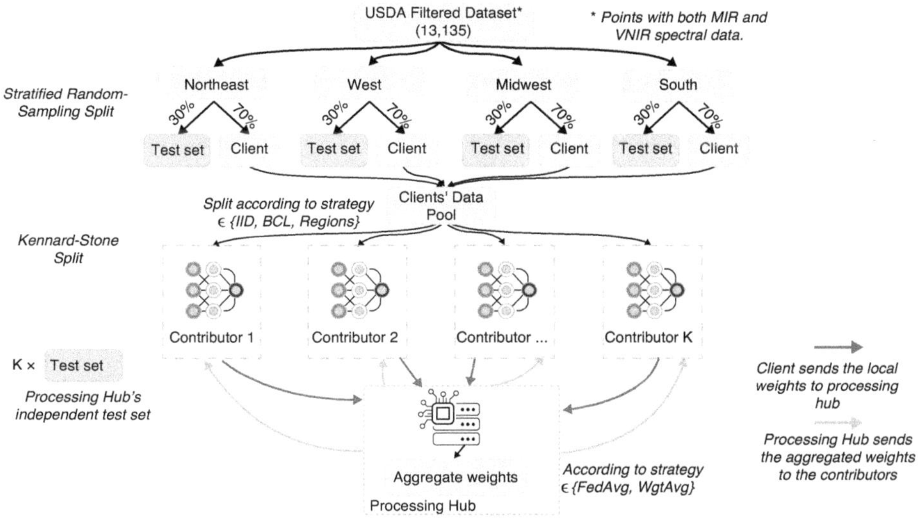

FIGURE 15.7 The FL framework implemented by Gallios et al. [18] to predict soil.

seed or chemical agriculture businesses may get access to and utilize the data without authorization, likely increasing prices or discriminating against farmers. Besides, when major agribusinesses are provided with such data and small-scale farms are not, it levels an unfair playing field, enabling larger players to manipulate market conditions to their advantage. Therefore, protecting data privacy ensures the autonomy of farmers, prevents their exploitation, and ensures fairness in the agricultural industry. Dembani et al. [19] summarizes how FL helps in data privacy preservation in agriculture.

15.4 FL CHALLENGES IN MODERN FARMING

FL, blockchain, and IoT are promising for sustainable agriculture but face significant challenges [20]. The existing research is largely theoretical or based on individual case studies, which prevents widespread adoption. Data privacy and security measures are also underdeveloped [21]. Overcoming these challenges is important to implement these technologies effectively and fulfill their potential in advancing agricultural productivity and sustainability.

15.4.1 Heterogeneous Data Sources

In FL, clients may exist in different geographical regions collecting different types of data at different frequency. These datasets are often non-independent and identically distributed (non-IID). This means that each participating farm or device may contribute data with different distributions, formats, and levels of quality, making it difficult to train a global model that performs well for all participants. This data heterogeneity in agriculture leads to slower model convergence, reduced accuracy, and fairness concerns, especially when local conditions differ significantly from the global average [22]. Additional research notes that this heterogeneity complicates model aggregation and can result in "client drift," where local models diverge from the global objective, further impacting performance and reliability [23]. Addressing these challenges requires the development of robust FL algorithms tailored to the peculiarities of agricultural data, such as personalized FL or hierarchical aggregation strategies, to ensure that collaborative models remain accurate, fair, and useful across diverse farming environments [19,13].

15.4.2 Connectivity Limitations in Remote Areas

FL experiences connectivity challenges in remote agricultural areas due to unreliable connectivity. Communication links between sensors-edges, edges-central server, and edges-edges experience frequent channel drops. The factors that challenges FL in remote areas are:

1. **Lack of Internet Infrastructure**: The remote areas lack stable internet infrastructure leading to frequent disconnections during FL training rounds and model parameter distribution rounds. This delays model updates and lacks in model synchronization. This results in "straggler" clients. "Straggler" clients are the client nodes that fail in timely sharing of local model parameters as well as in receiving of global parameter updates. "Straggler" clients result in reduced training efficiency [24].
2. **Low Bandwidth and High Latency**: The wireless connectivity in rural regions offers limited bandwidth, making it costly to transmit large model updates in FL. Moreover, in FL the model parameters are finalized after iterative parameter learning and sharing process. The repetitive exchange of parameters might overwhelm the limited network bandwidth and increase network latency.
3. **Power Outages**: Remote agriculture areas also lack continuous power supplies, affecting the operation of electricity-powered sensors and edge nodes. Power outages are a severe issue during bad weather conditions, resulting in a lag in data generated by sensors in the field. A power outage is a serious challenge in modern farming practices. However, it is the most underrated and needs to be addressed.

15.4.3 Computational Constraints on Edge Devices

The limited processing power at the sensor nodes and edge devices makes it challenging to calculate FL model gradient. These devices not only lack computational resources but are also equipped with limited memory. The restricted RAM at edge devices limits their capability to use large datasets for model training. The limited computations, storage, and connectivity create straggler nodes in the network, which affects the ultimate aim of using FL in modern farming.

15.5 FUTURE RESEARCH DIRECTIONS

Personalization and Adaption: Agricultural data has high data heterogeneity due to varying weather and geographical conditions. When FL is applied across multiple regions, there are high chances of non-IID data. For example, a pest prediction model trained on data from temperature climates may not generalize well to tropical regions. The convergence of global models is challenging in such scenarios. A personalized model that can address and utilize the heterogeneity of the data is still an open research problem. Adaptation of FL to local practices is another direction that can be utilized to address and utilize non-IID data. Moreover, personalization and adaption will help in fine-tuning the FL model to address the real-world constraints.

Hybrid AI: Merging FL with Explainable AI (XAI): XAI is a technology in trend which explains the trend and impact of an AI model. Use of XAI in contemporary farming will help in achieving trust and adoption of farmers in modern technology. For example, FL model may generate alarm for soil irrigation and XAI may provide details on which soil moisture readings or weather patterns influenced the irrigation decision. Using XAI together with FL will help in addressing biases in FL, inculcating human feedback into the FL loop using explanations as a feedback mechanism, improving personalization and trust.

Scalability of Large-Sensor Ecosystems*:* Modern agriculture depends on wide-reaching IoT device networks for real-time monitoring and data collection which makes the scalability of large sensor ecosystems an essential research area for FL. Through FL implementation, collaborative model training works with localized data storage to improve privacy and minimize communication costs and bandwidth usage required by extensive and varied sensor network datasets. The complexity of sensor network scalability creates technical difficulties through handling non-IID data together with maintaining secure and efficient communication and supporting devices with different capabilities and intermittent connections. The solution to these issues demands scalable FL architectures that combine hierarchical and edge-aggregated models and the creation of lightweight algorithms along with adaptive communication protocols for resource-limited agricultural systems. The advancement of FL scalability in large sensor networks serves as a critical step to establish robust context-aware privacy-preserving decision support systems that will boost precision and productivity and sustainability in smart agriculture.

REFERENCES

1. S. L. Ullo and G. R. Sinha, "Advances in IoT and smart sensors for remote sensing and agriculture applications," *Remote Sensing*, vol. 13, no. 13, p. 2585, 2021.
2. J. Villat and K. A. Nicholas, "Quantifying soil carbon sequestration from regenerative agricultural practices in crops and vineyards," *Frontiers in Sustainable Food Systems*, vol. 7, p. 1234108, 2024.
3. L. Schreefel, R. P. Schulte, I. De Boer, A. P. Schrijver, and H. Van Zanten, "Regenerative agriculture–the soil is the base," *Global Food Security*, vol. 26, p. 100404, 2020.
4. C. Sowmya, M. Anand, C. Indu Rani, G. Amuthaselvi, and P. Janaki, "Recent developments and inventive approaches in vertical farming," *Frontiers in Sustainable Food Systems*, vol. 8, p. 1400787, 2024.
5. M. Kannan, G. Elavarasan, A. Balamurugan, B. Dhanusiya, and D. Freedon, "Hydroponic farming–a state of art for the future agriculture," *Materials Today: Proceedings*, vol. 68, pp. 2163–2166, 2022.
6. B. M. Eldridge, L. R. Manzoni, C. A. Graham, B. Rodgers, J. R. Farmer, and A. N. Dodd, "Getting to the roots of aeroponic indoor farming," *New Phytologist*, vol. 228, no. 4, pp. 1183–1192, 2020.

7. S. Banabilah, M. Aloqaily, E. Alsayed, N. Malik, and Y. Jararweh, "Federated learning review: Fundamentals, enabling technologies, and future applications," *Information Processing & Management*, vol. 59, no. 6, p. 103061, 2022.
8. M. Dhanaraju, P. Chenniappan, K. Ramalingam, S. Pazhanivelan, and R. Kaliaperumal, "Smart farming: Internet of things (IoT)-based sustainable agriculture," *Agriculture*, vol. 12, no. 10, p. 1745, 2022.
9. W.-S. Kim, W.-S. Lee, and Y.-J. Kim, "A review of the applications of the internet of things (IoT) for agricultural automation," *Journal of Biosystems Engineering*, vol. 45, pp. 385–400, 2020.
10. C. Brewster, I. Roussaki, N. Kalatzis, K. Doolin, and K. Ellis, "IoT in agriculture: Designing a Europe-wide large-scale pilot," *IEEE Communications Magazine*, vol. 55, no. 9, pp. 26–33, 2017.
11. D. C. Nguyen, M. Ding, Q.-V. Pham, P. N. Pathirana, L. B. Le, A. Seneviratne, J. Li, D. Niyato, and H. V. Poor, "Federated learning meets blockchain in edge computing: Opportunities and challenges," *IEEE Internet of Things Journal*, vol. 8, no. 16, pp. 12806–12825, 2021.
12. T. Manoj, K. Makkithaya, and V. Narendra, "A blockchain-assisted trusted federated learning for smart agriculture," *SN Computer Science*, vol. 6, no. 3, p. 221, 2025.
13. R. Garro, C. Wilson, D. Swain, A. Pordomingo, and S. Wibowo, "A systematic literature review on the applications of federated learning and enabling technologies for livestock management," *Computers and Electronics in Agriculture*, vol. 234, p. 110180, 2025.
14. D. Mamba Kabala, A. Hafiane, L. Bobelin, and R. Canals, "Image-based crop disease detection with federated learning," *Scientific Reports*, vol. 13, no. 1, p. 19220, 2023.
15. P. Hari and M. P. Singh, "Adaptive knowledge transfer using federated deep learning for plant disease detection," *Computers and Electronics in Agriculture*, vol. 229, p. 109720, 2025.
16. A. Mukherjee and R. Buyya, "Federated learning architectures: A performance evaluation with crop yield prediction application," *Software: Practice and Experience*, vol. 55, no. 7, pp. 1165–1184, 2024.
17. R. Cowlishaw, N. Longépé, and A. Riccardi, "Balancing centralisation and decentralisation in federated learning for earth observation-based agricultural predictions," *Scientific Reports*, vol. 15, no. 1, p. 10454, 2025.
18. G. Gallios, N. Tsakiridis, and N. Tziolas, "Federated learning applications in soil spectroscopy," *Geoderma*, vol. 456, p. 117259, 2025.
19. R. Dembani, I. Karvelas, N. A. Akbar, S. Rizou, D. Tegolo, and S. Fountas, "Agricultural data privacy and federated learning: A review of challenges and opportunities," *Computers and Electronics in Agriculture*, vol. 232, p. 110048, 2025.
20. H. R. Hasan, A. Musamih, K. Salah, R. Jayaraman, M. Omar, J. Arshad, and D. Boscovic, "Smart agriculture assurance: IoT and blockchain for trusted sustainable produce," *Computers and Electronics in Agriculture*, vol. 224, p. 109184, 2024.
21. S. Mehta, V. Kukreja, and A. Gupta, "Revolutionizing maize disease management with federated learning CNNS: A decentralized and privacy-sensitive approach," in *2023 4th International Conference for Emerging Technology (INCET),* Belgaum, India, pp. 1–6, IEEE, 2023.
22. B. Mohammed, "A comprehensive overview of federated learning for next generation smart agriculture: Current trends, challenges, and future directions," *Informatica*, vol. 49, no. 1, pp. 117–126, 2025.
23. L. I. Barona L´opez and T. Borja Saltos, "Heterogeneity challenges of federated learning for future wireless communication networks," *Journal of Sensor and Actuator Networks*, vol. 14, no. 2, p. 37, 2025.
24. P. Patros, M. Ooi, V. Huang, M. Mayo, C. Anderson, S. Burroughs, M. Baughman, O. Almurshed, O. Rana, R. Chard and K. Chard, "Rural AI: Serverless-powered federated learning for remote applications," *IEEE Internet Computing*, vol. 27, no. 2, pp. 28–34, 2022.

16 Integration of Edge Computing and Artificial Intelligence for Precision Agriculture in the Era of Agriculture 4.0

Anamika Sharma and Deepshikha

16.1 INTRODUCTION

The food demand is increasing year by year, primarily due to the increase in the world's population. Conversely, the agricultural industry is affected by both implicit and explicit factors. Implicit factors include environmental constraints such as climate change, soil degradation, water scarcity, and the spread of pests, which threaten the sustainability and productivity of agriculture worldwide (Verma et al., 2025). Explicit factors are geopolitical tension, economic instability, and pandemics. Moreover, the traditional agricultural equipment requires intensive manual effort and elevates operational costs, which affect the efficiency and scalability goals of modern farming.

The adoption of Agriculture 4.0 is suffering from various challenges, particularly in terms of regional divergences. The European, North, and South American continents are adopting digital farming faster than the African continent (Abiri, Rizan, Balasundram, Shahbazi, & Abdul-Hamid, 2023). The Asian continent also participates in this adoption, as countries like India and China actively promote precision farming for sustainable food production. Notably, a country's economic strength plays a pivotal role in determining the pace and extent of new technological adoption (Benyam, Soma, & Fraser, 2021; Gaikwad, Vibhute, Kale, & Mehrotra, 2021). Balkrishna et al. (Balkrishna, Pathak, Kumar, Arya, & Singh, 2023) performed a comprehensive survey on Information and Communication Technology (ICT)-based innovations for smart agriculture in India. They identified 29 Indian applications focused on agricultural information management systems, all exhibiting similar functionalities. Agriculture 4.0 has travelled a long journey from the traditional and labour-intensive methods of agriculture in the early 20th century to smart agriculture in the present day. A detailed description of the evolution of Agriculture 4.0 is shown in Figure 16.1.

DOI: 10.1201/9781003637264-20

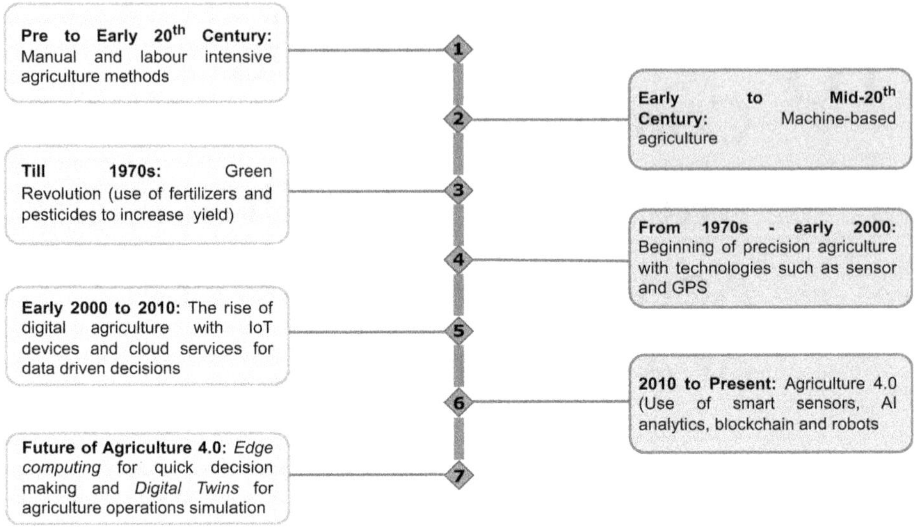

FIGURE 16.1 The evolution of Agriculture 4.0.

16.1.1 Agriculture 4.0

Agriculture 4.0 is a technology-driven transformation for food production, processing, and distribution. It has a broader scope, which is not only limited to technological adoption but also includes economic, environmental, and social dimensions, aiming for a future where agriculture is more productive and sustainable (Gyamfi, ElSayed, Kropczynski, Yakubu, & Elsayed, 2024). It combines four key components: Internet of Things (IoT) sensors, distributed storage, artificial intelligence (AI), and remote access to create an interconnected, intelligent, and precise agriculture system (Aceto, Persico, & Pescapé, 2019). The application domains of Agriculture 4.0 are water management, crop growth and health monitoring, land and soil monitoring, control of the usage of agrochemicals and fertilisers, indoor or vertical farming such as hydroponics and aquaponics systems, autonomous machinery and its navigation system, and logistics management(Araújo, Peres, Barata, Lidon, & Ramalho, 2021; Maffezzoli, Ardolino, Bacchetti, Perona, & Renga, 2022).

16.1.2 The Role of Digital Technologies in Agriculture 4.0

Digital technologies empower farmers with precise, controlled, and data-driven farming. These technologies can enhance both agricultural productivity and profitability when effectively used. The key roles of digital technologies in precision farming are listed below:

1. Control the use of fertilisers, pesticides, and water to the specific needs of the crop.
2. Real-time collection of soil nutrients, plant health, and weather data.

3. AI applications can precisely predict crop growth and detect and classify plant diseases based on historical and input data sets.
4. Enhance operational efficiency using automation and robotics for planting, fertiliser spray, irrigation and harvesting.
5. Collecting spatial data using satellites or drones to evaluate crop health, soil properties, and land utilisation.
6. Blockchain technology can transparently track the production and supply chain of agricultural products from the farm to customers.
7. The application of digital twins can create virtual replicas of farms to simulate scenarios and optimise farming decisions.
8. The applications of edge computing for quick and localised decision-making.

16.2 ENABLING TECHNOLOGIES IN AGRICULTURE 4.0

Agriculture 4.0 represents the integration of *data collection technologies*—such as sensor-enabled tools with *communication systems* like Bluetooth, 4G/5G, LoRaWAN, and satellite communication combined with *cloud platforms* including AWS and Azure, and *advanced data processing* using AI on high-performance computing systems. The key components of Agriculture 4.0 are shown in Figure 16.2.

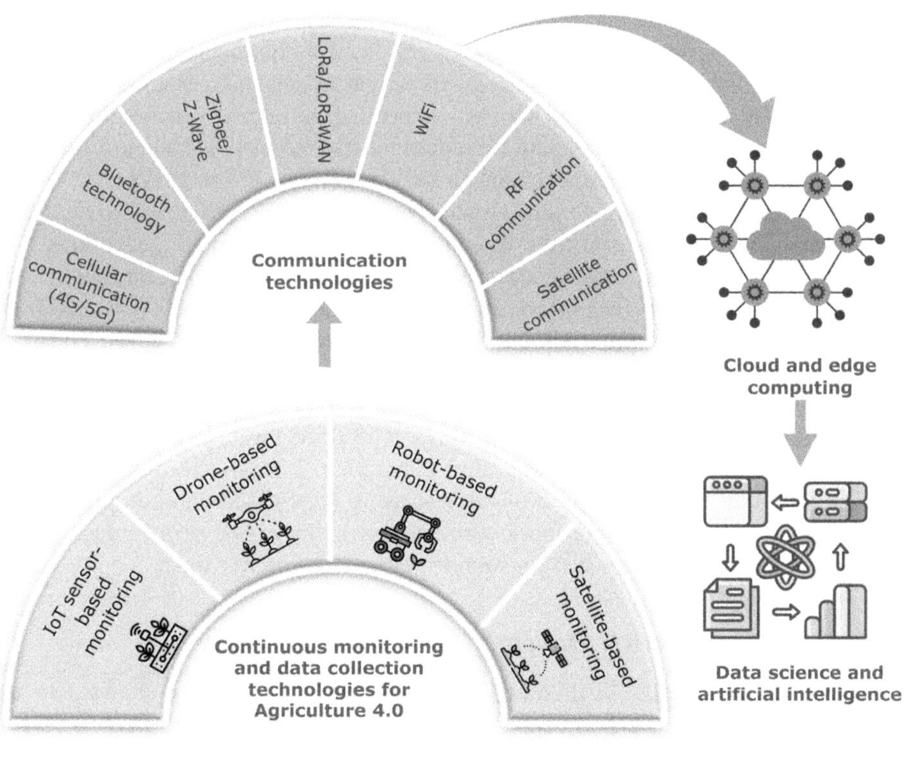

FIGURE 16.2 Components of Agriculture 4.0.

Figure 16.1 explains the integrated architecture of Agriculture 4.0. It emphasises the flow from data acquisition to intelligent decision-making through layered technologies. The continuous monitoring and data collection are achieved using heterogeneous systems such as IoT sensor-based monitoring for soil and crop conditions, drone-based platforms equipped with imaging systems for aerial surveillance, robot-based ground monitoring for plant-level assessment, and satellite-based remote sensing for large-scale environmental analysis. These data sources rely on a range of communication technologies for real-time transmission, including short-range protocols like Bluetooth and Zigbee for local sensor networks, Wi-Fi for medium-range data exchange, and long-range protocols such as LoRa/LoRaWAN, cellular (4G/5G), radio frequency (RF), and satellite communication. The long-range protocols, such as LoRa/LoRaWAN, are essential in remote or infrastructure-less agricultural zones. The collected data is then processed using a hybrid cloud and edge computing framework. Edge computing facilitates low-latency, energy-efficient analysis at the field level, while cloud computing offers scalable resources for complex analytics and long-term data storage. This computational pipeline supports advanced data science and AI applications, including machine learning, deep learning, predictive modelling, and decision support systems. These AI tools extract actionable insights from raw sensor data, enabling precise and autonomous farm management. This architecture represents the core of Agriculture 4.0, an ecosystem that integrates sensing, connectivity, computing, and intelligence to drive efficiency, sustainability, and data-driven decision-making in modern agriculture.

The future of agriculture is emphasising 5G-enabled communication for ultra-low latency and high-bandwidth tasks such as live drone videos and transmission of high-resolution videos. The researchers are also connecting the edge computing IoT devices with the LoRa communication module to reduce the dependency on cloud infrastructure to get real-time alerts or messages (Singh, Sobti, Jain, Malik, & Le, 2022).

Prakash et al. (Prakash, Singh, Gupta, & Lohan, 2023) proposed a four-layer architecture that enables real-time monitoring and automation in agriculture. The **Sensor Layer** comprises various deployed sensors (e.g., temperature, humidity, CO_2, radiation, moisture, and wind speed) and unmanned aerial vehicles (UAVs) equipped with cameras to collect environmental and crop data. The second layer, **Network Layer,** facilitates communication using technologies such as Zigbee, LoRaWAN, Bluetooth, Wi-Fi, GPS, and long term evolution (LTE). This layer ensures reliable data transmission from sensors to gateways. The third layer, **Service Layer,** performs data processing, analysis, and alert generation. It acts as a middleware connecting raw data from the region of interest using edge-cloud computing. The last layer, **Application Layer,** interfaces with end-users via mobile or web dashboards, enabling remote monitoring and decision-making based on live data.

16.2.1 Sensors and IoT Devices

Sensors play a crucial role in precision agriculture by detecting changes in the surrounding environment and transmitting the data. Smart sensors integrate computing capabilities for local data processing. The cost of agricultural sensors varies widely, impacting farmers' ability to invest in these technologies. While basic sensors are affordable, advanced systems require higher investments and have improved efficiency (Mansoor et al., 2025). IoT in agriculture connects sensors with communication devices to collect and generate large volumes of data, such as soil nutrients, crop health, climate, crop storage, and supply chain. This data is then given as input to analytics tools, which help farmers in automation, decision-making, real-time crop monitoring, and resource optimisation. In recent trends, drones have emerged to monitor agricultural activities. These drones have advanced sensors, high-resolution cameras, and integrated communication technologies to collect and analyse real-time data (Rajak, Ganguly, Adhikary, & Bhattacharya, 2023). Table 16.1 presents the list of sensors used for various agricultural application domains.

Data collection, automation, and connectivity in IoT are possible because of the IoT controller. Numerous IoT controllers are available in the market, and their integration is based on the application domain (Kumar et al., 2024). Arduino UNO and Arduino Mega 2560 are basic microcontrollers compatible with the sensors that can sense parameters such as temperature, pH level, moisture, Nitrogen, Phosphorus, and Potassium (NPK) level, rainfall level, etc. This microcontroller is enabled with cellular connectivity but does not have built-in cellular capabilities. Instead, it can be connected to a cellular module using external hardware such as a SIM800/SIM900 GSM module for data transmission.

The wireless connectivity is essential for data transmission in various application domains of smart agriculture. For this purpose, the ESP (Espressif Systems Platform) microcontroller unit is the best choice. The ESP has different variants, such as ESP8266 and ESP32, that are compatible with Wi-Fi (2.4 GHz), Bluetooth, BLE, and Zigbee technologies for data communication. For complex operations, such as image capture, image processing, and lightweight machine learning models, the Raspberry

TABLE 16.1
Types of Sensors Used in Various Agricultural Applications Domain

Application Domain	Type of Sensors
Soil nutrients monitoring	Soil temperature, moisture, pH, soil conductivity, NPK (Nitrogen, Phosphorus, and Potassium), optical sensors, electrochemical sensors, dielectric sensors and drones
Crop growth monitoring	Electrochemical sensors, dielectric sensors, NDVI sensors (Normalised Difference Vegetation Index), ultrasonic, LiDAR (Light Detection and Ranging), high-resolution cameras and drones
Plant disease monitoring	Hyperspectral sensors, gas sensors for stress detection, high-resolution camera, acoustic sensors, electrochemical sensors, biochemical and molecular sensors, and drones
Irrigation automation	Soil moisture, temperature and conductivity sensors, water flow and pressure sensor, water level sensors and UAV (Unmanned Aerial Vehicle)
Fertiliser spray automation	Flow meter, pressure sensors, GPS, NPK sensor, dielectric sensors, multi-spectral imaging and UAV
Fruit ripeness detection	High-resolution camera, LiDAR and electrochemical biosensors
Size and quality detection	High-resolution camera, Time-of-Flight (ToF) sensors, laser triangulation sensors, line-scan cameras and hyperspectral imaging

Pi Pico microcontroller is used. The advancement in microcontrollers, such as the Teensy 4.0/4.1 microcontroller, can perform high-speed data processing and vision processing. The STM32, a series of microcontrollers, is used for real-time environmental monitoring and drone control operations. The precise irrigation and automation are possible due to the PIC microcontrollers. The Xilinx Zynq (FPGA + ARM) microcontroller is used for AI operations such as plant growth monitoring, fruit ripeness detection, plant disease detection, and size and quality detection.

16.2.2 Communication Technologies

The variations in the transmission range and power consumption in various IoT communication technologies make them suitable for multiple applications in Agriculture 4.0. The IoT communication technologies are divided into three categories as listed below:

1. **Short-Range Communication**: The short-range communication or local communication technologies, such as Bluetooth, BLE (Bluetooth Low Energy), Wi-Fi (802.11), Zigbee, RFID, and near field communication (NFC), provide communication range within the farm or greenhouse. The technologies have a shorter communication range, ideally 10–100 m. Only Zigbee has a communication range of approximately 100–500 m. Bluetooth, Zigbee, and RFID have limited data rates. Bluetooth has a data rate of up to 1 Mbps, and Zigbee has a data rate of up to 250 Kbps. The Wi-Fi module provides a higher data rate, up to 1 Gbps within a 100 m communication range. The power consumption in the Wi-Fi module is higher than that of other modules. The short-range communication is best suited for smart irrigation, local soil nutrient monitoring, and local machine-to-machine communication.
2. **Long-Range Communication**: Long-range communication technologies such as cellular networks, LoRa/LoRaWAN, Narrowband-IoT (NB-IoT), LTE, Sigfox, and satellite communication enable reliable remote data transmission. Except for satellite communication, which offers global coverage, these modules typically support communication ranges up to 15 km. LoRa/LoRaWAN, NB-IoT, LTE, and Sigfox provide low data rates and are well-suited for applications with minimal data requirements, such as remote soil nutrient monitoring and remote actuation of smart agricultural devices (e.g., fertiliser sprayers, sprinklers, and drones). Cellular networks support higher data rates—up to 1 Gbps—making them suitable for bandwidth-intensive tasks such as high-resolution image transmission and executing machine learning or deep learning models. Satellite communication offers scalable data rates, ranging from low to high, depending on the application needs. It is particularly beneficial in scenarios requiring coverage in remote or disaster-affected areas with limited or unavailable infrastructure.
3. **Internet-Based Communication:** Internet-based communication protocols enable IoT sensor-to-cloud integration, web-based dashboards for monitoring agricultural and machine activities, real-time data streaming,

and device-to-device communication. The Message Queuing Telemetry Transport (MQTT) protocol is commonly employed for lightweight, efficient IoT-to-cloud communication. Hypertext Transfer Protocol (HTTP/HTTPS) and WebSocket protocols implement interactive, web-based dashboards that display real-time data feeds. The Constrained Application Protocol (CoAP) is designed for constrained environments and facilitates low-overhead, device-to-device communication. These protocols are optimised for low-power, resource-constrained IoT devices, making them suitable for smart agriculture applications.

The communication infrastructure for Agriculture 4.0 has the following requirements:

1. Requirement of sensors, base stations, gateways, and mesh networks.
2. Reliable connectivity and communication medium for sensor-to-edge and edge-to-cloud.
3. Requirement of a low-latency communication medium.
4. Ensuring network scalability, data integrity, and data security.

16.2.3 Cloud and Edge Computing

Cloud and edge computing enable intelligent and data-driven operations in Agriculture 4.0. Cloud computing provides centralised storage, high-performance processing, and advanced analytics capabilities using a scalable infrastructure-as-a-service (IaaS) architecture. It is particularly effective for computationally complex tasks such as machine learning and deep learning model training, big data analytics, and long-term data storage from distributed agricultural sensors.

The power consumption cost of heterogeneous data transfer is high. Moreover, the cloud uses powerful servers for computation and data analysis that consume large amounts of energy. Therefore, both bulk data transmission and computation on the cloud have an adverse impact on the environment, as large energy consumption leads to a large amount of carbon emissions. To overcome this issue, **edge computing** brings data processing closer to the source, i.e., near the IoT devices deployed in the region of interest. Enabling low-latency processing and real-time data analysis for decision-making at the edge significantly reduces communication delays and bandwidth consumption. Edge computing is ideal for time-constrained agricultural operations such as irrigation control, pest detection, and autonomous drone navigation.

The integration of cloud with edge/fog computing is a growing trend that enhances pervasive computing capabilities in smart agriculture. The AgriTalk platform addresses key challenges such as device management, AI provisioning, data privacy, and sensor calibration. It demonstrates that fog/edge computing can reduce delays by 50% compared to cloud solutions and enables farmers to build applications using a low-code/no-code interface (Lin, Chen, & Chang, 2023). Alharbi and Aldossary (Abiri et al., 2023) proposed an architecture that integrates edge, fog, and cloud layers to enable real-time processing of agricultural sensor data, reducing the load on the cloud and improving energy efficiency. Using Mixed-Integer Linear Programming (MILP), the system shows a 36% reduction in energy consumption, a 43% reduction in carbon emissions, and up to an 86% reduction in network traffic compared to traditional cloud-based models.

This transformation in edge and cloud computing facilitates a hybrid architecture, where edge nodes handle local processing and essential or aggregated data is transmitted to the cloud for deeper analysis. This distributed computing paradigm enhances system reliability, scalability, and efficiency in Agriculture 4.0.

16.2.4 Data Science and AI

According to a report by Facts and Factors, the global AI in agriculture market was valued at approximately USD 1,002.02 million in 2019 and is projected to reach around USD 3,984.5 million by 2026, growing at a compound annual growth rate (CAGR) of about 25.2% from 2020 to 2026. The report examines the key drivers, challenges, and their impact on market demand during the forecast period, while also highlighting emerging opportunities within the AI in agriculture sector (Factors, 2025). The increasing adoption of IoT, demand for higher crop yields, advancements in crop management, and drone-based pesticide application are key enablers of AI integration in agriculture. Population growth, limited fertile land, and the need for sustainable food production further drive the demand for AI-driven solutions. AI technologies, supported by IoT and data analytics, facilitate predictive and prescriptive analytics, plant disease identification, pest detection, and real-time soil monitoring. It is contributing to enhanced agricultural productivity and food security.

Agriculture 4.0 is the integration of digital and intelligent technologies to transform traditional farming into a more sustainable, efficient, and data-driven industry. The AI in agriculture significantly modernises the agricultural practices by providing precision agriculture, enhancing data analysis, automating processes, and increasing the accuracy of predictive analytics (Misra et al., 2020). AI uses sensors for real-time data collection with high accuracy. The algorithms of machine learning, deep learning, and computer vision are essential for automation tasks. Convolutional Neural Networks (CNNs) or the variants of YOLO (You Only Look Once) are used for image-based disease detection. Random forest and decision trees are best suited for recommendation systems. LSTM (Long Short-Term Memory) and RNNs (Recurrent Neural Networks) are time series analysis models and are used for yield forecasting. Along with this, 3D laser scanning and analysis of hyperspectral images are also key applications of AI (Javaid, Haleem, Khan, & Suman, 2023).

Data science provides the foundation for processing and interpreting large volumes of heterogeneous data collected from various agricultural sources. The data sources are IoT sensors (soil moisture, weather), drones, satellite imagery, market trends, and genomics. The key techniques used in data science are predictive modelling, descriptive analytics, big data integration, and data visualisation. These analytics are helpful in various applications such as yield prediction, disease outbreak forecasting, soil health assessment, and crop selection and rotation planning.

AI, especially machine learning and deep learning, is increasingly transforming agriculture by enhancing decision-making, resource efficiency, and environmental sustainability. Technologies like remote sensors, GPS-assisted irrigation, drones, and self-driving robots address key agricultural challenges, including irrigation, weed control, and plant monitoring. AI enables efficient data collection, early disease detection, and precision farming, improving crop yields and reducing manual labour. Pre-trained models using large datasets like PlantVillage and ImageNet can accurately identify plant diseases, though recognition accuracy and precision may vary based on image context and dataset quality. Overall, AI-driven approaches promote productivity and reduce environmental impact in Agriculture 4.0 (Shaikh, Rasool, & Lone, 2022).

16.3 EDGE COMPUTING IN AGRICULTURE

Edge computing is defined as processing the data near the data source (IoT sensors) to improve response times, reduce transmission costs, ensure low latency for critical tasks, and create a sustainable environment. A farm uses edge devices to process temperature, humidity, and soil moisture data in real-time to control the automatic irrigation pumps. This prevents overwatering, reduces water waste, and improves crop quality without relying on cloud connectivity. The algorithms of AI are used at local edge servers for decision-making. These servers are integrated with AI microcontrollers such as NVIDIA Jetson, Raspberry Pi, and Google Coral, as well as communication technologies such as LoRaWAN, Zigbee, NB-IoT, and 5G. The edge servers use AWS Greengrass, Azure IoT Edge, and EdgeX Foundry.

16.3.1 Edge Computing Architecture for Agricultural Systems

Cloud and edge computing are complementary technologies. Edge computing excels in real-time responsiveness and decentralised control, while cloud computing offers superior capacity for large-scale analytics and global coordination. A hybrid architecture combining both often generates optimal performance, cost, and scalability in smart agriculture systems.

Figure 16.3 explains the architecture of a smart agriculture framework integrating edge computing and cloud computing to automate and optimise various farming operations. At the core of the system is a cloud server that communicates with several local edge computing servers, each responsible for processing data from specific agricultural domains in real-time. These domains include automatic irrigation, pest and disease detection, greenhouse automation, drone automation, and agricultural robot control.

In the automatic irrigation system, edge servers gather data from soil moisture, rainfall, temperature, wind speed, and flow control sensors to make instant irrigation decisions without relying solely on the cloud. The pest and disease detection system uses optical sensors (RGB, thermal, or hyperspectral), electrochemical biosensors, acoustic sensors, and gas sensors to detect anomalies in crops; the data is processed locally for faster diagnostics and alerts. Similarly, in greenhouse automation, sensors like soil moisture, temperature, light, humidity, and CO_2 detection provide environmental feedback that the edge server uses to regulate conditions within the greenhouse automatically.

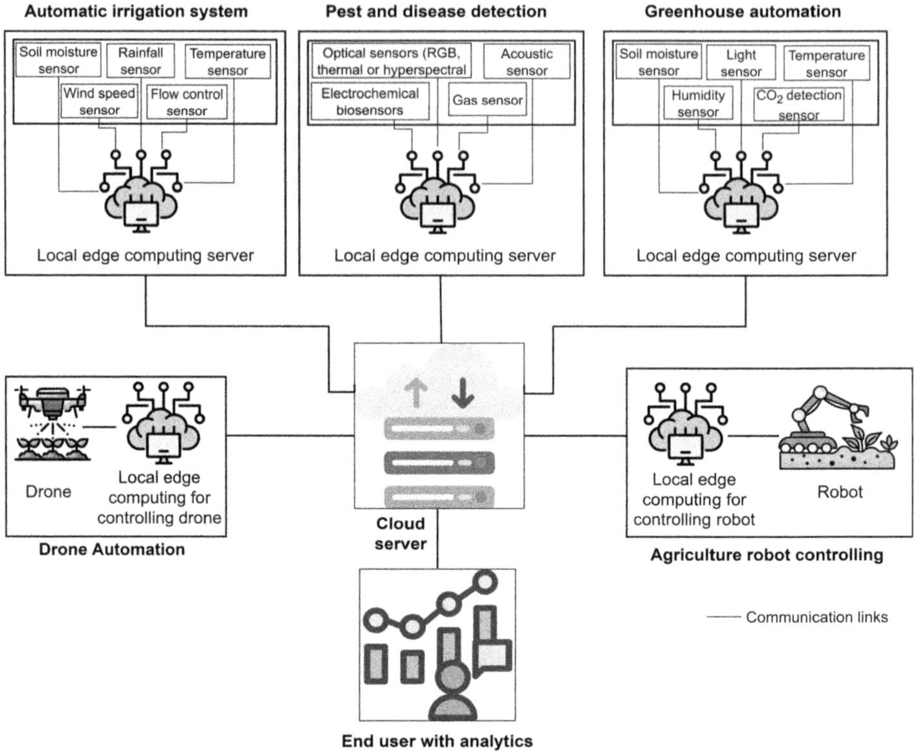

FIGURE 16.3 Architecture of edge computing in agriculture.

For mobility and task-specific automation, drone automation is handled by a dedicated edge server that controls drone operations such as spraying and field monitoring. Agriculture robots used for tasks like weeding, planting, or harvesting are also managed through local edge computing units, enabling precise control and adaptability to field conditions.

All these subsystems communicate with the cloud server for centralised data storage, advanced analytics, machine learning model updates, and global farm management strategies. The communication links ensure seamless data flow between the devices, edge nodes, and the cloud, enabling a resilient, scalable, and intelligent agriculture system. The cloud server transmits the data to the end-user for decision-making. The edge AI is used in the following use cases of precision agriculture:

1. Soil analysis and nutrient monitoring.
2. Fertiliser and pest spray management.
3. Automation of the irrigation system.
4. Real-time environmental monitoring.
5. Crop growth and health assessment.
6. AI-driven navigation and control for autonomous agricultural machinery such as tractors, drones, and robots.

16.3.2 Algorithms in Edge Computing

Edge computing requires lightweight, low-latency, and energy-efficient algorithms that can operate with limited hardware resources and often without a stable internet. These are classified as follows:

1. **Data Pre-processing Algorithms**: These algorithms are used to filter, clean, and reduce raw sensor data before transmission or analysis. The various algorithms used are low-pass filters (e.g., moving average), data compression (e.g., Huffman coding, LZW ((Lempel-Ziv-Welch)), feature extraction (e.g., PCA (Principal Component Analysis), edge detection in images), and outlier detection (e.g., Z-score, IQR) (Junior et al., 2022).
2. **Machine Learning Algorithms at the Edge**: The decision trees are simple to implement, interpretable, and fast for classification tasks such as crop diseases. Naive Bayes is a lightweight probabilistic model for classification tasks. The K-nearest neighbours (KNN) is used for image classification, e.g., plant species or pest identification. The Support Vector Machine (SVM) is efficient for binary classification problems in small datasets such as soil type detection (Junior et al., 2022; Rathore & Rajavat, 2024). A lightweight variant of neural networks known as TinyML optimises neural models such as MobileNet, SqueezeNet, or models converted via TensorFlow Lite for edge deployment. The applicability of TinyML is for disease detection and classification in various crops (Gookyi et al., 2024; Hayajneh et al., 2024; Kimutai & Förster, 2024).
3. **Deep Learning on Edge (TinyML)**: For time-series analysis in weather forecasting or irrigation scheduling, LSTM or RNN is used. CNNs are applied for plant disease detection, drone image processing, etc. YOLO is a fast object detection algorithm commonly used on edge devices for tasks such as livestock monitoring and pest detection from images (Rai et al., 2024; Upadhyay, Sunil, Zhang, Koparan, & Sun, 2024).

   In a recent study, Hayajneh et al. (2024) proposed a Tiny Machine Learning algorithm optimised for edge computing environments. This lightweight algorithm enables efficient inference on resource-constrained devices, delivering valuable outputs with minimal computational overhead. The research demonstrates explicitly the integration of TinyML with unmanned aerial vehicles to support precision tasks in smart farming applications. This research proposed a TinyML-based framework for UAV-assisted smart farming, specifically targeting soil moisture prediction through transfer learning (TL) using deep neural networks and long short-term memory models. The framework integrates IoT sensors and a custom-designed sensor board, built on an ESP32 microcontroller, to measure soil and environmental parameters. It enables the deployment of pre-trained models on edge devices for time-series forecasting, thereby supporting precision irrigation. The system demonstrates practical implementation on UAVs and evaluates performance using inference metrics such as inference time, accuracy, mean squared error (MSE), and the coefficient of determination (R^2). This work is among the first to explore UAV-assisted TL on edge IoT devices using TinyML and shows the feasibility of balancing prediction accuracy with computational efficiency in low-power environments. It also lays the foundation for extending TinyML applications to broader smart farming use cases.

4. **Reinforcement Learning**: It is applied in adaptive systems like robotic weeders or autonomous tractors (Goldenits, Mallinger, Raubitzek, & Neubauer, 2024).
5. **Routing Algorithm**: These algorithms are used between edge networks and sensor nodes. LEACH (Low-Energy Adaptive Clustering Hierarchy) is used for energy-efficient sensor node clustering (Vaiyapuri et al., 2022). AODV (Ad hoc On-demand Distance Vector) is applied for dynamic routing in edge-based IoT networks.

   In mobile edge computing within MANETs, frequent link failures occur due to node mobility and energy depletion, leading to increased latency in single path routing protocols. To mitigate this, a multi-path routing protocol, LLECPAOMDV (Link Lifetime and Energy Consumption Prediction-based AOMDV), is proposed. It incorporates an energy-aware algorithm during route discovery by excluding low-energy nodes and selects routes based on predicted link lifetime and minimal energy consumption. Performance evaluation shows that LLECPAOMDV achieves the best results in terms of energy efficiency, packet delivery ratio, end-to-end delay, and network lifetime compared to existing protocols, making it well-suited for mobile edge computing (Zhang et al., 2020).

16.3.3 Advantages of Edge over Cloud in Farming Environments

In the evolving technology, understanding the differences between edge and cloud computing is essential. While cloud computing is more established, edge computing is a newer paradigm offering complementary capabilities. Together, they are transforming data processing and storage by replacing traditional infrastructures like corporate data centres, expensive hardware, and frequent software upgrades. A detailed comparison of edge and cloud computing is shown in Table 16.2.

While cloud and edge computing differ in architecture and application, both aim to enhance modern computing through *scalability*, *efficient data management*, and *cost-effectiveness.* While cloud computing provides centralised, scalable resources ideal for handling large-scale analytics, long-term data storage, and compute-intensive tasks, edge computing offers localised, low-latency processing essential for real-time responsiveness and decentralised control. Cloud systems are well-suited for global applications with extensive user bases, whereas edge systems are optimised for time-sensitive operations in remote or bandwidth-constrained environments. Both architectures contribute to cost-effectiveness in different ways: cloud computing by reducing capital expenditures through shared infrastructure and edge computing by minimising bandwidth usage and enabling local data processing. Their strategic integration enables robust and flexible solutions in domains such as smart agriculture, healthcare, and smart cities, where both real-time and long-term data insights are important (Kurduban, 2024).

TABLE 16.2
Comparison of Edge Computing with Cloud Computing

Feature	Edge Computing	Cloud Computing
Definition	Local data processing on edge devices such as sensors, drones, or robots	Centralised processing on cloud platforms after data transmission
Latency	Very low: supports real-time actions like irrigation or pest control	High: suitable for non-real-time tasks like historical trend analysis
Bandwidth usage	Low: only processed or filtered data is transmitted	High: continuous raw data transmission, such as numerical, textual, images or videos
Reliability	Operates offline or in low-connectivity areas	Requires stable internet connectivity
Computation capacity	Limited: supports lightweight ML models, such as TinyML	High: suitable for complex models, AI training, and large-scale processing
Privacy and security	High: sensitive data remains on-site	Requires strong security for centralised storage and access
Scalability	Hardware dependent, scales per device	Cloud native scalability with virtual resources
Use cases	Real-time irrigation, pest detection, livestock monitoring	Crop yield prediction, weather-based forecasting, model training
Cost considerations	Higher device cost but reduced data cost	Lower device cost but higher transmission and storage cost
Data storage	Short-term, temporary storage and usage	Long-term archival, cross-season analysis

16.3.4 Operational Challenges of Integrating Edge Computing with Agriculture 4.0

The integration of edge computing in digital agriculture has the potential to revolutionise farming practices by enabling real-time data processing, localised decision-making, and reduced dependency on cloud infrastructure. However, the deployment of such systems in agricultural environments faces several operational barriers that can hinder scalability and long-term sustainability. These challenges include technical, financial, infrastructural, and human resource dimensions, as discussed below:

1. **High Infrastructure Initialisation Cost**: Deployment of edge computing infrastructure demands substantial initial investment, including the procurement and integration of IoT sensors, UAVs (drones), edge servers, and autonomous robotic systems.
2. **Data Security and Privacy**: Ensuring secure data acquisition, transmission, and storage at edge nodes remains a primary challenge, particularly in preventing unauthorised access and cyberattacks in distributed environments.
3. **Digital Literacy Gap**: A key limitation is the lack of digital awareness and technical proficiency among farmers, which hampers the adoption and effective use of edge-enabled technologies.

4. **Skilled Workforce Requirement**: The operation and management of edge-based agricultural systems require trained personnel with interdisciplinary expertise in agronomy, data science, and embedded systems.
5. **Power Supply Constraints**: Continuous and reliable access to electricity is essential for charging and operating drones, sensors, and robotic platforms, especially in remote and off-grid rural areas.
6. **Hardware Maintenance Overheads**: The maintenance of edge hardware (e.g., sensors, gateways, UAVs) introduces recurring operational costs and logistical complexities, especially under harsh agricultural field conditions.
7. **Network Connectivity**: A stable internet connectivity is essential for real-time data processing and communication between edge devices and cloud systems, which may be limited in rural and low-infrastructure regions.

16.4 THE ROLE OF EDGE AI IN PRECISION AGRICULTURE

Edge AI has emerged as a critical enabler of real-time decision-making and localised intelligence in precision agriculture. By processing data at the source, closer to sensors and actuators, Edge AI reduces dependency on cloud connectivity, lowers latency, and enhances operational efficiency in the field (El Jarroudi et al., 2024; Qian, Qian, & Yu, 2025). Its applications span several key domains in Agriculture 4.0, including soil analysis, fertiliser management, irrigation control, environmental monitoring, and crop health assessment.

1. **Soil Analysis**: Edge AI enables localised analysis of soil health through direct integration with deployed sensors. By deploying machine learning models at the edge, devices can infer parameters such as soil moisture, pH, electrical conductivity, and macronutrient concentrations (N, P, K) in real time. This eliminates the need for cloud transmission, reduces decision latency, and allows region-specific soil diagnostics. Moreover, CNNs applied to hyperspectral data collected from soil surfaces can classify soil types and detect contamination or compaction zones (Kasemi, Vincze, & Lammer, 2023).
2. **Fertiliser Management**: Edge AI facilitates variable rate application of fertilisers by processing real-time multi-sensor inputs, including crop nutrient status via leaf imaging, soil nutrient levels, and environmental conditions. Reinforcement learning or decision tree-based models at the edge compute the optimal nutrient dosage and timing per zone or plant. These edge-based decisions integrate with automated actuators or robotic platforms for precise application, minimising fertiliser waste and maximising the yield. Additionally, historical data fusion with real-time inputs enables predictive nutrient scheduling (Ramirez, 2025).
3. **Irrigation Control**: Edge computing nodes equipped with AI-based control logic process data from soil moisture sensors, ambient humidity, leaf water potential sensors, and local weather forecasts to implement dynamic irrigation scheduling. For instance, recurrent neural networks or fuzzy logic controllers at the edge can predict evapotranspiration rates and adjust irrigation frequency and volume accordingly. Edge-based irrigation systems enhance

temporal granularity, support zone-specific water delivery, and operate independently of unreliable rural internet connectivity, making them suitable for autonomous irrigation in remote fields. LoRaWAN is also providing the flexibility of autonomy for agricultural systems (Munir, Bajwa, Ashraf, Anwar, & Rashid, 2021; Ramirez, 2025; Y. Zhang et al., 2025).

4. **Real-Time Environment Monitoring**: Edge AI enables continuous monitoring of environmental variables such as temperature, relative humidity, solar radiation, wind speed, and atmospheric gases using embedded sensor networks. AI models (e.g., anomaly detection algorithms or spatiotemporal clustering methods) deployed at edge gateways detect microclimatic anomalies, predict frost or heat stress events, and provide alerts for real-time climate adaptation. This real-time responsiveness supports precision pest control, greenhouse management, and weather-sensitive crop operations such as spraying or harvesting (Prajapati, Kumar, Pramanik, Prajapati, & Saini, 2023).
5. **Crop Growth and Health Assessment**: Edge AI utilises machine vision systems integrated with RGB, thermal, and multi-spectral cameras on drones, robots, or stationary platforms to monitor plant phenotypes, growth stages, and health conditions. CNNs and transformer-based vision models at the edge can classify diseases, detect nutrient deficiencies, and estimate biomass or canopy coverage directly in the field. These systems offer on-site decision-making, enabling real-time precision interventions like spot treatment of diseased plants or adaptive scheduling of agronomic operations. Further, growth prediction models can be embedded at edge nodes for yield forecasting and stress diagnostics under varying environmental conditions (Huang, Chen, Lin, & Zheng, 2022; Natraj, Raj, Karpagam, & Gunanandhini, 2025; Pintus, Colucci, & Maggio, 2025).

16.5 FUTURE DIRECTIONS

The evolution of Agriculture 4.0 is characterised by the convergence of advanced computing architectures, intelligent modelling systems, and novel hardware paradigms designed to address the increasing complexity and scale of modern agricultural systems. As conventional methods struggle to cope with the demands of real-time data processing, adaptive decision-making, and resource optimisation, next-generation technologies such as cognitive digital twins, field-scale virtual replication, and predictive scenario modelling offer transformative capabilities for dynamic, data-driven farm management. Simultaneously, breakthroughs in neuromorphic edge processors, energy-efficient AI chipsets, and in-sensor computing paradigms are redefining the computational landscape by enabling ultra-low-power, low-latency intelligence directly at the edge. These innovations collectively aim to enhance precision, autonomy, and sustainability in agricultural practices, setting the foundation for fully integrated, cyber-physical farming ecosystems.

1. **Cognitive Digital Twins**: Future research developments will integrate AI-driven cognitive capabilities into digital twin frameworks, enabling autonomous learning, reasoning, and decision-making in real time.

These twins will evolve from static models to dynamic, adaptive systems capable of simulating crop physiology, predicting stress responses, and prescribing optimised interventions. Integration with multi-modal sensor data and cloud–edge fusion will enhance situational awareness and farm intelligence.

2. **Field-Scale Virtual Replication**: The digital agriculture advancements will focus on high-resolution, GIS-integrated, real-time virtual environments that digitally mirror entire farms down to micro-zones. These replicas will support remote diagnostics, autonomous machinery navigation, and climate impact simulations, making large-scale precision agriculture feasible. Incorporating augmented reality and digital soil–plant interfaces will further enhance farmer interaction and decision-making accuracy.
3. **Predictive Scenario Modelling**: The future research will use hybrid AI models combining deep learning with physics-based simulations to predict multiple agricultural scenarios under various climatic and resource constraints. These models will be embedded at the edge for real-time responsiveness and continuously refined using streaming data from IoT devices, enhancing farm resilience and sustainability.
4. **Neuromorphic Edge Processors**: Agriculture 4.0 will leverage brain-inspired, event-driven computing architectures for ultra-low-latency and energy-efficient edge intelligence. Neuromorphic processors enable real-time sensory fusion, anomaly detection, and autonomous control in edge devices such as UAVs and agro-bots. Research will focus on agriculture-specific neuromorphic models optimised for environmental variability and bio-physical signal interpretation.
5. **Energy-Efficient AI Chipsets**: Development will shift toward domain-specific AI accelerators that operate under ultra-low power constraints and are suitable for remote agricultural deployments. These chipsets will support edge-native learning, reduce reliance on cloud connectivity, and facilitate solar-powered smart nodes for long-term deployment in off-grid environments.
6. **In-Sensor Computing Paradigms**: The advancement in sensors will integrate compute capabilities directly at the pixel or data-capture level, allowing on-device pre-processing and decision-making. This paradigm, known as *compute-in-pixel* or *sensor fusion at the source,* will reduce data transmission loads and latency. Applications will include real-time leaf disease detection, soil nutrient monitoring, and autonomous sprayer control, making the system highly efficient and scalable.

16.6 CONCLUSION

The integration of edge computing and AI in Agriculture 4.0 represents a transformative shift in modern farming practices, enabling real-time, data-driven, and sustainable solutions to address global food production challenges. By deploying IoT sensors, communication technologies, and advanced computing systems, precision agriculture achieves localised decision-making, reduced latency, and enhanced operational efficiency. Edge AI plays a pivotal role in applications such as soil analysis, irrigation control, crop health assessment, and environmental monitoring, while hybrid architectures combining edge and cloud computing optimise scalability and resource utilisation. Despite operational challenges like high infrastructure costs, digital literacy gaps, and connectivity issues, advancements in neuromorphic processors, energy-efficient AI chipsets, and cognitive digital twins promise a future of fully integrated, cyber-physical agriculture ecosystems. This convergence of technologies enhances productivity and sustainability and sets the foundation for innovative, autonomous agricultural systems capable of meeting the demands of a growing global population.

REFERENCES

Abiri, R., Rizan, N., Balasundram, S. K., Shahbazi, A. B., & Abdul-Hamid, H. (2023). Application of digital technologies for ensuring agricultural productivity. *Heliyon*, *9*(12), e22601.

Aceto, G., Persico, V., & Pescapé, A. (2019). A survey on information and communication technologies for industry 4.0: State-of-the-art, taxonomies, perspectives, and challenges. *IEEE Communications Surveys & Tutorials*, 21(4), 3467–3501.

Araújo, S. O., Peres, R. S., Barata, J., Lidon, F., & Ramalho, J. C. (2021). Characterising the agriculture 4.0 landscape—emerging trends, challenges and opportunities. *Agronomy*, 11(4), 667.

Balkrishna, A., Pathak, R., Kumar, S., Arya, V., & Singh, S. K. (2023). A comprehensive analysis of the advances in Indian digital agricultural architecture. *Smart Agricultural Technology*, 5, 100318.

Benyam, A. A., Soma, T., & Fraser, E. (2021). Digital agricultural technologies for food loss and waste prevention and reduction: Global trends, adoption opportunities and barriers. *Journal of Cleaner Production*, 323, 129099.

El Jarroudi, M., Kouadio, L., Delfosse, P., Bock, C. H., Mahlein, A.-K., Fettweis, X., … Hamdioui, S. (2024). Leveraging edge artificial intelligence for sustainable agriculture. *Nature Sustainability*, 7(7), 846–854.

Factors, F. (2025). AI in agriculture market size, share global analysis report, 2021–2026. https://www.fnfresearch.com/global-ai-in-agriculture-market-by-technology-machine-1122 (Accessed: 2025-07-04).

Gaikwad, S. V., Vibhute, A. D., Kale, K. V., & Mehrotra, S. C. (2021). An innovative IoT based system for precision farming. *Computers and Electronics in Agriculture*, 187, 106291.

Goldenits, G., Mallinger, K., Raubitzek, S., & Neubauer, T. (2024). Current applications and potential future directions of reinforcement learning-based digital twins in agriculture. *Smart Agricultural Technology*, 8, 100512.

Gookyi, D. A. N., Wulnye, F. A., Arthur, E. A. E., Ahiadormey, R. K., Agyemang, J. O., Agyekum, K. O.-B. O., & Gyaang, R. (2024). Tinyml for smart agriculture: Comparative analysis of tinyml platforms and practical deployment for maise leaf disease identification. *Smart Agricultural Technology*, *8*, 100490.

Gyamfi, E. K., ElSayed, Z., Kropczynski, J., Yakubu, M. A., & Elsayed, N. (2024). Agricultural 4.0 leveraging on technological solutions: Study for smart farming sector. In *2024 IEEE 3rd International Conference on Computing and Machine Intelligence (ICMI)* (pp. 1–9), Michigan, USA.

Hayajneh, A. M., Aldalahmeh, S. A., Alasali, F., Al-Obiedollah, H., Zaidi, S. A., & McLernon, D. (2024). Tiny machine learning on the edge: A framework for transfer learning empowered unmanned aerial vehicle assisted smart farming. *IET Smart Cities*, 6(1), 10–26.

Huang, C.-H., Chen, B.-W., Lin, Y.-J., & Zheng, J.-X. (2022). Smart crop growth monitoring based on system adaptivity and edge AI. *IEEE Access*, 10, 64114–64125.

Javaid, M., Haleem, A., Khan, I. H., & Suman, R. (2023). Understanding the potential applications of artificial intelligence in agriculture sector. *Advanced Agrochem*, 2(1), 15–30.

Junior, F. M. R., Bianchi, R. A., Prati, R. C., Kolehmainen, K., Soininen, J.-P., & Kamienski, C. A. (2022). Data reduction based on machine learning algorithms for fog computing in IoT smart agriculture. *Biosystems Engineering*, 223, 142–158.

Kasemi, R., Vincze, M., & Lammer, L. (2023). Edgesoil: Low-cost soil sensor and edge AI fusion in agricultural robotics. In *# Placeholder Parent /Metadata Value#* (pp. 42–46).

Kimutai, G., & Förster, A. (2024). Domain-adaptive tinyml model for efficient pest and disease detection in domestic crops: A practical approach for developing countries. In *Proceedings of the 2024 International Conference on Information Technology for Social Good* (pp. 44–55), Bremen, Germany.

Kumar, V., Sharma, K. V., Kedam, N., Patel, A., Kate, T. R., & Rathnayake, U. (2024). A comprehensive review on smart and sustainable agriculture using IoT technologies. *Smart Agricultural Technology*, 8, 100487.

Kurduban, V. (2024). Edge computing vs cloud computing: Differences and relationship. https://www.digi.com/blog/post/edge-computing-vs-cloud-computing. (Accessed: 2025-07-05)

Lin, Y.-B., Chen, W.-E., & Chang, T. C.-Y. (2023). Moving from cloud to fog/edge: The smart agriculture experience. *IEEE Communications Magazine*, 61(12), 86–92.

Maffezzoli, F., Ardolino, M., Bacchetti, A., Perona, M., & Renga, F. (2022). Agriculture 4.0: A systematic literature review on the paradigm, technologies and benefits. *Futures*, 142, 102998.

Mansoor, S., Iqbal, S., Popescu, S. M., Kim, S. L., Chung, Y. S., & Baek, J.-H. (2025). Integration of smart sensors and IoT in precision agriculture: Trends, challenges and future prospectives. *Frontiers in Plant Science*, 16, 1587869.

Misra, N., Dixit, Y., Al-Mallahi, A., Bhullar, M. S., Upadhyay, R., & Martynenko, A. (2020). IoT, big data, and artificial intelligence in agriculture and food industry. *IEEE Internet of Things Journal*, 9(9), 6305–6324.

Munir, M. S., Bajwa, I. S., Ashraf, A., Anwar, W., & Rashid, R. (2021). Intelligent and smart irrigation system using edge computing and IoT. *Complexity*, *2021*(1), 6691571.

Natraj, N., Raj, P., Karpagam, M., & Gunanandhini, S. (2025). A study on edge AI application in crop monitoring. In S. Mahajan, S. Munirathinam, & P. Raj (Eds.), *Edge of Intelligence: Exploring the Frontiers of AI at the Edge* (pp. 51–72). Wiley.

Pintus, M., Colucci, F., & Maggio, F. (2025). Emerging developments in real-time edge aiot for agricultural image classification. *IoT*, 6(1), 13.

Prajapati, J. B., Kumar, A., Pramanik, J., Prajapati, B. G., & Saini, K. (2023). Edge AI for real-time and intelligent agriculture. In P. Raj, K. Saini, & V. Pacheco (Eds.), *Applying Drone Technologies and Robotics for Agricultural Sustainability* (pp. 215–244). IGI Global.

Prakash, C., Singh, L. P., Gupta, A., & Lohan, S. K. (2023). Advancements in smart farming: A comprehensive review of IoT, wireless communication, sensors, and hardware for agricultural automation. *Sensors and Actuators A: Physical*, 362, 114605.

Qian, M., Qian, C., & Yu, W. (2025). Edge intelligence in smart agriculture cps. In W. Yu (Ed.), *Edge Intelligence in Cyber-Physical Systems* (pp. 265–291). Elsevier.

Rai, N., Zhang, Y., Villamil, M., Howatt, K., Ostlie, M., & Sun, X. (2024). Agricultural weed identification in images and videos by integrating optimised deep learning architecture on an edge computing technology. *Computers and Electronics in Agriculture*, 216, 108442.

Rajak, P., Ganguly, A., Adhikary, S., & Bhattacharya, S. (2023). Internet of things and smart sensors in agriculture: Scopes and challenges. *Journal of Agriculture and Food Research*, 14, 100776.

Ramirez, J. G. C. (2025). Optimising water and fertiliser use in agriculture through aidriven IoT networks: A comprehensive analysis. *Artificial Intelligence and Machine Learning Review*, 6(1), 1–7.

Rathore, N., & Rajavat, A. (2024). Smart farming based on IoT-edge computing: Applying machine learning models for disease and irrigation water requirement prediction in potato crop using containerised microservices. In N. Khatri, A. K. Vyas, C. Iwendi, & P. Chatterjee (Eds.), *Precision Agriculture for Sustainability* (pp. 399– 424). Apple Academic Press.

Shaikh, T. A., Rasool, T., & Lone, F. R. (2022). Towards leveraging the role of machine learning and artificial intelligence in precision agriculture and smart farming. *Computers and Electronics in Agriculture*, 198, 107119.

Singh, D. K., Sobti, R., Jain, A., Malik, P. K., & Le, D.-N. (2022). Lora based intelligent soil and weather condition monitoring with internet of things for precision agriculture in smart cities. *IET Communications*, 16(5), 604–618.

Upadhyay, A., Sunil, G., Zhang, Y., Koparan, C., & Sun, X. (2024). Development and evaluation of a machine vision and deep learning-based smart sprayer system for site-specific weed management in row crops: An edge computing approach. *Journal of Agriculture and Food Research*, 18, 101331.

Vaiyapuri, T., Parvathy, V. S., Manikandan, V., Krishnaraj, N., Gupta, D., & Shankar, K. (2022). A novel hybrid optimisation for cluster-based routing protocol in informationcentric wireless sensor networks for IoT based mobile edge computing. *Wireless Personal Communications*, 127(1), 39–62.

Verma, K. K., Song, X.-P., Kumari, A., Jagadesh, M., Singh, S. K., Bhatt, R., … Li, Y.-R. (2025). Climate change adaptation: Challenges for agricultural sustainability. *Plant, Cell & Environment*, 48(4), 2522–2533.

Zhang, D.-G., Chen, L., Zhang, J., Chen, J., Zhang, T., Tang, Y.-M., & Qiu, J.-N. (2020). A multi-path routing protocol based on link lifetime and energy consumption prediction for mobile edge computing. *IEEE Access*, 8, 69058–69071.

Zhang, Y., Wang, X., Jin, L., Ni, J., Zhu, Y., Cao, W., & Jiang, X. (2025). Research and development of an IoT smart irrigation system for farmland based on lora and edge computing. *Agronomy*, 15(2), 366.

Index

R

S

T

U

V

W

Y

For Product Safety Concerns and Information please contact our EU
representative GPSR@taylorandfrancis.com
Taylor & Francis Verlag GmbH, Kaufingerstraße 24, 80331 München, Germany

www.ingramcontent.com/pod-product-compliance
Ingram Content Group UK Ltd.
Pitfield, Milton Keynes, MK11 3LW, UK
UKHW020145200926
475702UK00005B/928
9781041068273